SCIENTIFIC METHOD IN PRACTICE

As the gateway to scientific thinking, an understanding of the scientific method is essential for success and productivity in science. Whereas research techniques among the sciences clearly are unique to the individual problems they approach, fundamentally the various disciplines are alike in their reliance on evidence and their use of inductive logic, probability, parsimony, and hypothesis testing – that is, on the scientific method. Yet excessive specialization can lead to neglect of these general principles, often at the cost of productivity and perspective.

Between the covers of a single book, Hugh Gauch has compiled the first synthesis of the practice and the philosophy of the scientific method. This book will enable scientists to become better scientists by offering them a deeper understanding of the underpinnings of the scientific method. For instance, parsimonious models often increase accuracy and thereby accelerate progress and increase returns on research investments. An examination of each of the principles noted earlier and a selective discussion of history and philosophy will demonstrate that the synergy of specialized techniques with general principles can enhance scientific productivity and experimental outcome. Once science is understood in its historical and philosophical context, scientists also can achieve a more accurate perspective on the rationality of the scientific approach and the role of science in society. Beginning with a discussion of today's "science wars" and science's presuppositions, the book then looks at deductive and inductive logic, probability, statistics, and parsimony and concludes with an examination of science's powers and limits and a look at science education. Topics relevant to a variety of disciplines are treated, and clarifying figures, case studies, and chapter summaries enhance the pedagogy.

This adeptly executed, comprehensive, yet pragmatic work yields a new synergy suitable for scientists and instructors as well as for graduate students and advanced undergraduates.

Hugh G. Gauch, Jr., is Senior Research Specialist in Crop and Soil Sciences at Cornell University.

SCIENTIFIC METHOD IN PRACTICE

Hugh G. Gauch, Jr.

Cornell University

CAMBRIDGE
UNIVERSITY PRESS

PUBLISHED BY THE PRESS SYNDICATE OF THE UNIVERSITY OF CAMBRIDGE
The Pitt Building, Trumpington Street, Cambridge, United Kingdom

CAMBRIDGE UNIVERSITY PRESS
The Edinburgh Building, Cambridge CB2 2RU, UK
40 West 20th Street, New York, NY 10011-4211, USA
477 Williamstown Road, Port Melbourne, VIC 3207, Australia
Ruiz de Alarcón 13, 28014 Madrid, Spain
Dock House, The Waterfront, Cape Town 8001, South Africa

http://www.cambridge.org

First published 2003

Printed in the United Kingdom at the University Press, Cambridge

Typefaces Times Ten 9.75/12.5 pt. and Helvetica Neue Condensed   *System* LaTeX 2_ε [TB]

A catalog record for this book is available from the British Library.

Library of Congress Cataloging in Publication Data

Gauch, Hugh G., 1942–
Scientific method in practice / Hugh G. Gauch.
    p.   cm.
Includes bibliographical references and index.
ISBN 0-521-81689-0 – ISBN 0-521-01708-4 (pbk.)
1. Science – Philosophy.   2. Science – Methodology.   I. Title.
Q175 .G337   2002
501 – dc21   2002022271

ISBN 0 521 81689 0 hardback
ISBN 0 521 01708 4 paperback

This book is dedicated

to a wise philosopher and fine scientist,

my dear Godson,

Jonathan Xavier.

CONTENTS

FOREWORD

Approximately halfway through her Ph.D. program in the biological sciences, my first doctoral student requested that we meet to review progress toward the degree. Knowing that both the research and course work of this student were progressing very nicely, I entered the appointment confident of a glowing report both for graduate student and major professor. However, the discussion took an unanticipated twist as we finished talking about the topics on my agenda.

When and where, this student asked, do we get to the more philosophical part of this Doctorate of Philosophy degree in science? Was it really true that this program that we guidance-committee members had so carefully designed for her was not going to include even one advanced course in philosophy or history of science? If not in formal course work, when would we as major professor and graduate student deal with the logical underpinnings and processes of science at the level of basic principles? The final question had the greatest unintended sting – something to the effect of "Will I graduate feeling worthy of more than a technical degree?"

Stunned and somewhat befuddled, I sent this student on her way with lame explanations: There simply wasn't sufficient time in a modern science education for students to become renaissance scholars as well as well-published researchers capable of competing successfully for federal grant dollars. Moreover, to venture into science philosophy required an unhealthy tolerance for time wasted in silly, perfectionist arguments over whether or not the sun would rise tomorrow. The better path to becoming a successful scientist, I argued, was to function as an apprentice to successful researchers and to get on generating results from real-world experiments. After all, I concluded, the quality of your Ph.D. degree will be at least equal to my own. Had not that Ph.D. landed me a great postdoctoral experience and an enviable tenure-stream Assistant Professor position at a major research-intensive university?

To this day, my former student does not realize how that conversation awakened my conscience to the awesomeness of the responsibility that educators shoulder in passing the scientific torch across intellectual generations. Although it came too late for my first Ph.D. student, that conversation precipitated my sustained appetite for suitable teaching materials to put some Ph. back into the Ph.D. degree in science. For two decades, I have been teaching a graduate seminar course entitled "The Nature and Practice of Science." This course seeks to leaven the minds of College of Natural Science graduate students with sufficient science philosophy, logic, and best practices for enriched and unusually productive careers in science and education. Having found no suitable text for such a course, we have relied on diverse readings, including John Platt, Karl Popper, Thomas Kuhn, Ronald Giere, and others, including university and governmental documents intended for graduate-student science education. Unfortunately, these authors use differing and confusing terminologies, and significant gaps are left that cause instructors and students to struggle in building a unified whole. Moreover, our coverage is introductory; the student or faculty member wishing to seek further is left wondering where to go for more and how to do so efficiently.

Thus it was with enthusiasm and relief that I as a reviewer read this book on *Scientific Method in Practice* by Hugh G. Gauch, Jr. Here at last is a comprehensive and up-to-date treatise on the fundamentals of science philosophy and method between the covers of one book and written from the pragmatic perspective of a credible science practitioner with whom researchers can identify. Here is the book that my graduate student and I needed on the day of our discussion about injecting appropriate Ph. into her Ph.D. degree in science.

The scope and depth of *Scientific Method* are truly amazing. Hugh Gauch has read, distilled, and integrated the contents of literally hundreds of books and articles on science history, philosophy, and practice. Readers treated to the "intellectual gold" emerging from this mammoth "smelting operation" are guaranteed to emerge with a deep sense of appreciation for the efficiency inherent in a well-executed and comprehensive scholarly review yielding a new synthesis. While practicing scientists cannot be expected to read hundreds of texts on science history and philosophy, they certainly can and should read this seminal book.

This is not to say that *Scientific Method* is an easy, comfortable read throughout. Questions of what we humans know, how we can know, with what certainty we can know, and what are the best ways to proceed in efficiently acquiring new knowledge sufficiently reliable to guide practical work and living have absorbed the minds of the greatest thinkers since antiquity. Readers hardly need forewarning that an understanding of a modern philosophical and scientific perspective on these profound questions will require significant work and, in some areas like approaches to statistical analyses, appreciable technicality. However, readers of this book will be rewarded with substantial answers.

Hugh Gauch has done a good job in keeping this important material interesting, sensible, and pragmatic. The book lives up to its billing – it will increase the productivity and perspective of scientists. Thus, in addition to practicing scientific researchers, this book should be of great interest to managers of scientific research. Also, science education has always sought to be grounded in an accurate understanding of science philosophy and practice. So it seems appropriate that science educators should be at the front of the line of folks benefiting from Gauch's work. For that reason, as one involved in science education, I consider it a special honor to participate in the discovery and promotion of this important book.

Finally, it should not be overlooked that *Scientific Method* is a substantial philosophical contribution in its own right. In addition to all of its aforementioned benefits to science practitioners, this book is also Hugh Gauch's attempt to rescue science from the clutches of postmodernist philosophers who argue that the credibility of scientific knowledge is undermined by inherent logical and procedural weaknesses. Science from that weakened perspective becomes nothing special. I find convincing Gauch's rebuttal to the postmodernists and therefore believe that this book makes an important statement to the anti-science political movement currently afoot. Clearly, this book is a major intellectual work establishing and defending a new philosophical position that any mainstream philosopher of science must take seriously.

I boldly predict that in the field of scientific method, the work of Hugh Gauch will stand as a contribution similar in magnitude to the work of a Francis Bacon, Karl Popper, or Thomas Kuhn. Moreover, I would like to think that at the turn of the twenty-first century, this work represents the beginning of a trend away from schism and toward a meaningful reuniting of science practice with philosophy. Each of these major intellectual disciplines has much to offer the other. With the production of this book, Gauch has taken a significant step in recapturing a desirable synergy between science philosophy and science practice, upon which we should capitalize.

Dr. James R. Miller
Michigan State University
East Lansing, Michigan
March 2002

PREFACE

The thesis of this book, as set forth in Chapter 1, is that there exist general principles of scientific method that are applicable to all of the sciences, but excessive specialization often causes scientists to neglect the study of these general principles, even though they undergird science's rationality and greatly influence science's efficiency and productivity. These general methodological principles involve the use of deductive and inductive logic, probability, parsimony, and hypothesis testing. Neither specialized techniques nor general principles can substitute for one another, but rather the winning combination for scientists is mastery of both.

The primary purpose of this book is to help scientists become better scientists, more creative and more productive, by fostering a deeper understanding of the general principles of scientific method. For instance, parsimonious models often can lead to greater accuracy and thereby improve decision-making, accelerate progress, and increase returns on research investments. Also, scientists can improve the statistical analyses of their data by understanding how the Bayesian and frequentist paradigms relate to different research questions and technological objectives.

The secondary purpose is to help scientists gain perspective on science's rationality and role. Every conclusion of science, when fully disclosed, involves components of three kinds: presuppositions, evidence, and logic. Accordingly, an explanation of scientific method amounts to disclosing and securing these three inputs. Also, clearly understood methods beget realistic expectations and legitimate claims. Then scientists can defend science's legitimate claims from influential attacks with a measure of sophistication and confidence, while also perceiving the proper domain and real limits of science, and thereby avoiding excessive claims for science or diminished roles for the humanities. A humanities-rich version of science is more beneficial and engaging than a humanities-poor version.

Understandably, some readers may have greater interest in one of these projects than the other. Those interested in higher productivity should focus on Chapters 5–9. On the other hand, those pursuing philosophical and historical perspective should focus on Chapters 2–4 and 10. Both projects are addressed in Chapters 1 and 12. Science educators will find Chapter 11 particularly relevant. But that said, readers are encouraged not to be overly hasty in judging what is or is not interesting or useful. To master scientific method for the purpose of increasing productivity is commendable, but the ensuing day-to-day labor is arduous and sometimes tedious, so an occasional philosophical joy along the way is not to be despised. On the other hand, to master scientific method for the purpose of gaining perspective on science is commendable, but this turns on understanding scientific practice in technical detail. In a word, this book's two purposes – productivity and perspective – complement each other.

This book's intended audience includes primarily professional scientists and graduate and advanced undergraduate students in the sciences, and it can be used for either individual study or classroom instruction. This book is also for science educators at all levels. Because methods precede results, scientific method is the gateway to all scientific thinking. It is becoming increasingly common for universities to offer courses on the nature and practice of science that are team-taught by scientists and philosophers (and historians, sociologists, and ethicists). Accordingly, some philosophers of science and others in the humanities may be interested to see which topics in their literature are found by scientists to be particularly interesting and helpful.

Because of some unfamiliar historical and philosophical content, at first glance this book may seem somewhat demanding. But in fact, were it added to scientists' bookshelves, it usually would be the least technical book there. Accordingly, although it is more advanced than typical undergraduate texts on science's method and philosophy, motivated juniors and seniors should find it within their grasp.

This book is addressed to a general audience of scientists, rather than being customized for the specific interests of one scientific specialty. To serve those diverse needs, the strategy adopted here is to treat numerous topics fairly thoroughly. Understandably, one reader may be fascinated by the historical contributions of Robert Grosseteste and Albertus Magnus but may be uninterested in the axioms of predicate logic, whereas another reader may have the opposite reaction. Therefore, readers are encouraged to study or skim various sections as their interests dictate, especially when this book is used for individual study.

There are several older books on scientific method that still have much merit, but obviously they cannot address current debates and recent advances (Ritchie 1923; Wilson 1952; Ackoff 1962; Burks 1977; Grinnell 1987).

There is also a more recent book aimed at undergraduates with little or no background in the sciences (Carey 1994). Although admirable for its intended audience and stated objectives, it cannot be expected to benefit professional scientists in their practice. The books by Giere (1984) and Howson and Urbach (1993) are particularly insightful. Other recent books have emphasized historical and philosophical aspects of scientific method (Gower 1997; Rosenberg 2000).

As there did not seem to be a recent book on scientific method aimed at professional scientists and university science majors, this book was written to meet that need. Given the importance of science and technology in contemporary society, and given the inherent beauty and interest of scientific method, it is astonishing that this topic has been so neglected.

Because the literature on scientific method is so underdeveloped, the ideas presented here had to be gleaned from diverse sources far and wide, especially those on the philosophy of science. My role has been largely that of an importer, on behalf of my fellow scientists, gathering useful ideas from numerous books about the history, philosophy, and logic of science. This book is rather different from its sources, however, because there are substantial differences between the needs of philosophers and those of scientists. Philosophers like to study the philosophy of science to become better philosophers, but scientists need to study the philosophy of science to become better scientists. This is a book on scientific method by scientists and for scientists.

Having said what this book is, a few words on what it is not: It is not a systematic or conventional text on the philosophy of science, although it draws substantially from that intriguing literature. Likewise, it is not a comprehensive survey of all topics that might be included in a course on the nature and practice of science, particularly because it does not explore science's ethics and priorities. However, scientific method could well be a core topic occupying a sizable fraction of a more broadly conceived curriculum on the nature and practice of science. Regrettably, this book is not a comprehensive history of scientific method, although a few gems are included. Philosophy, ethics, and history are important, but this book's topic is method, and that poses sufficient challenge.

Finally, in the wars against disease and hunger, as well as poverty and ignorance, scientists and technologists have wonderful opportunities and hence substantial responsibilities. In its larger role alongside others of the liberal arts, science contributes to our picture of the world and life, implying the necessity to capture and communicate a valid picture. The main intention here is to spark a realization that the general principles of scientific method are more difficult and yet more beautiful than some readers may previously have recognized, and thereby to stimulate within the scientific community greater attention to this neglected topic that has tremendous potential to enhance productivity and perspective.

I appreciate helpful suggestions on earlier drafts of various chapters from several scientists, philosophers, and statisticians, including James O. Berger, Mark A. Case, Gary W. Fick, Malcolm R. Forster, William H. Jefferys, James R. Miller, Roger E. Steele, and Martin T. Wells. I am particularly grateful to Gregory J. Velicer, whose wise counsel did much to guide, shape, and encourage this work. Of course, all remaining deficiencies are my sole responsibility. I thank Millard Baublitz, Jr., and P. Andrew Karplus for contributing fascinating case studies to Chapter 9. I also thank James R. Miller for writing the Foreword. I am grateful to my parents, who both were scientists, for instilling in me a love of learning, an interest in science, and a respect for truth. I also appreciate the sustained enthusiasm for this research project shown by my sister and brothers, Susan, Jonathan, Christopher, Gary, and Ken, and their families. Cornell University provided a wonderfully favorable environment for writing this book, especially by virtue of its superb library system.

Hugh G. Gauch, Jr.
Cornell University
Ithaca, New York
March 2002

CHAPTER ONE

INTRODUCTION

This book explores the general principles of scientific method that pervade all of the sciences, focusing on practical aspects. The implicit contrast is with specialized techniques for research that are used in only certain sciences. The structure of science's methodology envisioned here is depicted in Figure 1.1, which shows individual sciences, such as astronomy and chemistry, as being partly similar and partly dissimilar in methodology. What they share is a core of the general principles of scientific method. This common core includes such topics as hypothesis generation and testing, deductive and inductive logic, parsimony, and science's presuppositions, domain, and limits. Beyond methodology as such, some practical issues are shared broadly across the sciences, such as relating the scientific enterprise to the humanities and implementing effective science education.

The general principles that are this book's topics are shown in greater detail in Figure 1.2. These principles are of three kinds: (1) Some principles are relatively distinctive of science itself. For instance, the ideas about Ockham's hill that are developed in Chapter 8 on parsimony have a distinctively scientific character. If occasionally lawyers or historians happen to use those ideas, they will not be reprimanded. Nevertheless, clearly those ideas are used primarily by scientists and technologists. (2) Other principles are shared broadly among all forms of rational inquiry. For example, deductive logic is squarely in the province of scientists, and it is explored in Chapter 5. But deductions are also important in nearly all undertakings. (3) Still other principles are so rudimentary and foundational that their wellsprings are in common sense, such as the principle of noncontradiction. Also, science's presuppositions, which are discussed in Chapter 4, have their roots in common sense. Naturally, the boundaries among these three groups are somewhat fuzzy, so they are shown with dashed lines. Nevertheless, the broad distinctions among these three groups are clear and useful.

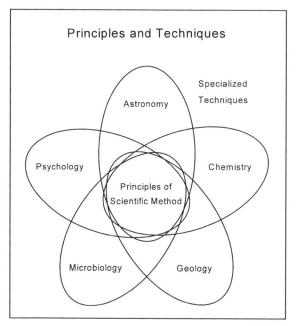

Figure 1.1. Science's methodology depicted for five representative scientific disciplines, which are partly similar and partly dissimilar. Accordingly, scientific methodology has two components. The general principles of scientific method pervade the entire scientific enterprise, whereas specialized techniques are confined to particular disciplines or subdisciplines.

There is a salient difference between specialized techniques and general principles in terms of how they are taught and learned. Precisely because specialized techniques are specialized, each scientific specialty has its own more or less distinctive set of techniques. Because there are hundreds of specialties and subspecialties, the overall job of communicating these techniques requires millions of instructional courses, books, and articles. But precisely because general principles are general, the entire scientific community has a single, shared set of principles, and it is feasible to collect and communicate the main information about these principles within the scope of a single course or book. Whereas a scientist or technologist needs to learn new techniques when moving from one project to another, the pervasive general principles need be mastered but once. Likewise, whereas specialized techniques and knowledge have increasingly shorter half-lives, given the unprecedented and accelerating rate of change in science and technology, the general principles are refreshingly enduring.

The central thesis of this book is that scientific methodology has two components, the general principles of scientific method and the specialized

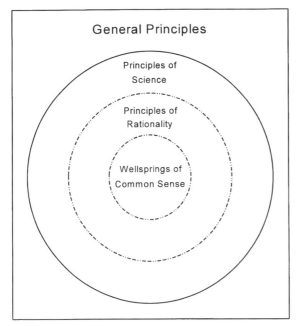

Figure 1.2. Detailed view of the general principles, which are of three kinds: principles that are relatively distinctive of science itself, broader principles found in all forms of rational inquiry, and foundational principles with their wellsprings in common sense.

techniques of a given specialty, and the winning combination for scientists is strength in both. Neither basic principles nor research techniques can substitute for one another. This winning combination can enhance productivity and perspective.

A CONTROVERSIAL IDEA

The mere idea that there exist such things as general principles of scientific method is controversial. The objections are of two kinds, philosophical and scientific. But first, a potential misunderstanding needs to be avoided. The scientific method "is often misrepresented as a fixed sequence of steps," rather than being seen for what it truly is, "a highly variable and creative process" (AAAS 2000:18). The claim here is that science has general principles that must be mastered to increase productivity and enhance perspective, not that these principles provide a simple and automated sequence of steps to follow.

Beginning with the philosophical objection, it is fashionable among some skeptical, relativistic, and postmodern philosophers to say that there are no principles of rationality whatsoever that are reliably or impressively truth-conducive. For instance, in an interview in *Scientific American*, the noted philosopher of science Paul Feyerabend insisted that there are no objective standards of rationality, so naturally there is no logic or method to science (Horgan 1993). Instead, "Anything goes" in science, and it is no more productive of truth than "ancient myth-tellers, troubadours and court jesters." From that dark and despairing philosophical perspective, the concern with scientific method would seem to have nothing to do distinctively with science itself. Rather, science would be just one more instance of the pervasive problem that rationality and truth elude us mere mortals, forever and inevitably.

Such critiques are unfamiliar to most scientists, although some may have heard a few distant shots from the so-called science wars. Scientists typically find those objections either silly or aggravating, so rather few engage such controversies or bother to contribute in a sophisticated and influential manner. But in the humanities, those deep critiques of rationality are currently quite influential. Anyway, by that reckoning, Figure 1.1 should show blank paper.

Moving along to the scientific objection, many scientists have claimed that there is no such thing as a scientific method. For instance, a Nobel laureate in medicine, Sir Peter Medawar, pondered this question: "What methods of enquiry apply with equal efficacy to atoms and stars and genes? What *is* 'The Scientific Method'?" He concluded that "I very much doubt whether a methodology based on the intellectual practices of physicists and biologists (supposing that methodology to be sound) would be of any great use to sociologists" (Medawar 1969:8, 13). In this regard, consider a little thought experiment. Suppose that an astronomer, a microbiologist, and an engineer were each given a grant of $500,000 to purchase research equipment. What would they buy? Obviously they would purchase strikingly different instruments, and each scientist's new treasures would be quite useless to the others (apart from the universal need for computers). By that reckoning, Figure 1.1 should show the methodologies of the individual sciences dispersed, with no area in which they would all overlap.

What of these objections? Is it plausible that, contrary to Figure 1.1, the methodologies of the various disciplines and subdisciplines of science have no overlap, no shared general principles? Asking a few concrete questions should clarify the issues and thereby promote an answer. Do astronomers use deductive logic, but not microbiologists? Do psychologists use inductive logic (including statistics) to draw conclusions from data, but not geologists? Are probability concepts and calculations used in biology, but not in sociology? Do medical researchers care about parsimonious models and explanations, but not electrical engineers? Does physics have presuppositions about the existence and comprehensibility of the physical world, but not genetics? If the answers to such questions are no, then Figure 1.1 stands as a plausible picture of science's methodology.

THE AAAS VISION OF SCIENCE

Beyond such brief and rudimentary reasoning about science's methodology, it merits mention that the thesis proposed here accords with the official position of the American Association for the Advancement of Science (AAAS). The AAAS is the world's largest scientific society, the umbrella organization for almost 300 scientific organizations and publisher of the prestigious journal *Science*. Accordingly, the AAAS position bids fair as an expression of the mainstream opinion.

The AAAS views scientific methodology as a combination of general principles and specialized techniques, as depicted in Figure 1.1.

> Scientists share certain basic beliefs and attitudes about what they do and how they view their work.... Fundamentally, the various scientific disciplines are alike in their reliance on evidence, the use of hypotheses and theories, the kinds of logic used, and much more. Nevertheless, scientists differ greatly from one another in what phenomena they investigate and in how they go about their work; in the reliance they place on historical data or on experimental findings and on qualitative or quantitative methods; in their recourse to fundamental principles; and in how much they draw on the findings of other sciences.... Organizationally, science can be thought of as the collection of all of the different scientific fields, or content disciplines. From anthropology through zoology, there are dozens of such disciplines.... With respect to purpose and philosophy, however, all are equally scientific and together make up the same scientific endeavor. (AAAS 1989:25–26, 29)

Regarding the general principles, "Some important themes pervade science, mathematics, and technology and appear over and over again, whether we are looking at an ancient civilization, the human body, or a comet. They are ideas that transcend disciplinary boundaries and prove fruitful in explanation, in theory, in observation, and in design" (AAAS 1989:123).

Accordingly, "Students should have the opportunity to learn the nature of the 'scientific method'" (AAAS 1990:xii; also see AAAS 1993). That verdict is affirmed in official documents from the National Academy of Sciences (NAS 1995), the National Commission on Excellence in Education (NCEE 1983), the National Research Council of the NAS (NRC 1996, 1997, 1999), the National Science Foundation (NSF 1996), the National Science Teachers

Association (NSTA 1995), and the counterparts of those organizations in many other nations (Matthews 2000:321–351).

An important difference between specialized techniques and general principles is that the former are discussed in essentially scientific and technical terms, whereas the latter inevitably involve a wider world of ideas. Accordingly, for the topic at hand, the "central premise" of one AAAS (1990:xi) position paper is extremely important, namely, that "Science is one of the liberal arts and . . . science must be taught as one of the liberal arts, which it unquestionably is." Many of the broad principles of scientific inquiry are not unique to science, but also pervade rational inquiry more generally, as depicted in Figure 1.2. "All sciences share certain aspects of understanding – common perspectives that transcend disciplinary boundaries. Indeed, many of these fundamental values and aspects are also the province of the humanities, the fine and practical arts, and the social sciences" (AAAS 1990:xii; also see p. 11).

Likewise, the continuity between science and common sense is respected, which implies productive applicability of scientific attitudes and thinking in daily life. "Although all sorts of imagination and thought may be used in coming up with hypotheses and theories, sooner or later scientific arguments must conform to the principles of logical reasoning – that is, to testing the validity of arguments by applying certain criteria of inference, demonstration, and common sense" (AAAS 1989:27). "There are . . . certain features of science that give it a distinctive character as a mode of inquiry. Although those features are especially characteristic of the work of professional scientists, everyone can exercise them in thinking scientifically about many matters of interest in everyday life" (AAAS 1989:26; also see AAAS 1990:16).

Because science's general principles involve a wider world of ideas, many vital aspects cannot be understood satisfactorily by looking at science in isolation. Rather, they can be mastered properly only by seeing science in context, especially in philosophical and historical context. Therefore, this book's pursuit of the principles of scientific method will occasionally range into discourse that has a distinctively philosophical or historical or sociological character. There is a natural and synergistic traffic of great ideas among the liberal arts, including science.

The brief remarks in this and the previous sections are not offered as a rigorous defense of the (controversial) thesis that some general principles are vital components of scientific reasoning. Only a whole book, such as what follows, can aspire to such an ambitious goal. Rather, these preliminary remarks are offered as evidence that the idea that there is a scientific method has enough plausibility and backing to merit careful consideration, not breezy dismissal.

PRIMARY AND SECONDARY BENEFITS

Whatever else may be controversial, one thing that is certain is that mastery of the subject matter proposed and presented here, the principles of scientific method, will require some time and effort. Accordingly, it is natural to ask about the purposes and benefits that will result from this study.

Two general kinds of benefits are expected, namely, increased productivity and enhanced perspective. The primary benefit will be to help scientists become better scientists, more creative and more productive, by providing a deeper understanding of the basic principles of scientific method. A secondary benefit will be to cultivate a humanities-rich version of science, rather than a humanities-poor version, so that scientists can gain perspective on their enterprise.

Regarding the primary benefit, what a scientist or technologist needs in order to function well can be depicted by a resources inventory, as in Figure 1.3. All items in this inventory are needed. The first three items address the obvious physical setup that a scientist needs. The last two items are intellectual rather than physical, namely, mastery of the specialized techniques of a chosen specialty and mastery of the general principles of scientific method.

A common concern is that frequently the weakest link in a scientist's inventory is an inadequate understanding of science's principles. This weakness in understanding the scientific method has just as much potential to

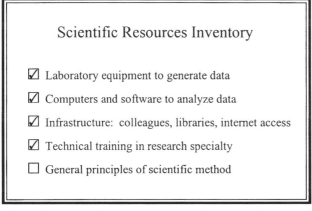

Figure 1.3. A typical resources inventory for a research group. The scientists in a given research group often have excellent laboratory equipment, computers, infrastructure, and technical training, but inadequate understanding of the general principles of scientific method is the weakest link. Ideally, a research group will be able to check off all five boxes in this inventory, and there will be no weak link.

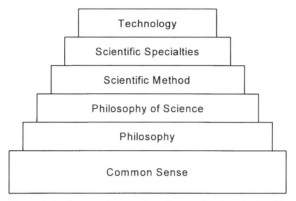

Figure 1.4. Perspective on the place and role of scientific method. The foundations of scientific method are provided by the philosophy of science, which depends more generally on philosophy, which is grounded ultimately by common sense. In turn, scientific method supports scientific specialties and technology.

retard progress as does, say, inappropriate laboratory equipment or inadequate training in some research technique.

Moving along to the secondary benefit, an initial perspective on the place and role of scientific method is offered in Figure 1.4. The scientific specialties and technological accomplishments that emerge from applying the scientific method are obvious. But a humanities-rich version of science will reveal science's roots in the philosophy of science and more generally in philosophy, which is grounded ultimately in common sense. Accordingly, scientific method will be better integrated and more interesting when presented in its philosophical and historical context. Such perspective will also facilitate realistic claims, neither timid nor aggrandized, about science's powers and prospects. A humanities-rich perspective will preclude imperialistic claims about science's domain, with all of the attendant false promises that could only disappoint and alienate.

The topic of science's basic principles has been around for millennia, even before Aristotle (Losee 1993). So naturally many opinions have been expressed about the value of this topic, especially for a scientist's ordinary, practical, day-to-day work. Scientists themselves have written much, as have philosophers, though most of that literature has been rather speculative or anecdotal.

Fortunately, scholars in a different field, the science educators, have done the work of conducting hundreds of careful empirical studies to characterize and quantify the benefits that can result from learning scientific method. Many of those studies have involved impressive sample sizes and carefully controlled experiments to quantify educational outcomes and scientific competencies for students who either have or else have not received instruction in science's general principles. Consequently it has been educators, rather than scientists or philosophers, who have provided the best information on these benefits.

Incidentally, among educators, what here goes under such labels as "scientific method" and "general principles" is most frequently termed the "nature of science." Because Chapter 11 will review the literature in science education, here only brief remarks without documentation will be presented, by way of anticipation. Empirical studies by educators have provided overwhelming evidence for six specific claims.

(1) Better Comprehension. The specialized techniques and subject knowledge that so obviously make for productive scientists are better comprehended when the underlying principles of scientific method are understood – somewhat like the way that calcium is better absorbed by the digestive system when accompanied by vitamin D.

(2) Greater Adaptability. It is facility with the general principles of science that contributes the most to a scientist's ability to be adaptable and to transfer knowledge and strategies from a familiar context to new ones, and that adaptability will be necessary for productivity as science and technology continue to experience increasingly rapid and pervasive changes.

(3) Greater Interest. Most people find a humanities-rich version of science, with its wider perspective and big picture, much more engaging and interesting than a humanities-poor version, so including something of science's method, history, and philosophy in the science curriculum results in higher rates of retention of students in the various sciences (and it especially helps those ranked near the bottom, so that educational outcomes can become more nearly equal).

(4) More Realism. An understanding of the scientific method leads to a realistic perspective on science's powers and limits, and more generally to balanced views of the complementary roles of the sciences and the humanities.

(5) Better Researchers. Researchers who master science's general principles gain productivity because they can make better decisions about whether or not to question an earlier interpretation of their data as a result of new evidence, whether or not there is a need to repeat an experiment, where to look for other scientific work related to their project, and how certain or accurate their conclusions are.

(6) Better Teachers. Teachers and professors who master science's general principles prove to be better at communicating science content, in part because they are better at detecting and correcting students' prior mistaken notions and logic, and hence such teachers can better equip the next generation of scientists to be productive.

The facts of the case are clear, having been established by hundreds of empirical studies involving various age groups, nations, and science subjects: Understanding the principles of scientific method does increase productivity and enhance perspective. But why? Why does mastery of these principles help scientists to become better scientists? The most plausible explanation is simply that the central thesis of this book is true: It really is the case that scientific methodology has two components, the general principles of scientific method and the specialized techniques of a chosen specialty, and the winning combination for scientists is strength in both. Therefore, adequate understanding of scientific method is essential for an astronomer, botanist, chemist, dietitian, engineer, floriculturalist, geologist, . . . , or zoologist.

BEYOND THE BASICS

Do scientists typically have an adequate understanding of the scientific method? Is it sufficient to yield the expected benefits of productivity and perspective? Unfortunately, the current state of affairs seems rather dismal. "Ask a scientist what he conceives the scientific method to be, and he will adopt an expression that is at once solemn and shifty-eyed: solemn, because he feels he ought to declare an opinion; shifty-eyed, because he is wondering how to conceal the fact that he has no opinion to declare" (Medawar 1969:11). Furthermore, countless recent studies by science educators have confirmed that verdict.

The cause of the current situation is no mystery. Scientists are not born already knowing the principles of scientific method, and neither are they taught those principles in a vigorous and systematic manner. "The hapless student is inevitably left to his or her own devices to pick up casually and randomly, from here and there, unorganized bits of the scientific method, as well as bits of *un*scientific methods" (Theocharis and Psimopoulos 1987).

Just where do scientists get what meager bits they do have? Because few science majors ever take a course in scientific method, logic, or the history and philosophy of science, their exposure to any focused attention to science's principles usually is limited to the occasional science textbook that begins with an introductory chapter on scientific method. Figure 1.5 lists typical contents for such chapters.

Despite the perhaps scandalous resemblance of such accounts to the antiquated view of science offered by Francis Bacon in the early 1600s (Urbach 1987; Peltonen 1996), it may well be that elementary ideas along those lines provide the most suitable picture of science to convey to an eighth-grade student working on a science-fair project. Also, it may well be that such a rudimentary cartoon of science is much closer to the mark than is the

---

Elementary Scientific Method

- Hypothesis formulation
- Hypothesis testing
- Deductive and inductive logic
- Controlled experiments; replication and repeatability
- Interactions between data and theory
- Limits to science's domain

Figure 1.5. Typical topics in an elementary presentation of scientific method intended for college freshmen and sophomores. Introductory science texts often start with several pages on scientific method, discussing the formulation and testing of hypotheses, collection of data from controlled and replicated experiments, and so on. They are unlikely, however, to include any discussion of parsimony or any exploration of the history of scientific method beyond a passing mention of Aristotle.

currently fashionable postmodern take on science, as reviewed and criticized by Koertge (1998). Whatever its merit in terms of simplicity, such an elementary view of scientific method is wholly inadequate for science professionals and professors. In pursuit of increased productivity and sophisticated perspective, science professionals *must* go beyond the basics.

In some sense, the basic principles of scientific method, or at least their wellsprings in common sense, are obvious and compelling. But these principles are also difficult, challenging, and exacting. They yield their secrets and benefits only to those who pay their dues and do their work. Surely it is sobering to realize in historical perspective that civilizations rose and fell around the globe for millennia before anything recognizable as the scientific method emerged (say around A.D. 1200 to 1600 by various accounts). So there must be some limit to that view about scientific principles being obvious.

Indeed, learning enough about deductive logic to avoid common fallacies is no easy task. Learning enough about inductive logic and statistics to analyze data properly and vigorously is difficult, especially given the internal debate in statistics between the frequentist and Bayesian paradigms. Mastering the principle of parsimony or simplicity in order to gain efficiency and increase productivity is anything but simple, requiring precise distinctions, subtle concepts, and complex calculations. Acquiring some philosophical and historical perspective on science is challenging. And developing the habit of applying general principles to daily scientific work with creativity and effectiveness requires considerable mentoring and practice.

Moreover, these principles work in concert, so to understand their connections and interrelations in a functioning whole, they must be gathered
together and presented in a book or in a course in an integrated fashion.
Because the principles of concern here are general principles, they do indeed
emerge repeatedly in science and technology, but that alone is not sufficient
to guarantee that scientists will perceive and grasp their generality. For instance, the case studies in Chapter 9 will reveal parsimony at work in diverse
applications in genetics, agriculture, drug design, and electrical engineering.
The great risk is that a neophyte may see the material involving parsimony as
being just one more of the technicalities needed to accomplish some specialized task. Therefore, it is imperative to present general principles as being
truly *general* principles!

If the wide generality and applicability of a principle are taught *explicitly*
and *near the beginning* of a scientist's training, then subsequent instances of
that principle at work will reinforce the lesson and will promote adaptability
and productivity. But if relevant instances involving that principle are merely
encountered repeatedly, but with no larger story ever being told, then only
the rare student can be expected to assemble the big picture without proper
mentoring.

However, despite the deficiencies of the current situation and despite
the inherent challenges of this topic of science's principles, two factors are
encouraging. They imply that rapid and dramatic improvement certainly is
possible:

First, students are receptive, finding this subject matter quite interesting.
For example, Albert Einstein observed that "I can say with certainty that the
ablest students whom I met as a teacher were deeply interested in the theory
of knowledge. I mean by 'ablest students' those who excelled not only in
skill but in independence of judgment. They liked to start discussions about
the axioms and methods of science" (Frank 1957:xi). Likewise, Machamer
(1998) remarked, "Now part of the fun of science, as in most interesting
human activities, lies in thinking about how and why it is done, and how it
might be done better." Also recall the story about the graduate student in
this book's Foreword who eagerly wanted to know "the logical underpinnings
and processes of science at the level of basic principles" so that she could
graduate with a doctorate in science and feel "worthy of more than a technical
degree." Again, countless studies by science educators have shown decisively
that students are interested in a humanities-rich account of the principles of
scientific method.

Second, the AAAS vision for science includes a vigorous and sustained call
for students and scientists to understand these principles well, and it is reasonable to suppose that this leadership will be influential. So, on two counts,
there are good prospects for scientists to acquire their winning combination,

the principles and techniques that can enhance productivity and perspective. The bottom line is that if this book's central thesis is true, then it will not be possible to keep this winning combination a secret for much longer.

A TIMELY OPPORTUNITY

Again, the central thesis of this book is that science's methodology involves both general principles and specialized techniques, and these principles and techniques together constitute the winning combination for scientists that will enhance their productivity and perspective. However, such views have suffered considerable neglect during the past century or so.

A major cause of that neglect has been the common perception that even if there are such things as the principles of scientific method, the study of those principles would confer no benefit to scientists. For instance, writing in his usual witty and engaging style, Medawar (1969:8, 12) mused, "If the purpose of scientific methodology is to prescribe or expound a system of enquiry or even a code of practice for scientific behavior, then scientists seem to be able to get on very well without it. Most scientists receive no tuition in scientific method, but those who have been instructed perform no better as scientists than those who have not. Of what other branch of learning can it be said that it gives its proficients no advantage; that it need not be taught or, if taught, need not be learned?" (Medawar 1969:8; also see p. 12). In fairness to Medawar, he also remarked that "Of course, the fact that scientists do not consciously practice a formal methodology is very poor evidence that no such methodology exists" (Medawar 1969:9), and he did go on to offer some positive comments.

In any case, the sentiment that scientists get along just fine without probing science's philosophical and methodological foundations is at least common. Such sentiments are mistaken. They bespeak a lamentable and dangerous complacency. Indeed, on three counts, it is time for serious consideration of this book's central thesis.

(1) Science Education. The AAAS has stated with confidence and enthusiasm its vision that a humanities-rich understanding of science makes for better scientists. During the past decade, science educators have generated an enormous literature that provides a wealth of compelling empirical evidence in support of the AAAS vision, as will be reviewed in Chapter 11.

Because this literature is so recent, however, one cannot blame scientists and philosophers in the past for not having taken into account the findings of educators when they offered their anecdotes and speculations about the relevance or irrelevance of science's principles for day-to-day research. But this does mean that earlier assessments, such as that by Medawar, now need to be taken with a grain of salt. More important, any future opinions and reflections from scientists and philosophers about the value of mastering science's principles can gain in realism and interest by incorporating the factual findings of science educators.

(2) Recent Developments. In many respects, the topic of science's general principles is an ancient story with refreshingly enduring content. In some other respects, however, it is a living and growing topic that includes exciting recent developments.

The foremost instance of recent advances is that wonderfully subtle but surprisingly practical application of parsimony, as will be explored in Chapter 8. Briefly, experiments and investigations in many scientific specialties produce large amounts of relatively noisy or inaccurate data. This situation is especially common in some applied sciences, such as agriculture and medicine. Remarkably, parsimonious models of such data can yield findings considerably more accurate than are indicated by the original data. Frequently a parsimonious model can improve accuracy, prompt better decisions, and increase productivity as much as can the collection of several times as much data. Yet the cost of a few seconds of computer time to fit a parsimonious model is minute compared with the cost of collecting much more data, so parsimonious models offer a remarkably cost-effective means to increase productivity.

Sadly, however, that opportunity to increase productivity is one of science's best-kept secrets. Apart from scientists and technologists in a few specialties, such as signal processing, that option is all but unknown. Nor is such an application of parsimony something so simple and obvious that scientists are likely to stumble across it without intentional study and mentoring. The underlying concepts are unfamiliar and even somewhat counterintuitive. Though in typical applications the calculations require only seconds on an ordinary computer, that represents the millions or billions of arithmetic steps in a highly structured algorithm that earlier someone mastered and programmed. Obviously such powerful methods were unknown prior to the general availability of computers beginning in the 1960s, as well as the advent, at around the same time, of some critical developments in statistical theory.

Recent decades and even recent years have seen great advances in deductive logic, probability, and inductive logic (or statistics), as will be explored in Chapters 5–7. So, on many fronts, the recent advances in understanding and implementing science's general principles have been so spectacular that earlier accounts are outdated, and nothing but a contemporary evaluation of the possibilities merits serious consideration.

(3) Appropriate Focus. The AAAS vision of a humanities-rich version of science has breathtaking sweep. The history, philosophy, and methodology of science compose an enormously broad and involved interdisciplinary field.

Consequently, when approaching this field for an audience of science students and professionals, one needs to focus on those specific portions of the information that are of greatest interest and benefit to scientists. Otherwise the result could be more dissipative than beneficial.

Too often the curriculum in science's principles that has reached scientists has been developed along the path of least resistance by borrowing wholesale from the literature in the philosophy of science. But the goal of that literature has been to make philosophers better philosophers, not to make scientists better scientists, and similar remarks could be made about the literature in the history of science or the sociology of science.

Obviously, any book or course for scientists regarding science's principles must seriously engage the companion literatures from historians, philosophers, sociologists, and educators, but the borrowing must be selective and focused if the goal is to help scientists become better scientists. Consequently, any assessment of the value of having scientists study science's basic principles will be inaccurate if that study was based on materials without the proper focus. A scientist could study an enormous number of pages directed at making philosophers better philosophers or making historians better historians and still not become a better scientist. The real issue is whether or not a scientist will benefit from mastering material properly focused for scientists.

There has been little such focus in the past. However, now is the time to put the AAAS vision to the test in a manner that is properly focused and fair. Besides the AAAS (1990) vision, relatively recent position papers that set forth specific recommendations for curricula in the nature of science include those by the AAAS (1989), NAS (1995), NRC (1996, 1999), and NSF (1996). In all of those reports, the principles of scientific method hold a prominent position in the proposed curricula.

So, on three counts, there is now a new day for the thesis that mastering science's general principles can help scientists. The time for complacency is past. The time is right for rapid progress.

PERSONAL EXPERIENCE

Thus far, this introductory chapter has drawn on the insights of others, especially those of the AAAS and science educators, to illustrate and support this book's central thesis. As this chapter approaches its close, perhaps some readers would be interested in the personal experience that has prompted my interest in the principles of scientific method.

My research specialty at Cornell University during the past three decades has been the statistical analysis of ecological and agricultural data. A special focus in this work has been agricultural yield trials. Worldwide, billions of

Figure 1.6. A soybean yield trial conducted in Aurora, New York. The soybean varieties here varied in terms of numerous traits. For example, the variety in the center foreground matured more quickly than the varieties to its left and right, making its leaves light yellow rather than dark green as the end of the growing season approached. Yield is a particularly important trait. (Reprinted from Gauch, 1992:3, with kind permission from Elsevier Science.)

dollars are spent annually to test various cultivars, fertilizers, insecticides, and so on. For instance, Figure 1.6 shows a soybean yield trial to determine which cultivars perform best. The objective of yield-trial research is, of course, to increase crop yields.

From studying the philosophy and method of science, but not from reading the agricultural literature, I came to realize that a parsimonious model could provide a more accurate picture than could its raw data. So I tried that concept on yield-trial data and found that the resulting gain in accuracy could be assessed empirically and exactly by data splitting using replicated data (Gauch 1988). It worked. The parsimonious model, which required but a few seconds of computer time, typically produced findings as accurate as would have been achieved using averages over replications based on two to five times as much data. Such additional data would have cost tens to hundreds of thousands of dollars, in various instances, so those gains in accuracy and efficiency were spectacularly cost-effective. Furthermore, statistical theory was able to explain that surprising phenomenon, which was demonstrated repeatedly for many crops in diverse locations and agroecosystems.

Accordingly, I submitted a manuscript to a prestigious statistics journal. One reviewer flatly rejected my manuscript, complaining that my results were "magical" and too good to be true – the ideas involving parsimony, one of the principles of scientific method, were just too unfamiliar. But fortunately the editor understood my work better and accepted the paper. Subsequently I published a paper in *American Scientist* that provided a broad philosophical and scientific perspective for understanding the relationship between parsimony and accuracy (Gauch 1993). Meanwhile, the groundbreaking idea (within the agricultural literature) that parsimonious models could increase accuracy and efficiency has now become rather common, and it has made no small contribution to yield increases for many crops in many nations.

The salient feature of that story is that the requisite parsimonious models and computers had been available to agronomists and breeders for a couple of decades, but no one had capitalized on that opportunity until 1988. What has been the opportunity cost? Standard practices in agricultural research today are increasing the yields for most of the world's major crops by about 0.5% to 1.5% per year. An exact projection is impossible, but a conservative estimate is that parsimonious models of yield-trial data often can support an additional increment of about 0.4% per year (Gauch 1992:184–185).

In other words, for a typical case, if ordinary data analysis supports an average annual yield increase of 1%, whereas a parsimonious model supports 1.4%, then something like 30% of the information in the data is wasted when researchers fail to put parsimony to work. As will be reviewed in Chapter 8, over the past several years there has been compelling evidence from numerous plant-breeding projects that parsimonious models can routinely support that additional yield increment of 0.4% per year.

If an additional increment of 0.4% per year had been achieved after that window of opportunity was opened around 1970, then today's crop yields would be 12.7% higher. But to be conservative, suppose that just half of that advantage had been transferrable from research plots to farmers' fields. Even then, putting parsimony to work could have increased crop yields by 6% over the past 30 years. That may not seem like much, but given that the world's population today is about 6 billion people, that 6% increase would feed 360 million people, more than the population of North America.

But tragically, that window of opportunity from 1970 to 2000 has come and gone. Thus far, only a fraction of agricultural researchers have learned to use parsimonious models to gain accuracy, and even that limited application did not take place until the last of those three decades. The resultant loss of 6% is now irretrievable, because breeding is an incremental process. Each year's efforts begin where the last year left off as breeding stocks are gradually improved. Even though parsimony can be put to work in greater measure in the future, that does not change the historical fact that the ongoing process of plant breeding already has built into it a 6% opportunity cost, caused by neglecting parsimony from 1970 to 2000, and that loss can never be erased. Opportunities come and go; they do not linger forever.

What was missing? What caused that now permanent 6% reduction in crop productivity? It was not the lack of specialized techniques. Nor was it inability to easily perform billions of arithmetic steps. Nor was it lack of funding. It was lack of understanding, or, better yet, mastery, of parsimony, one of the principles of scientific method. What was missing was the last of the critical resources listed in Figure 1.3.

Needless to say, during the past three decades, countless additional measures could have been taken to strengthen agricultural research and thereby to boost farm productivity. The lost 6% could have been regained in many different ways. What is so remarkable about that particular lost opportunity to exercise parsimony, however, is how cost-effective it would have been. Besides having low cost, it also would have involved low risk, because that approach to data analysis had already been tried and proved, unlike many other possible measures that were unproved and risky. Nor would any new or different or expensive requirements have been imposed on data-collection procedures. Rather, the missed opportunity was failure to apply parsimonious models to extract more accurate and more useful information from data already in hand using computers already in place.

That loss is analogous to buying a bag of oranges and then squeezing out only half of the juice. Such waste doesn't make sense. Regardless of which particular opportunities could have been used to change agricultural research for the better, one factor remains the same: Whatever data are obtained, it makes sense to extract all or nearly all of the useful information from the data. Getting only half of the juice is deplorable. And the only way to get all or nearly all of the juice is to master not only research techniques but also general principles.

The bottom line is that for lack of mastering the principles of scientific method, crop yields worldwide are now considerably less (about 6% less) than they could have been had the value of parsimony been appreciated three decades earlier. The principles of scientific method matter.

The larger issue that this example raises is that many other scientific and technological specialties present us with tremendous opportunities that cannot be realized until some specialist in a given discipline masters and applies a critical general principle, be it parsimony or another principle in a given instance. Precisely because these are *general* principles, my suspicion is that my own experience is representative of what can be encountered in countless specialties, as the diverse case studies in Chapter 9 will clearly indicate.

Finally, these reflections on my own experience have focused thus far on only one of the two proposed benefits, namely, productivity. On balance, something might be added about the other benefit, namely, the perspective gained from a humanities-rich perception of science. My own experience resonates with the AAAS (1990:xi) expectation that broad experience of science as a liberal art is worthwhile for "the sheer pleasure and intellectual satisfaction of understanding science."

Like the graduate student mentioned in this book's Foreword, I also had a restless curiosity and deep interest regarding the basic principles of scientific thinking. And also like her, that spark of curiosity had received no stimulus or encouragement whatsoever from the courses and ideas presented in my university education. Nevertheless, such curiosity is normal and common, as Aristotle observed in the opening words of his *Metaphysics*: "All men by nature desire to know" (McKeon 1941:689). More recently, the AAAS (1990:xi) has reaffirmed Aristotle's observation that there is great satisfaction and pleasure in "the human desire to know and understand."

In a campus bookstore, I stumbled across a book by Burks (1977) not long after it was first published in 1963. Arthur Burks was a professor of both philosophy and computer science. His book was quite long, about 700 pages, and frequently was rather repetitious and tedious. But it had the content that I had been seeking and had not yet found anywhere else. There at last I had found an intellectually satisfying account of the underlying principles and rationality of scientific thinking. That book immediately became a great favorite of mine. Subsequently I sought and occasionally found additional books to nourish my continuing interest in the principles of scientific method, most notably that by Jeffreys (1983), which was first published in 1961.

Thus my interest in science's principles dates to about 1965. For the next two decades, my primary motivation for that interest was – to echo the AAAS – the "sheer pleasure" that accompanies "the human desire to know and understand." Grasping the big ideas that are woven throughout the fabric of the entire scientific enterprise generates delight and confidence.

Because of youth and bad company, however, the idea that mastery of those principles could also promote productivity was an idea that would slumber in my mind for a couple of decades! It was not until 1982 that some scattered thoughts began to be reawakened and to coalesce (Gauch 1982), eventually resulting in the *Biometrics* article mentioned earlier (Gauch 1988). Since then, I have been keenly aware that the principles of scientific method can enhance not only perspective but also productivity. Whether at present my interest in these principles is motivated more by a desire for intellectual perspective or for scientific productivity I am not able to say.

SUMMARY

This book takes as its subject matter the general principles of scientific method that pervade all of the sciences, as contrasted with specialized techniques that occur only in some sciences. These basic principles include hypothesis generation and testing, deductive and inductive logic, parsimony, and science's presuppositions, domain, and limits.

The primary benefit to be expected from understanding these principles is increased productivity. A secondary benefit will be a humanities-rich version of science that will promote a wider perspective on the scientific enterprise. To obtain these benefits, however, scientists must go beyond the basics of scientific method. They must master the principles of scientific method as an integrated, functioning whole.

The central thesis of this book is that scientific methodology has two components, the general principles of scientific method and the specialized techniques of a chosen specialty, and the winning combination for scientists is strength in both. Neither basic principles nor research techniques can substitute for one another. This winning combination will enhance both productivity and perspective.

On three counts, this thesis merits serious consideration. First, that science has a scientific method with general principles and that these principles can benefit scientists is the official, considered view of the AAAS (and the NAS, NRC, NSF, and other major scientific organizations in the United States, as well as similar entities in numerous other nations). Second, science educators have demonstrated in hundreds of empirical studies, often involving sizable samples and controlled experiments, that learning science's general principles can benefit students and scientists in several specific, quantifiable, important respects. Third, my own research experience, primarily involving agricultural yield-trial experiments, confirms the two expected benefits. Mastery of the principles of scientific method promotes vital scientific productivity and technological progress. In addition, there is intellectual pleasure in gaining a humanities-rich perspective on how scientific thinking works.

CHAPTER TWO

SCIENCE IN PERSPECTIVE

This is the first of four chapters directed mainly at this book's secondary goal of enhancing perspective (the others being Chapters 3, 4, and 10). The particular kind of perspective sought here is intellectual perspective, as contrasted with, say, perspective on science's monetary costs, technological benefits, or sociological role. The focus here is on the traffic of ideas between science and the humanities, especially philosophy and history. In pursing perspective on science, this chapter tackles three rudimentary questions: What does it mean to say that science is a liberal art? What claims does science make for its methods and conclusions? What is the meaning of truth in science, and how has that meaning developed historically?

SCIENCE AS A LIBERAL ART

Is science a liberal art? What would it mean to affirm that it is, or to deny that it is? How has this conception of science waxed and waned over the centuries, and what are its current prospects? To address such issues about science's intellectual identity, a good point of departure is the position of the world's largest scientific organization, the American Association for the Advancement of Science. The official, energetic position of the AAAS (1990:xi) is that "Science is one of the liberal arts and . . . must be taught as one of the liberal arts, which it unquestionably is." Lest there be any confusion, the AAAS clarifies that what is meant here by "science" is the natural sciences, as contrasted with mathematics and engineering (and presumably other applied sciences, such as agriculture and medicine) (AAAS 1990:vii).

The AAAS suggests several practical advantages from placing science within the liberal-arts tradition:

> Without the study of science and its relationships to other domains of knowledge, neither the intrinsic value of liberal education nor the practical

benefits deriving from it can be achieved. Science, like the other liberal arts, contributes to the satisfaction of the human desire to know and understand. Moreover, a liberal education is the most practical education because it develops habits of mind that are essential for the conduct of the examined life. Ideally, a liberal education produces persons who are open-minded and free from provincialism, dogma, preconception, and ideology; conscious of their opinions and judgments; reflective of their actions; and aware of their place in the social and natural worlds. The experience of learning science as a liberal art must be extended to all young people so that they can discover the sheer pleasure and intellectual satisfaction of understanding science. In this way, they will be empowered to participate more fully and fruitfully in their chosen professions and in civic affairs.... Education in science is more than the transmission of factual information: it must provide students with a knowledge base that enables them to educate themselves about the scientific and technological issues of their times; it must provide students with an understanding of the nature of science and its place in society; and it must provide them with an understanding of the methods and processes of scientific inquiry. (AAAS 1990:xi–xii)

A mosaic in a Cornell University chapel beautifully depicts the integration of all learning (Figure 2.1). The Arts and the Sciences are located to the right and left of the central figure, Philosophy, flanked by Truth and Beauty (not shown).

The essence of liberal-arts education specifically in the sciences as described by Matthews is in agreement with the AAAS (1989, 1990):

Contributors to the liberal tradition believe that science taught ... and informed by the history and philosophy of the subject can engender understanding of nature, the appreciation of beauty in both nature and science, and the awareness of ethical issues unveiled by scientific knowledge and created by scientific practice.... The liberal tradition maintains that science education should not just be an education or training *in* science, although of course it must be this, but also an education *about* science. Students educated in science should have an appreciation of scientific methods, their diversity and their limitations. They should have a feeling for methodological issues, such as how scientific theories are evaluated and how competing theories are appraised, and a sense of the interrelated role of experiment, mathematics and religious and philosophical commitment in the development of science.... Students doing and interpreting experiments need to know something of how data relies upon theory, how evidence relates to the support or falsification of hypotheses, how real cases relate to ideal cases in science, and a host of other matters which all involve philosophical or methodological concerns. (Matthews 1994:2–4)

Figure 2.1. The Arts and the Sciences. The Arts are represented by Literature, Architecture, and Music, and the Sciences by Biology, Astronomy, and Physics. These details are from the mosaic *The Realm of Learning*, in Sage Chapel, Cornell University, that was designed by Ella Condie Lamb. (These photographs by Robert Barker of Cornell University Photography are reproduced with his kind permission.)

He offers a specific example: "To teach Boyle's Law without reflection on what 'law' means in science, without considering what constitutes evidence for a law in science, and without attention to who Boyle was, when he lived, and what he did, is to teach in a truncated way. More can be made of the educational moment than merely teaching, or assisting students to discover that for a given gas at a constant temperature, pressure times volume is a constant" (Matthews 1994:3).

Among the many disciplines in the humanities that have vital interactions with science, philosophy is especially prominent:

> Philosophy is not far below the surface in any scientific investigation. At a most basic level any text or scientific discussion will contain terms such as "law," "theory," "model," "explanation," "cause," "truth," "knowledge," "hypothesis," "confirmation," "observation," "evidence," "idealization," "time," "space," "fields," "species." Philosophy begins when students and teachers slow down the science lesson and ask what these terms mean and what the conditions are for their correct use. All of these concepts contribute to, and in part arise from, philosophical deliberation on issues of epistemology and metaphysics: questions about what things can be known and how we can know them, and about what things actually exist in the world and the relations possible between them. (Matthews 1994:87)

A Living Tradition

Despite the AAAS verdict that science unquestionably is one of the liberal arts, that has not been generally and clearly appreciated in recent decades. Furthermore, the liberal arts have a living, changing, developing tradition, so partnering with them is not entirely straightforward. Consequently, turning the AAAS vision into reality will require some effort: "In spite of the importance of science and the ubiquity of its applications, science has not been integrated adequately into the totality of human experience. . . . Understanding science and its influence on society and the natural world will require a vast reform in science education from preschool to university" (AAAS 1990:xi).

What went wrong? To understand the huge discrepancy between the AAAS vision of a humanities-rich science and the current reality of a humanities-poor science, some rudimentary historical perspective is needed. By A.D. 500 the classical liberal arts had already become well codified. Figure 2.2 lists these seven arts, with three arts in the lower division, the trivium, and four arts in the higher division, the quadrivium. Note that several sciences are included among these arts.

Around A.D. 1200, and coincident with the founding of the earliest universities, there was a great influx of new knowledge into western Europe, the "renaissance of the twelfth century" (Haskins 1923:4). From that time until now, the liberal arts have continued to expand, although without universal consensus on precise boundaries. For instance, an early accretion was the addition of geography to geometry. Plainly, at the present time, the AAAS takes such accretions to the original liberal arts to have encompassed all of the natural sciences.

Ironically, it was science's maturation – its development of powerful methods and amazing discoveries – that initiated the rift between the sciences and the humanities: "Only in the seventeenth century, in the course of what

```
┌─────────────────────────────────────┐
│                                     │
│        The Liberal Arts             │
│                                     │
│   Trivium      Quadrivium           │
│                                     │
│   Grammar      Arithmetic           │
│   Logic        Geometry             │
│   Rhetoric     Astronomy            │
│                Music                │
│                                     │
└─────────────────────────────────────┘
```

Figure 2.2. The seven liberal arts of medieval times.

historians were much later to dub 'the scientific revolution,' did achievements in the study of the natural world come to be widely regarded as setting new standards for what could count as genuine knowledge, and therefore the methods employed by the 'natural philosophers' (as they were still termed) enjoyed a special cultural authority" (Stefan Collini, in Snow 1993:x). Gradually the rift widened between the two cultures, the sciences and the humanities.

Around 1850, at about the time when many of the great universities in the United States and elsewhere beyond Europe were being founded, there were two revealing developments: invention of the new word "scientist," and acquisition of a new meaning for the word "science." The term "scientist" was coined in 1834 by members of the British Association for the Advancement of Science to describe students of nature, by analogy with the previously existing term "artist." Subsequently that new word was established securely in 1840 through William Whewell's popular writings. Somewhat later, in the 1860s, the *Oxford English Dictionary* (OED) recognized that "science" had come to have a new meaning as "physical and experimental science, to the exclusion of theological and metaphysical," and the 1987 supplement to the OED remarked that "this is now the dominant sense in ordinary use" (http://dictionary.oed.com). Those new or modified words certified science's coming of age, with its own independent intellectual identity. Increasingly since 1850, science has also had its own institutional identity.

The rift between the sciences and the humanities reached its maximum around the 1920s and 1930s, with logical positivism discounting human factors in science and disdaining philosophy, especially metaphysics (Frank 1957). A turning point came in 1959 with the publication of two books destined to have enormous influence: In 1959 Karl Popper called for a human-sized account of science, with significant philosophical, historical, and sociological content (Popper 1968), and C. P. Snow drew attention to the divide between "the two cultures" and the resulting lamentable intellectual fragmentation (Snow 1993). So the connection between science and the humanities has varied somewhat during the past century, but on the whole there has been a considerable rift between the two cultures.

Two Camps

Certainly the depictions by the AAAS (1989, 1990) of productive inter-actions between science and the other liberal arts are decidedly convivial and promising, but it must be acknowledged that science's recommended partners, the humanities, currently are in a state of tremendous turmoil and controversy. This is evident in recent books by specialists in the liberal-arts tradition (Farnham and Yarmolinsky 1996; Glyer and Weeks 1998).

"Serious scholars disagree over many high-level topics in the history and philosophy of science" (Matthews 1994:8). "Unfortunately, the nature of sci-ence finds its place in the curriculum just when academic wars are erupting over the very subject. It is an interesting postmodern time for educationalists to embrace the history, sociology, and philosophy of science: Each of these subdisciplines is riven with internal dispute over fundamentals" (Matthews 1998b). "If experts disagree about the nature of science, how should we de-cide what to teach students? . . . Unfortunately, there is . . . a lack of consensus about the nature of science. . . . What then is a teacher to do?" (Smith and Scharmann 1999).

With keen insight, Matthews (1994:9) discerns that there are "two broad camps" in the history and philosophy of science (HPS) literature, "those who appeal to HPS to support the teaching of science, and those who appeal to HPS to puncture the perceived arrogance and authority of science." This second camp stresses "the human face of science" and argues for pervasive "skepticism about scientific knowledge claims." Matthews's sensible reaction is to "embrace a number of the positions of the second group: science does have a human, cultural, and historical dimension, it is closely connected with philosophy, interests and values, and its knowledge claims are frequently tentative," and yet, "none of these admissions need lead to skepticism about the cognitive claims of science."

Given the humanities' profound internal controversies, to suggest that science could gain strength by partnering with the humanities might seem like suggesting that a sober person seek support from a staggering drunk! But that would be an unfortunate overreaction. True, there are enough troubles in the humanities that a wanton relationship could weaken science. But much more importantly, there are enough insights and glories in the humanities that a discerning relationship could greatly strengthen science. "This is just to say that science is more complex, and more interesting, than many simple-minded accounts might have us, and science students, believe" (Matthews 1994:9).

Chapter 3 and, to a lesser extent, Chapter 11 will discuss the troubling controversies that this second camp emphasizes. By way of anticipation, the attitude taken there is that disturbing controversies are more likely to lead to clear and correct ideas than is slumbering complacency. So scientists have nothing to fear and everything to gain from some hard thinking prompted by the opposition.

For the present, however, the foregoing rather cheerful and innocent account of science as a liberal art provides a fitting point of departure. Unquestionably and wonderfully, science is a liberal art.

Apart from limited attention given to the second camp, mainly in Chapter 3, this book's primary attention is directed toward contributors from the first camp. That is, our emphasis is on the ideas of those who perceive the potential for a humanities-rich version of science to enhance the productivity and perspective of scientists. The reason for this selectivity is simple: It is not the naysayers who invent the wheel and the silicon chip, discover distant galaxies and atomic structures, and find cures for diseases.

Future Prospects

The heritage of today's scientists, ranging from the time of Aristotle to the present, includes twenty-two centuries of humanities-rich investigations and one century of humanities-poor contributions. To have a dark age of just one in twenty-three centuries is rather good! But it is that one most recent century that has had, for better or for worse, the dominant influence on contemporary scientists.

Looking toward the future, a return to a humanities-rich science seems virtually certain because of the articulate and energetic call from so many leading scientific organizations, as well as the potential benefits that have already been demonstrated by so many science educators. Accordingly, we can expect that the verdict of history will be that humanities-poor science was an aberration that occurred in just one century out of more than two millennia, namely, the twentieth century.

FOUR BOLD CLAIMS

Exactly what do scientists claim for science's methods and results? Or, in different words, just what about science is good and worthy of respect? Imagine that we ask a scientist "Why do you respect science?" The scientist replies "I respect science because it is scientific." Obviously, that reply is empty, circular, and stupid. An interesting reply must point to some merit of science other than merely its being scientific. It must point to some other goods that are more fundamental.

So what is it that scientists say is good about science? This question of the value of science can apply to scientific results in general, but it is more tangible when asked of specific, individual claims. For the sake of concreteness, any little exemplar of scientific thinking will suffice. So envision a scientist declaring that "Table salt is composed of sodium and chlorine." What claims attend this statement?

The answer offered here is that scientists make four principal claims: rationality, truth, objectivity, and realism. Of course, other terms or plaudits could also be voiced to praise science, but basically they would be variations on, and preconditions and implications of, the four basic terms already offered. For example, science could be said to be coherent, but that is a necessary precondition for truth. Also, science could be praised as pragmatic or successful, but that is an implication of truth, because truth works in a way that error does not. The broad interpretation and application of truth pursued here will include its preconditions and implications. So these four terms suffice to praise science.

This section defines four words: rationality, truth, objectivity, and realism. But first a few remarks about the very concept of "definition" itself: There are several kinds of definitions, so, more specifically, this section provides conceptual definitions of these four words. A conceptual definition specifies the meaning of a word. By contrast, an operational definition provides methods for deciding to which instances the word does or does not apply. For example, a conceptual definition of an acid could say that it is a substance that produces free hydrogen ions in water solution, whereas an operational definition could say that an acid in water solution turns litmus paper red or gives a reading below 7 from a pH meter. Both of these kinds of definitions are needed so that a word can apply a specific meaning selectively to appropriate instances.

In contrast to this section's conceptual definitions, much of the remainder of this book on scientific method will comprise, in essence, an operational definition of truth for statements about the physical world. Naturally, the degree of certainty accorded to a given conclusion will vary from case to case, depending on the ambitiousness of the hypotheses, the availability of data, the kind of reasoning employed, and other factors.

In Chapter 5, on deduction, truth appears when we apply rules to axioms in order to derive theorems or perform calculations. So, given the axioms of standard arithmetic, "$2 + 2 = 4$" is an irrefutable, absolutely true statement. On the other hand, in Chapter 7, on induction, science's resource is imperfect data, rather than assumed axioms, so truth claims are only probable. For instance, a diagnostic test may indicate, with a probability of 80%, that a patient has lung cancer (and, unfortunately, subsequent surgery may change that tentative conclusion to a certainty).

So, on an operational or methodological level, scientific truth has diverse appearances, using various resources to generate certainties or else only probabilities. What gives unity to science's diverse strategies and methods for pursuing knowledge, however, is its underlying conceptual definition of truth. The kinds of data and reasoning may differ from one scientific investigation to another, and the degrees of certainty may also differ, but the meaning of truth will not change.

Conceptual and operational definitions must be kept distinct for another reason: The conceptual definition of truth plays the important role of making scientific hypotheses meaningful even before an operational definition or method leads to the collection of data and the drawing of conclusions. For example, the hypothesis that "a carbon atom contains nine protons" is meaningful precisely because it is understood as an attempt at truth, although in this particular case experimental data would result in rejection of that hypothesis.

Another feature of definitions is worth noting. Individual definitions of words are somewhat sterile, because meaning emerges from webs of words (Hayes 1999). For instance, the meaning of "acid" is inseparable from other words such as "water," "solution," "hydrogen," "base," and "salt." Likewise, "truth" and "realism" are related. Not surprisingly, definitions of "truth" often use the word "realism" or "reality," and the reverse also occurs. Consequently, overly precise boundaries between related words are pointless. What merits emphasis here, however, is that although each of science's four claims is a strong claim, it is the simultaneous assertion of all four of these claims that fully expresses science's boldness.

Finally, to ward off insidious complacency, a reply may be offered to the potential objection that scientists already know full well what "truth" and related words mean, so this section is needless. Of course, scientists have the basic idea of truth, which is also found in the realms of common sense, law, and so forth. Accordingly, what this section offers are precising definitions intended to refine the somewhat unreflective although basically correct concepts. Also, the concept of truth that circulates among scientists can be enriched by opening the discourse to the humanities, especially philosophy. Much of what follows is drawn from relevant philosophical literature, which tends not to be as familiar to scientists as is their own literature.

Rationality

Rationality is good reasoning. The traditional concept of rationality in philosophy, which is also singularly appropriate in science, is that reason holds a double office: regulating belief and guiding action. Rational beliefs have appropriate evidence and reasons that support their truth, and rational actions promote what is good. Rational persons seek true beliefs to guide good actions. "Pieces of behaviour, beliefs, arguments, policies, and other exercises of the human mind may all be described as rational. To accept something as rational is to accept it as making sense, as appropriate, or required, or in accordance with some acknowledged goal, such as aiming at truth or aiming at the good" (Blackburn 1994:319).

Of science's four bold claims, rationality is discussed first because it is so integral to this book's topic, the scientific method. Although beliefs, persons, and other things can be the objects of a claim of rationality, the principal target here is method. First and foremost, rationality is needed for science's method. Method precedes and produces results, so claims of rationality for science's conclusions are derivative from more strategic claims of rationality for science's method. Rational methods produce rational beliefs.

A rational-knowledge claim follows this formula: I hold belief X for reasons R with level of confidence C, where inquiry into X is within the domain of competence of method M that accesses the relevant aspects of reality. A rational belief is not an arbitrary guess, but rather is a justified conclusion based on specific reasons and evidence. The first-order belief X is accompanied by a second-order belief that assesses the strength of the evidence and hence the appropriate level of confidence, which may range from low probability to high probability to certainty. Besides supporting X, some of the evidence may also be directed at discrediting various alternative hypotheses. Lastly, the reasons and evidence have meaning and force from a third-order appeal to an appropriate method that accesses the aspects of reality that are relevant for an inquiry into X. For example, the scientific method is directed at physical reality, and its domain of competence includes reaching a confident belief, based on compelling evidence, about the composition of table salt.

This business of giving reasons R for belief X must eventually stop somewhere, however, so not quite all knowledge claims can follow this formula. Rather, some must follow an alternative formula: I hold belief X because of presuppositions P. This is a story, however, that is better deferred to Chapter 4. The important story at present is just that methods underlie reasons, which in turn underlie beliefs.

Reason's double office, of regulating belief and guiding action, means that true belief goes with good action. When belief and action do not agree, which is a moral problem rather than an intellectual problem, the result is insincerity and hypocrisy. When reason is wrongfully demoted to the single office of only regulating belief, thus severing belief from action, the inevitable consequence is sickly beliefs deliberately shielded from reality. That pathology causes persistent errors and tremendous losses.

The traditional opponent of reason was passion, as in Plato's picture of reason as a charioteer commanding unruly passions as the horses. So a rational person is one who sincerely intends to believe the truth, even if occasionally strong desires go against reason's dictates.

The claim to be defended here, that science is rational, should not be misconstrued as the different and imperialistic claim that only science is rational. To the contrary, science is a form of rationality applied to physical objects, and science flourishes best when integrated with additional forms of rationality, including common sense and philosophy. "The method of natural science is not the sole and universal rational way of reaching truth; it is one version of rational method, adapted to a particular set of truths.... By studying science and becoming familiar with that form of rational activity, one is helped to understand rational procedure in general; it becomes easier to grasp the principles of all rational life through practice of one form of it, and so to adapt those principles to other studies and to life in general" (Caldin 1949:134–135).

Likewise, the claim to be defended here, that science is rational, should not be conflated to the different and undefendable claim that science is always beneficial. It is unfair to confuse science's beliefs and actions, or to pit one against the other in order to malign science's rationality, as in deeming that atomic weapons and carcinogenic insecticides count against science's rationality. Obviously, the simple truth is that knowledge of physical reality can be used for good or for ill. Science in the mind is like a stick in the hand: It increases one's ability to work one's will, regardless whether that will is good or bad, or informed or careless. Hence, unwise and even malicious technology can count against science being beneficial, and yet at the same time can count for science being rational in the limited sense of providing the objective truth about physical reality that is required for technological innovations.

Truth

Truth is a property of a statement, namely, that the statement corresponds with reality. This correspondence theory of truth goes back to Aristotle, who wrote that "To say of what is that it is not, or of what is not that it is, is false, while to say of what is that it is, and of what is not that it is not, is true" (McKeon 1941:749). This definition has three components: a statement declaring something about the world, the actual state of the world, and the relationship of correspondence between the statement and the world. For example, if I say "This glass contains orange juice" and the state of affairs is that this glass does contain orange juice, then this statement corresponds with the world and hence it is true. But if I say that it contains orange juice when it does not, or that it does not contain orange juice when it does, then

Figure 2.3. The correspondence concept of truth, with priority of nature over belief. Here the state of nature is a flower with five petals, and the person's belief is that the flower has five petals, so nature and belief correspond, and consequently this brilliant scientist's belief is true. It is the flower's petals, not the scientist's beliefs, that control the right answer. Beliefs corresponding with reality are true. (This drawing by Carl R. Whittaker is reproduced with his kind permission.)

such statements are false. Truth claims may be expressed with various levels of confidence, such as "I am certain that 'Table salt is sodium chloride' is true" or "The doctors believe that 'The tumor is not malignant' with 90% confidence" or "There is a 95% probability that the sample's true mass is within the interval 1,072 ± 14 grams."

The correspondence theory of truth grants reality priority over beliefs: "the facts about the world determine the truth of statements, but the converse is not true," and this asymmetry is nothing less than "a defining feature of truth about objective reality" (Irwin 1988:5; also see Chakrabarti 1995:79). "In claiming that truth is correspondence to the facts, Aristotle accepts a biconditional; it is true that p if and only if p. But he finds the mere biconditional inadequate for the asymmetry and natural priority he finds in the relation of correspondence; this asymmetry is to be captured in causal or explanatory terms" (Irwin 1988:5–6). Again, "Truth is accuracy or representation of an independent world – a world that, while it includes us and our acts of representing it, decides the accuracy of our representations of it and is not constructed by them" (Leplin 1997:29). Figure 2.3 depicts Aristotle's correspondence concept of truth.

For example, a weary chemistry student might blurt out that "Table salt is magnesium sulfate." But this belief will not turn all the world's salt into magnesium sulfate nor change even a gram of salt in the student's hand. Rather, truth is to be regained by correcting a false belief, not by changing an objective reality. Table salt was sodium chloride, is now, and will remain so. Our birth, life, thoughts, and death will leave this matter altogether unaltered. Scientific truth is neither defined nor determined by opinions or votes, but by the objective state of nature. When it comes to scientific truth, reality has all of the votes, and scientists have none. This is the essential humility that fosters the pursuit of scientific truth.

In the correspondence definition of truth, notice that the bearers of truth are statements, not persons. Persons are the bearers of statements, but statements are the bearers of truth. Accordingly, truth is not affected by who does or does not say it. For example, "This glass contains orange juice" is not made true because Mary says it nor made false because Jim says it. This mistake of judging the truth of a statement by its sponsor has an official name in logic, the genetic fallacy. It is also called an *ad hominem* fallacy, appealing to prejudices rather than reason by attacking one's opponent rather than debating the issue.

A claim of truth can be opposed in three ways. The obvious opposite is a claim of falsity, such as that "It is false that 'Table salt is composed of sodium and chlorine.'" Another alternative is a claim of ignorance, that "We do not possess substantial evidence and reasons for asserting either that 'Table salt is composed of sodium and chlorine' or that it is composed of something else." More generally, even if one accepts a truth claim itself, one may still think that it should be offered with a higher or lower level of confidence in view of the strength or weakness of the evidence. Still another alternative is a claim of nonsense, that "It is meaningless nonsense to say that 'Table salt is composed of sodium and chlorine.'"

For better or for worse, philosophers have proposed numerous definitions of truth besides the correspondence theory advocated here. What is valid in those other definitions is best regarded as routine elaboration of the correspondence definition, which alone can serve science as the core concept of truth.

For example, the coherence theory says that truth consists in coherence (agreement) among a set of beliefs. The valid element here is that coherence is crucial. Thus, if I say that "Table salt is sodium chloride" and at the same time also blithely voice the contrary that "Table salt is not sodium chloride," then I lose credit for this first statement because of the incoherence and insincerity caused by the second statement. Likewise, to be either true or false, a statement must at least make sense; "big it run brown" is neither true nor false, but nonsense.

For another example, the pragmatic theory of truth says that truth consists in showing that something works. The valid element here is that truth does have practical value for doing business with reality. Thus, if your doctor puts you on a low-sodium diet, then there is practical value in understanding the truth that table salt is sodium chloride. Truth has both theoretical and practical importance. Accordingly, reason holds the double office of regulating belief and guiding action. The danger here would be to let pragmatic actions replace true beliefs, rather than complement them, in a theory of truth.

As the correspondence theory of truth is elaborated, it is clear that meaningfulness and coherence are necessary preconditions for truth and that truth has practical implications and value. To abandon the correspondence theory and to substitute coherence, pragmatism, or anything else as the core concept or the entirety of truth would be disastrous. If the definition of truth becomes defective or vague, science languishes.

When the correspondence, coherence, pragmatic, and other theories of truth are all considered seriously and respected equally, in practice none of them wins the day. Rather, the only winner would seem to be a "mystification theory" of truth, saying that it is beyond humans to understand or define truth. Is this *your* theory of truth? There is a simple test: Your mother asks this question: "Did you eat the last cookie? Now tell the truth!" If you are capable of answering that question, then someone else may be mystified about what truth is, but you are not. The mystification theory of truth is just bad philosophy.

The definition of truth is one easy little bit of philosophy that scientists must get straight before their enterprise can make meaningful claims. A true statement corresponds with reality. The most characteristic feature of antiscientific and postmodern views is to place the word "truth" in scare quotes or else proudly to avoid the big t-word altogether. Indeed, *every* kind and variety of anti-scientific philosophy has, as an essential part of its machinery, a defective notion of truth that assists in the sad task of rendering truth elusive. Scientists must take warning from the words of Leplin (1997:28) that "All manner of truth-surrogates have been proposed" by some philosophers "as what science *really* aims for," and scientists must reject all substitutes.

Because true statements correspond with objective reality, a theory of truth should be complemented by theories of objectivity and realism. Accordingly, the next two subsections discuss these two related concepts.

Objectivity

In its primary usage, the concept of objectivity often appears in adjectival form as objective belief, objective knowledge, or objective truth. This concept is complex and somewhat subtle, having three interrelated aspects. Objective knowledge is about an object, rather than a subject or knower; it is achievable by the exercise of ordinary endowments common to all humans, so agreement among persons is possible; and it is not subverted and undone by differences between persons in their worldview commitments, at least for nearly all worldviews.

The first of the three interrelated aspects of objectivity is that objective knowledge is about an object. For example, "Table salt is sodium chloride" expresses an objective claim about an object, table salt, while expressing nothing about persons who do or do not hold this belief. Because objective beliefs are about objects themselves, not the persons expressing beliefs, the truth or falsity of an objective belief is determined by the belief's object, such as table salt. This thinking reflects and respects the correspondence theory of truth and its priority of reality over beliefs.

In Aristotle's terms, an objective truth about nature is a truth "known by nature," meaning that it expresses a real feature of the physical world, not just an opinion suited to our cognitive capacities or our questionable theories (Irwin 1988:5). Indeed, "As one physicist remarked, physics is about how atoms appear to atoms," and "in science the ultimate dissenting voice is nature itself, and that is a voice which even an entrenched scientific establishment cannot silence for ever" (O'Hear 1989:229, 215). Science's goal is "observer-independent truths about a world independent of us," and "The truths science attempts to reveal about atoms and the solar system and even about microbes and bacteria would still be true even if human beings had never existed" (O'Hear 1989:231, 6). For instance, nothing that anyone or everyone has believed or does believe or will believe about table salt can affect its chemical composition in the slightest. To think otherwise would be to impute to humans quite extravagant powers. There is an essential scientific humility in understanding our position in the world, that we are observers of a world not created by ourselves.

The second aspect of objectivity is that objective knowledge is achievable by the exercise of ordinary endowments common to all humans, so agreement among persons is possible. Consequently, science's claims are public and verifiable (Bugliarello 1992). For example, a chemist might say that "Table salt is sodium chloride." But the basis for that knowledge is not some privileged position deriving from, say, the chemist's superior race, nationality, or intelligence. Rather, that knowledge arrives through observant eyes and ears, manipulative hands, and a contemplative mind – through ordinary human endowments. Scientists are humans. Indeed, all chemists have come to the same conclusion and report without exception that table salt is sodium chloride. Furthermore, if you wanted to satisfy yourself, you could repeat the relevant experiments to determine table salt's composition. Accordingly, the scientific attitude is not "I am a superior and unique person who alone knows fact X," but rather "I know X, and so can any other human who cares to make the effort required to learn it." There is an essential humility in the understanding that science is public and shared.

The third and final aspect of objectivity is immunity to worldview differences. A major reason why science is respected is that it cuts across political, cultural, and religious divisions.

> The impartiality of nature to our feelings, beliefs, and desires means that the work of testing and developing scientific theories is insensitive to the ideological background of individual scientists.... [Indeed,] science does cut through political ideology, because its theories are about nature, and made true or false by a nonpartisan nature, whatever the race or beliefs of their inventor, and however they conform or fail to conform to political or religious opinion.... There is no such thing as British science, or Catholic science, or Communist science, though there are Britons, Catholics, and Communists who are scientists, and who should, as scientists, be able to communicate fully with each other. (O'Hear 1989:6–7, 2, 8)

There is an essential humility, openness, and generosity of spirit in realizing that not only your own worldview supports science but also most other worldviews allow science to make sense.

Having just emphasized that science rises above worldview divisions, on balance it must also be said that this immunity to worldview differences is substantial and satisfactory, but not total. Although held by only a small minority of the world's population, there are some worldviews that are so deeply skeptical or relativistic that they do not and cannot support anything recognizable as science's ordinary claims. And those worldview commitments have a deeper role and greater influence than any and all of science's evidence. But that is a story better told in Chapter 4, on science's presuppositions. For the present, it suffices to acknowledge that science is for almost everyone, but not quite everyone.

"Objectivity" also has a secondary usage that applies to persons, rather than to beliefs as in its primary usage. When formulating their beliefs, objective persons are willing to allow facts and truth to overrule prejudices and desires. Science "forbids a man to sink into himself and his selfish claims, and shifts the centre of interest from within himself to outside" (Caldin 1949:135–136). Objective inquirers welcome truth.

These primary and secondary senses of objectivity, applied to beliefs and persons respectively, have a strong link. Envision a class of chemistry students performing experiments to determine table salt's composition. There is a shared object of investigation (some table salt) and a shared human nature (allowing all students to participate, although non-humans such as rocks and trees cannot participate); that shared object and shared nature lead to shared evidence, and consequently there will emerge shared knowledge and subjective agreement. Objective truth expresses beliefs "toward which all inquirers of good will are destined to converge" (Allan Megill, in Megill 1994:1). Truth draws objective inquirers.

This link between objective truth and intersubjective agreement is so strong that the former is difficult to defend when the latter fails. For example, if all of those chemistry students reached different conclusions, who could believe that the class had found the truth? A claim of objective truth without public agreement is worrisome.

Furthermore, it must be emphasized that objective knowledge is claimed or possessed by human subjects, for otherwise, unrealistic and undefendable versions of objectivity would emerge. Scientists, as human beings, "must inevitably see the universe from a centre lying within ourselves and speak about it in terms of a human language shaped by the exigencies of human intercourse. Any attempt rigorously to eliminate our human perspective from our picture of the world must lead to absurdity" (Polanyi 1962:3). The untenable alternative has been described disparagingly and aptly as the view from nowhere and as a God's-eye view. Science cannot be implemented simply as a robotic method of gathering facts and reaching infallible conclusions, with complete disregard for the powers and limits of the humans who are the scientists.

Objective knowledge that is shared among numerous persons gives science a convivial social aspect, the scientific community. Scientific life is shared in community: "Articulate systems which foster and satisfy an intellectual passion can survive only with the support of a society which respects the values affirmed by these passions.... [Thus,] our adherence to the truth can be seen to imply our adherence to a society which respects the truth, and which we trust to respect it. Love of truth and of intellectual values in general will ... reappear as the love of the kind of society which fosters these values, and submission to intellectual standards will be seen to imply participation in a society which accepts the cultural obligation to serve these standards" (Polanyi 1962:203).

But having acknowledged the subjective and social aspects of objectivity, a grave pathology develops if subjectivity supplants rather than complements objectivity. For example, Theodore R. Schatzki discussed aspects of objectivity concerning beliefs about objects and attitudes of subjects, here called the primary and secondary senses of objectivity, respectively, but he went on to treat objectivity exclusively as "a property of persons and their communities" (Natter, Schatzki, and Jones 1995:137). Hence, objectivity was shifted away from physical objects and toward human persons; philosophy was replaced by sociology. Why? Because prior commitment to relativistic philosophy has already declared dead any attempt to really get at physical reality. So there was nothing left for objectivity to be about, except for nice attitudes such as "openness to learning and readiness to revise preunderstandings" (p. 145), with deliberate avoidance of any mention of truth or reality or rationality in descriptions of ideal scientific attitudes.

Regrettably, such elevation of the knower over the known demeans the personal aspect of knowing because it leaves scientists with nothing for their beliefs to be about. That outcome illustrates the principle that every excess becomes its own punishment. To destroy either the objective aspect or the personal aspect of knowledge is to destroy both. Any attempt to eliminate physical objects from science's picture of the world and any attempt to eliminate human persons from science's picture of the world must alike lead to absurdity.

Realism

Realism, as regards the physical world, is the philosophical theory that both human thoughts and independent physical objects exist and that human endowments render the physical world substantially intelligible and reliably known. Scientific realism embodies the claim that the scientific method provides rational access to physical reality, generating much objective knowledge. Realistic beliefs correspond with reality. Realistic persons welcome reality.

> We are trying to refer to reality whenever we say what we think exists. Some may wish to talk of God, and others may think matter is the ultimate reality. Nevertheless, we all talk about tables and chairs, cats and rabbits. They exist, and are real, and do not just depend in some way on our thought for their existence.... Man himself is part of reality, and causally interacts with other segments of reality. He can change things, and even sometimes control them. He does not decide what is real and what is not, but he can make up his mind what he thinks real. This is the pursuit of truth. Man's attempt to make true assertions about the self-subsistent world of which he is a part may not always be successful, and may not always prove easy or straightforward. The repudiation of it as a goal would not only destroy science, but would make human intellectual activity totally pointless. (Trigg 1980:200)

Reality does not come in degrees, because something either does or does not exist. Thus, one little potato is fully as real as is the entire universe. It is not as big, not as important, and not as enduring, but it is just as real. Likewise, one little potato that exists fleetingly now is completely real regardless of whatever ultimate reality may be invoked to explain or cause or sustain its existence. Science claims to deal with reality – with real reality – but clearly some humility is in order regarding the extent of science's reach. Scientists can agree that a little potato is real even while there is disagreement, uncertainty, or even ignorance about the deep philosophical or physical explanation of its existence.

Common-sense belief in reality is practically universal. For example, a child may say "I am patting my dog." What does this mean? Manifestly, the philosophical story, too obvious to be elaborated in ordinary discourse, is that the child feels and sees and enjoys the dog by virtue of having hands and eyes and brain in close proximity to the furry quadruped. And science's realism is the same. "The simple and unscientific man's belief in reality is fundamentally the same as that of the scientist" (Max Born, quoted by Nash 1963:29). For example, on the basis of numerous conversations, Rosenthal-Schneider (1980:30) summarized Einstein's view: "Correspondence to the real physical universe, to nature, was for him the essential feature, the only one which would give 'truth-value' to any theory."

The opposite of realism is anti-realism, in any of its many variants. Recall from this subsection's opening definition that realism combines two tenets: the existence of objects and minds, and the intelligibility of objects to minds. Idealism denies the first tenet. It says that only minds exist and that "objects" are just illusions imagined by minds. Constructivism claims that the physical world is a projection of the mind, so we construct rather than discover reality. Instrumentalism denies that external physical objects should be the targets of our truth claims, substituting internal perceptions and thoughts as the material for analysis. Skepticism denies the second tenet. It does not deny that the physical world exists, but it denies that we do have or could have any reliable knowledge about the physical world. Relativism accepts personal truth-for-me but not public truth-for-everyone, so there is no objective and shared knowledge about the world such as the scientific community claims.

Ordinary science is so thoroughly tied to realism that realism's competitors seem to scientists to be somewhat like the philosophical joke expressed well in a little story by Wittgenstein: "I am sitting with a philosopher in the garden; he says again and again 'I know that that's a tree', pointing to a tree that is near us. Someone else arrives and hears this, and I tell him: 'This fellow isn't insane. We are only doing philosophy'" (Anscombe and von Wright 1969:61e). Without realism, ordinary science perishes.

The debate between realists and anti-realists is more philosophical than scientific: "Realism is . . . partly metaphysical and partly empirical; it has implications beyond experience but is testable by experience" (Jarrett Leplin, in Leplin 1984:7). Similarly, "Faith in the lawfulness of nature, the endeavor to find general universally valid laws, and the hope – or even expectation – of approaching 'truth' about 'reality' are grounded in a distinctive personal philosophy which transcends the sphere of all sciences and all scholarly philosophical systems" (Rosenthal-Schneider 1980:26).

The ground of a scientist's realism is the belief that we have informative contact with physical reality, that nature constrains our beliefs. This belief underlies a scientist's pursuit of empirical observations and experimental data. But some intellectuals hold the opposite, as carefully documented by Pinnick (1992). For example, Barnes (1991) boldly declares that "Reality will tolerate alternative descriptions without protest. We may say what we will of it, and it will not disagree. Sociologists of knowledge rightly reject realist epistemologies that *empower* reality." Of course, a scientist may feel that reality has already empowered itself, without our help. Anyway, such a skeptical view leads ultimately to the triumphant declaration that "Realism is dead" (Arthur Fine, in Leplin 1984:83; also see Arthur Fine, in Galison and Stump 1996:231–254).

Accordingly, despite this debate's intense battles over a thousand and one technicalities, the pivotal issues are mercifully few. Chapter 4, on science's presuppositions, argues that the realist and anti-realist debate turns ultimately on the acceptance or rejection of a scrap of rudimentary common sense, and that choice turns ultimately on faith. Faith underlies reason. But again, this section's task is only to define science's four claims, leaving their defense for later.

The full force of science's claims results from the joint assertion of all four: rationality, truth, objectivity, and realism. Science claims to have a rational method that provides humans with objective truth about physical reality. The meanings of science's four claims are summarized in Figure 2.4.

Naturally, science's four claims are deeply interconnected and partially similar. For example, Ellis (1990:1) remarked that "Our theories of truth and reality should be complementary." Brown (1987:205) said that "whatever claim to truth the results of science have derives from the role of objective procedures in science." Likewise, the intent of rationality is to pursue truth and realism. Not surprisingly, there are many books with titles such as "Truth and Objectivity," "Realism and Truth," and "Rationality and Reality." Nevertheless, there is something basic and strategic about the concept of truth, so its history is explored in the next section.

A BRIEF HISTORY OF TRUTH

This history of the conceptions of truth covers twenty-three centuries in about as many pages. Such extreme brevity allows only four stops, each separated by several centuries: Aristotle, around 350 B.C.; Augustine, around A.D. 400; several scholars in the fledgling medieval universities of Paris and Oxford in the 1200s; and philosopher-scientists of the past several centuries. This material focuses specifically on truth about the physical world, that is, scientific truth.

Science's Four Claims

Rationality

Rational methods of inquiry use reason and evidence correctly to achieve substantial and specified success in finding truth, and rational actions use rational and true beliefs to guide good actions.

Truth

True statements correspond with reality:

Correspondence

External Physical World   Internal Mental World
of Objects and Events   of Perceptions and Beliefs

Objectivity

Objective beliefs concern external physical objects; they can be tested and verified so that consensus will emerge among knowledgeable persons; and they do not depend on controversial presuppositions or special worldviews.

Realism

Realism is correspondence of human thoughts with an external and independent reality, including physical objects.

Figure 2.4. Science's claims of rationality, truth, objectivity, and realism.

Before starting, a preliminary word may be said about the importance of history. It is fair to say that for most scientists, their research frontiers are moving so rapidly that almost all relevant work comes from the past several years. Of course, astronomers know of Copernicus, and microbiologists hear of Pasteur. But usually the details are not crucial in current research programs, so scientists rarely emphasize history.

Readers of this book must recognize, however, that our topic here – scientific method – is different from routine scientific research in having a far greater debt to history and benefit from history. Concepts of truth, objectivity, rationality, and method have been around for quite some time. Consequently, great minds from earlier times still offer us diverse perspectives and penetrating insights that can significantly improve our chances of arriving at rich and productive solutions. Also, current thinking and debates about scientific method can be better understood in light of science's intellectual history.

There is an overarching theme regarding scientific method that runs throughout this entire history, and by being alerted to that theme from the outset, a reader is likely to gain twice or thrice as much insight. That overarching theme is the subtle and indecisive struggle over the centuries among empiricism, rationalism, and skepticism, caused by an underlying confusion about how to integrate science's evidence, logic, and presuppositions.

How do evidence, logic, and presuppositions work together to support our scientific conclusions about the physical world? That is the central question that can help readers assemble their own intact understanding out of the bits and pieces offered by others. By taking the best from an intellectual history that has been both brilliant and blundering and by integrating correct pieces into a working whole, contemporary scientists can surpass their predecessors. They can make their own a profound and beneficial understanding of scientific method.

Besides this overarching question, science's meandering intellectual history raises countless intriguing questions about deduction, induction, and other principles of scientific method. However, my own answers to these questions are postponed to later chapters, starting with Chapter 4. The intent is to give readers time and liberty to struggle on their own for a while with science's great questions, with the hope that that will stimulate readers' own thoughts about the way ahead. Only as the scientific community as a whole wrestles with science's most fundamental ideas can real progress be made.

Aristotle

Aristotle (384–322 B.C.) was enormously important in science's early development. He was a student of Plato, who was a student of Socrates, and he became the tutor of Alexander the Great. Aristotle established a school of philosophy called the Lyceum in Athens, Greece. He wrote more than 150 treatises, of which about 30 have survived. A bearded Plato and youthful Aristotle are depicted in Figure 2.5 in the perennial posture of philosophers – debating with each other.

Aristotle defined truth by the one and only definition that fits common sense and benefits science and technology: the correspondence concept of truth. A statement is true if it corresponds with reality; otherwise it is false. This definition of truth is obvious and easy, despite the temptation to think that a philosophically respectable definition must be difficult, mysterious, and elusive. As Adler (1978:151) said, "The question 'What is truth?' is not a

Figure 2.5. Plato and Aristotle in philosophical debate. Plato was the founder of the Academy in Athens. Aristotle was an important contributor to early science and logic and the founder of another school of philosophy, the Lyceum, also in Athens. (This photograph of a relief by Luca Della Robbia is reproduced with kind permission from Art Resource.)

difficult question to answer. After you understand what truth is, the difficult question, as we shall see, is: How can we tell whether a particular statement is true or false?" The problem is how to find truth, not how to define it. Scientists must make short work of defining truth in order to allow for the long work of finding a scientific method that can sort true statements from false statements.

English translations of Aristotle's works use the word "science," from *scientia* in the Latin translations, or ἐπιστήμη in the Greek original, meaning

Figure 2.6. The meaning of *scientia* in the work of Aristotle. The target of a belief or truth claim could be mere appearance or actual reality, and the level of confidence could be mere opinion, moderate probability, or absolute certainty. *Scientia* is true and certain knowledge about reality.

"knowledge." But the current use and original use of this term are rather different. Currently, science points to a particular kind or domain of knowledge, namely, knowledge about the natural world. But originally, *scientia* pointed to a particular quality of knowledge, namely, demonstrative knowledge about reality having absolute certainty of truth. Essentially, *scientia* was certain knowledge resulting from valid deductive logic applied to self-evidently true axioms and unassailable first principles. Because *scientia* was a quality rather than a domain of knowledge, it was sought in every domain, including mathematics, natural science, ethics, and philosophy.

Incidentally, the ancient term corresponding to our current use of "science" was not *scientia*, but rather "natural philosophy," which was that branch of philosophy that studied the natural world. The current view of science as an independent discipline in its own right, rather than as a branch of philosophy, is of relatively recent origin. The best one-word translation of *scientia* into English probably is not "science," but rather "certainty."

The meaning of *scientia* is further clarified in Figure 2.6. *Scientia* is distinctive in two respects. Its target is actual reality, not just the superficial appearance of something. And the level of confidence is absolute certainty, not merely probability or opinion. *Scientia* is true and certain knowledge about reality.

For comparison, the contemporary conception of science is much broader, encompassing all six boxes in this figure. Science offers probable as well as certain knowledge and recognizes that even opinions or hypotheses can constitute real contributions to scientific research and progress. Likewise, sometimes the goal is only to fit a mathematical model to data, without any claim that the terms in the model refer to specific things that really exist.

For Aristotle, knowledge was justified true belief, that is, belief that is true and that is known to be true on the basis of compelling reasons and evidence supplied by a rational method of inquiry. Truth is the fruit of good method.

Aristotle had a deductivist vision of scientific method, at least for a mature or ideal science (Losee 1993:5–15). The implicit golden standard behind that vision was geometry. Ancient philosophers were quite taken with geometry's clear thinking and definitive proofs. Consequently, geometry became the standard of success against which all other kinds of knowledge were judged. Indeed, the motto of Plato's Academy, where Aristotle had been a student, was "Let no one enter here without geometry." To restate that sentiment in contemporary terms, Geometry 101 was the prerequisite for Philosophy 101.

Naturally, despite that ideal of deductive certainty, Aristotle's actual method in the natural sciences featured careful observations of stars, plants, animals, and other objects, as well as inductive generalizations from the data. The axioms that seemed natural and powerful for geometry had, of course, no counterpart in the natural sciences. For example, no self-evident axioms could generate knowledge about a star's location, a plant's flowers, or an animal's teeth. Rather, one had to look at the world to see how things are.

Accordingly, Aristotle devised an inductive-deductive method that used inductions from observations to infer general principles, deductions from those principles to check the principles against further observations, and additional cycles of induction and deduction to continue the advance of knowledge (Losee 1993:6–9, 29–44). So Aristotle's deductivist vision of science served as a stirring ideal of certainty, but not as an adequate method for science. Evidently, the pursuit of truth about the natural world could not rest on philosophers' or mathematicians' axioms and deductions alone, but rather scientific method must also involve observations and inductions. But in a philosophical climate in which deductions were considered unproblematic but observations and inductions were deemed precarious, recourse to observations and inductions raised thorny problems.

Science's reliance on observations raised questions, especially from skeptical philosophers, about the reliability of sense perceptions and the legitimacy of related presuppositions. Consequently, scientists needed to identify, examine, and legitimate science's presuppositions about the reliability of sense perceptions, the existence of the physical world, and so on. That business seemed much more precarious than geometry's safe appeal to several self-evidently true axioms.

Science's presuppositions were also problematic in another way. Especially in the clearer light of hindsight it seemed that presuppositions sometimes became unreasonably influential, thereby blocking available data from leading to true conclusions. For example, the planets had long been believed to travel in circles, despite growing observational evidence to the contrary, because of a philosophical presupposition declaring that revolution in circles is uniquely the perfect motion befitting planets.

Likewise, science's reliance on inductions raised a thorny question: How can the natural sciences be as certain as geometry? That is, how can natural sciences, which depend on shaky observations and uncertain inductions (as well as unproblematic deductions), be as secure as geometry, which depends only on pristine deductions? Inductive reasoning seemed weak compared with deductive reasoning. Indeed, looking ahead, induction was destined for a mighty hard fall that would become one of the most troublesome threats to scientific method.

For philosopher-scientists in antiquity, grounding science (or "natural philosophy" in their terminology) in philosophy was natural – just as familiar and automatic as is the opposite stance now. Philosophy provided science's concept of realism and goal of truth, philosophy's logic empowered scientific thinking, and philosophical presuppositions legitimated science's appeal to sense perceptions and observational data. And Aristotle's philosophy, with its no-nonsense realism and confident expectations, provided an especially supportive environment for science.

But – and this is an enormous caveat – Aristotle's philosophy was not the only offering in Athens, the educational center of the ancient world. There were four main schools: the Academy of Plato, the Lyceum of Aristotle, the Stoic school of Zeno, and the Garden of Epicurus, as depicted in Figure 2.7. There were also several smaller movements, including Pythagoreans, Cynics, Sophists, and Skeptics. The relative influences of those schools waxed and waned over the centuries. The Stoics and Epicureans were especially influential in early Greek philosophy, as was Plato in late antiquity and in the early medieval period, and Aristotle in the late medieval period, when his works were rediscovered, universities were founded, and scientific method flourished.

Such philosophical pluralism, however, posed a serious problem for scientists. Scientists understood clearly the need to ground science in philosophy. But which philosophy? Obviously not skepticism! Even the comparatively tame Platonism was problematic. Platonism diminished the reality or significance of the visible, physical world to an illusion – a derivative, fleeting shadow of the eternal, unreachable "Forms" that, Plato thought, composed true reality. So how could philosophy help science? To say the least, philosophy's pluralism and interminable debates meant that science's pursuit of a philosophical grounding would require a little care and sophistication.

As empires rose and fell, the philosophical pluralism of Athens would be replicated in Alexandria and then Rome. The royal patronage of the Roman emperors was extended uncritically to all major philosophies. Naturally, many thinkers responded by not aligning themselves with any particular school. For example, Cicero (106–43 B.C.) received a pluralistic education under Stoic, Epicurean, and Platonist philosophers, but he concluded that philosophy could not deliver any certainty, but rather could provide only probable opinions at best. Consequently, science's search for a philosophical

Figure 2.7. Ancient Athens with its four schools of philosophy: Plato's Academy, Aristotle's Lyceum, Zeno's Stoic school, and Epicurus's school. (This drawing by Candace H. Smith is reproduced with her kind permission. © Candace H. Smith.)

grounding had to become more delicate, recognizing that many intellectuals had no specific or traditional philosophical stance.

Aristotle gave natural science a tremendous boost. Ironically, his achievements are not often appreciated by contemporary scientists, because his greatest contributions lay in influencing certain raging debates about fundamentals that we now take for granted. It must be emphasized – even though modern readers can scarcely grasp how something so obvious to them could ever have been hotly debated – that Aristotle advanced science enormously and strategically simply by insisting that the physical world is real. Aristotle, Plato's student, rejected his teacher's theory of a dependent status for physical things, claiming rather an autonomous and real existence. "Moreover, the traits that give an individual object its character do not, Aristotle argued, have a prior and separate existence in a world of forms, but belong to the object itself. There is no perfect form of a dog, for example, existing independently and replicated imperfectly in individual dogs, imparting to them their attributes. For Aristotle, there were just individual dogs" (Lindberg 1992:48). For Plato, the distance between appearance and reality was great; for Aristotle, it was small.

Another of Aristotle's immense contributions to science was to improve deductive logic. Aristotle's syllogistic logic was the first branch of mathematics to be based on axioms, predating Euclid's geometry. It is difficult to give a specific and meaningful number, but I would say that Aristotle got 70% of scientific method right. His contribution is impressive, especially for a philosopher-scientist living more than two millennia ago.

The greatest general deficiency of Aristotle's science was confusion about the integration and relative influences of philosophical presuppositions, empirical evidence, and deductive and inductive logic. How do all of these components fit together in a scientific method that can provide humans with considerable truth about the physical world? Aristotle's choice of geometry as the standard of success and truth for the natural sciences amounts to asking deduction to do a job that can be done only by a scientific method that combines presuppositions, observational evidence, deduction, and induction. Aristotle never reconciled and integrated the deductivism in his ideal science and the empiricism in his actual science. Furthermore, the comforting notion that logic and geometry had special, self-evidently true axioms was destined to evaporate two millennia later with the discovery of nonstandard logics and non-Euclidean geometries. Inevitably, the natural sciences could not be just like geometry. The study of physical things and the study of abstract ideas could not proceed by identical methods.

The greatest specific deficiency of Aristotle's science was profound disinterest in manipulating nature to carry out experiments. For Aristotle, genuine science concerned undisturbed nature, rather than dissected plants or manipulated rocks. Regrettably, his predilection to leave nature undisturbed greatly impeded the development of experimental science for a millennium and a half. Even at that early date there was something about rudimentary experimental methods that could have been learned from the trial-and-error procedures that had already been used to improve agriculture and medicine. But for Aristotle, reflection on the practical arts was beneath the dignity of philosophers, so philosophy gained nothing from that prior experience with experimentation in other realms.

Epicurus (341–271 B.C.), a contemporary of Aristotle, developed an interesting scientific method with distinctive emphasis on multiple working hypotheses (Asmis 1984; Chakrabarti 1995:89–91). The brevity of this section, however, does not allow exploration of other ancient philosopher-scientists.

Augustine

Skipping forward seven centuries in this brief history of truth, from Aristotle to Augustine (A.D. 354–430), the standard of truth and grounds of truth had shifted considerably. Augustine is *the* towering intellect of Western civilization, the one and only individual whose influence dominated an entire millennium. He is remembered primarily as a theologian and philosopher – as a church father and saint. Yet his contribution to science was also substantial (Crombie 1962:11–13; Marrone 1983:10–13; Lindberg 1992:150–151, 197–198; Trundle 1999). Augustine's treatise on logic, *Principia dialecticae*, adopted Aristotle's logic (rather than its main competitor, Stoic logic), thereby ensuring great influence for Aristotle in subsequent medieval logic.

A burning issue for Augustine was how to reconcile Greek philosophy and Christian theology, for both were taken quite seriously. Integrating philosophy and theology may seem irrelevant to some contemporary scientists. As will be seen momentarily, however, Augustine's solution for philosophy and theology became the model, centuries later, for the medieval integration of philosophy and science.

Augustine proposed the view that philosophy was the handmaiden of religion, and that became an enduring metaphor for Christian scholars.

> Philosophy, in Augustine's influential view, was to be the handmaiden of religion – not to be stamped out, but to be cultivated, disciplined, and put to use.... Augustine, who did so much to determine medieval attitudes, admonished his readers to set their hearts on the celestial and eternal, rather than the earthly and temporal. Nonetheless, he acknowledged that the temporal could serve the eternal by supplying knowledge about nature that would contribute to the proper interpretation of Scripture and the development of Christian doctrine. In his own works, including his theological works, Augustine displayed a sophisticated knowledge of Greek natural philosophy. (Lindberg 1992:150–151)

Lindberg (1992:151) has nicely summarized the relationship between science and religion in antiquity: "If we compare the early church with a modern research university or the National Science Foundation, the church will prove to have failed abysmally as a supporter of science and natural philosophy. But such a comparison is obviously unfair. If, instead, we compare the support given to the study of nature by the early church with the support available from any other contemporary social institution, it will become apparent that the church was one of the major patrons – perhaps the major patron – of scientific learning."

For Augustine, the foremost standard of rationality and truth was not Euclid's geometry. Rather, it was Christian theology, revealed by God in Holy Scripture. Theology had the benefit of revelation from God, the All-Knowing Knower. Accordingly, theology replaced geometry as queen of the sciences and the standard of truth. But Augustine's view of how humans acquire even ordinary scientific knowledge relied heavily on divine illumination, particularly as set forth in *The Teacher* (King 1995:94–146). His theory of divine illumination "claimed that whatever one held to be true even in knowledge attained naturally – that is to say, without the special intervention of God as in prophecy or in glorification – one knew as such because God's light, the light of Truth, shone upon the mind" (Marrone 1983:5).

With beautiful simplicity and great enthusiasm, Augustine saw that truth is inherently objective, public, communal, and sharable: "We possess in the truth ... what we all may enjoy, equally and in common; in it are no defects or limitations. For truth receives all its lovers without arousing their envy. It is open to all, yet it is always chaste. . . . The food of truth can never be stolen. There is nothing that you can drink of it which I cannot drink too. . . . Whatever you may take from truth and wisdom, they still remain complete for me" (Benjamin and Hackstaff 1964:69). He also extolled our enjoyment of truth: "Men exclaim how happy they are when, with throats parched from the heat, they come to a flowing and healthful spring; or when they are hungry and find a plentiful supper or dinner prepared. Shall we deny that we are happy when we are given the food and drink of truth? We usually hear voices of men declaring that they are happy if they rest among roses and other flowers or if they enjoy fragrant perfumes. What is more fragrant or more pleasant than the breath of truth?" (Benjamin and Hackstaff 1964:67–68).

Augustine is also notable for his book against skepticism, *Against the Academicians* (King 1995:1–93; Curley 1996). Plato's successors in the New Academy, and Cicero more recently, had argued that nothing could be known, and accordingly assent to any proposition should always be withheld. To the contrary, Augustine argued that skepticism was incoherent and that we can possess several kinds of knowledge impervious to skeptical doubts. Augustine defended the general reliability of sense perception. He appealed to common sense by asking if an influential skeptic, Carneades, knew whether he was a man or a bug! Augustine's concerns, however, were more than just intellectual. Augustine saw skepticism as a threat not only to knowledge but also to the wisdom, right action, and happiness that could proceed from secure truth. More specifically, Augustine wrote *Against the Academicians* while on a retreat to prepare for Christian baptism by Bishop Ambrose, and to take that step, he needed to shed his own former despair of finding truth. Curley (1996:2–9) has discussed that book's influence from medieval to modern times. Conventional views of science's method and success continue to be challenged by skepticism and relativism, so Augustine's analysis remains relevant.

Medieval Scholars

Moving forward another eight centuries in this brief history of truth, from Augustine, around A.D. 400, to the beginnings of medieval universities in the 1200s, the standard and grounds of scientific truth faced the most perplexing, exciting, and productive shift in the entire history of the philosophy of science. Some leading figures were Robert Grosseteste (c. 1168–1253) and later William of Ockham (c. 1285–1347) at the university in Oxford, William of Auvergne (c. 1180–1249), Albertus Magnus, or Albert the Great (c. 1200–1280), Henry of Ghent (c. 1217–1293), and Thomas Aquinas (c. 1225–1274) at the university in Paris, and Roger Bacon (c. 1214–1294) and John Duns Scotus (c. 1265–1308) at both universities. The rise of universities happened to coincide with the rediscovery and wide circulation of Aristotle's books and their Arabic commentaries.

The immensely original contribution of those medieval scholars was to ask a new question about science's truth, a question that may seem ordinary now, but it had not previously been asked or answered. Indeed, after Augustine, eight centuries would pass before the question would be asked clearly, and still another century would pass before it would be answered satisfactorily. It can be expressed thus: What human-sized and public method can provide scientists with real truth about the physical world? Here "real truth" means correspondence-with-reality truth, lest there be any confusion with other concepts of truth, such as the coherence or pragmatic concept.

The problem with Aristotle's vision of deductive certainty was that it concerned logical and mathematical ideas, not physical objects. And the problem with Augustine's vision of divine illumination was that it did not provide the detailed, human-sized account of scientific method and discovery called for by this new question. The aforementioned medieval scholars were all Christians, but they were busily interacting with Greek, Jewish, and Muslim scholars, so they wanted to develop an account of science that would be public, respecting all competent scientists. Indeed, "in 1341 new masters of arts at Paris were required to swear that they would teach 'the system of Aristotle and his commentator Averroes, and of the other ancient commentators and expositors of the said Aristotle, except in those cases that are contrary to the faith'" (Lindberg 1992:241).

Marrone (1983:3) has described the intellectual climate and challenge facing medieval scholars: "Philosophers have always puzzled over the nature of truth; indeed, the matter has generally fascinated intellectuals of any sort. Yet there have been times when the need to define truth carefully has loomed more important, or perhaps seemed more problematic, than usual. One such time came among intellectuals in Europe during the later Middle Ages. . . . Scholastics of the thirteenth and fourteenth centuries wanted to know how to identify that true knowledge which any intelligent person could have merely by exercising his or her natural intellectual capabilities."

The diverse answers given by the Scholastics ranged from pessimistic to optimistic verdicts on science. Henry of Ghent, the Somber Doctor, was pessimistic largely because of his extremely high standard for what he would be willing to call real truth, which placed impossible demands on science. He maintained that scientific knowledge (*scientia*) in the strict sense had to fulfill four exacting conditions: "First, it must be certain, i.e., exclusive of deception and doubt; secondly, it must be of a necessary object; thirdly, it must be produced by a cause that is evident to the intellect; fourthly, it must be applied to the object by a syllogistic reasoning process" (Vier 1951:117; also see Marrone 1985:69–92 and Adams 1987:552–571). However, such a high standard of certainty, inherited from Aristotle's definition of science in his *Posterior Analytics*, could never be met by the natural sciences, for they deal with truths about contingent physical things like stars and teeth. Henry's unyielding conditions severely reduced the domain of knowledge.

Other Scholastics responded optimistically to the challenging new question: What human-sized and public method can provide scientists with real truth about the physical world? They skillfully crafted a reinforced scientific method incorporating five great new ideas.

First, despite Aristotle's disinterest in manipulated nature, experimental methods were finally being developed in science, greatly expanding the opportunities to collect the specific data that could be used to discriminate effectively between competing hypotheses. Grosseteste was "the principal figure" in bringing about "a more adequate method of scientific inquiry" by which "medieval scientists were able eventually to outstrip their ancient European and Muslim teachers" (Dales 1973:62). He initiated a productive shift in science's emphasis, away from presuppositions and ancient authorities, and toward empirical evidence, controlled experiments, and mathematical descriptions. He combined the logic from philosophy and the empiricism from practical arts into a new scientific method. In Grosseteste's science, methods clearly precede results; philosophy of science precedes science. "He stands out from his contemporaries ... because he, before anyone else, was able to see that the major problems to be investigated, if science was to progress, were those of scientific method. ... He seems first to have worked out a methodology applicable to the physical world, and then to have applied it in the particular sciences" (A. C. Crombie, in Callus 1955:99, 101).

In scientific works written between 1220 and 1235, Grosseteste developed his scientific method and elaborated its philosophical import in commentaries on Aristotle's *Physics* and *Posterior Analytics* and in *On Lines, Angles and Figures* and *The Nature of Places*. His scientific findings regarding stars, comets, tides, heat, sound, color, the rainbow, and so on, flow from his scientific methodology. His thinking influenced Roger Bacon, who spread Grosseteste's ideas from Oxford to the University of Paris during a visit there in the 1240s. From the prestigious universities in Oxford and Paris, the new experimental science spread rapidly throughout the medieval universities: "And so it went to Galileo, William Gilbert, Francis Bacon, William Harvey, Descartes, Robert Hooke, Newton, Leibniz, and the world of the seventeenth century" (Crombie 1962:15). So it went to us also.

Roger Bacon, the Admirable Doctor, expressed the heart of the new experimental science in terms of three great prerogatives. The first prerogative of experimental science was that conclusions reached by induction should be submitted to further experimental testing; the second prerogative was that experimental facts had priority over any initial presuppositions or reasons and could augment the factual basis of science; the third prerogative was that scientific research could be extended to entirely new problems, many with practical value. The Admirable Doctor conducted numerous experiments in optics. He was eloquent about science's power to benefit humanity.

Second, an army of brilliant medieval logicians greatly extended the deductive and inductive logic needed by science. That stronger logic, combined with the richer data coming from new experiments with manipulated objects, as well as traditional observations of unmanipulated nature, brought data to bear on theory choices with new rigor and power. The details of those advances in logic, however, are better deferred to Chapters 5 and 7, on deductive and inductive logic.

Third, medieval philosopher-scientists enriched science's criteria for choosing a theory. The most obvious criterion is that a theory must fit the data. Ordinarily a theory is in trouble if it predicts or explains one thing but something else is observed. But awareness was growing that theories also had to satisfy additional criteria, such as parsimony.

William of Ockham, the Venerable Inceptor, is probably the medieval philosopher who is best known to contemporary scientists through the familiar principle of parsimony, often called "Ockham's razor" (Sober 1975;

Jefferys and Berger 1992; Gauch 1993). From Aristotle to Grosseteste, philosopher-scientists had valued parsimony, but Ockham advanced the discussion considerably. In essence, Ockham's razor advises scientists to prefer the simplest theory among those that fit the data equally well. Ockham's rejection, on grounds of parsimony, of Aristotle's theory of impetus paved the way for Newton's theory of inertia. Chapter 8 will explore parsimony in greater detail.

Fourth, science's presuppositions were handled with exquisite finesse. Chapter 4, on presuppositions, will look carefully at some issues introduced here only briefly.

Albertus Magnus, the Universal Doctor, gave Aristotle the most painstaking attention yet, writing more than 8,000 pages of commentary. He further developed Aristotle's suppositional reasoning. Following Albertus's example, the statement that "You are sitting" has suppositional truth given that "I see you sitting," given the presupposition of business as usual between us and the world (William A. Wallace, in Weisheipl 1980:116). In other words, "seeing is believing" is the ordinary presupposition of science, so seeing someone sitting justifies the belief that such is the truth.

But Albertus realized that the connection between perception and truth was not a self-evident or guaranteed principle, but rather was potentially subject to various philosophical or theological worries. The worry that comes to mind first for contemporary thinkers is skeptical doubt. But the Universal Doctor and his contemporaries had in mind a different worry, namely, that to respect God's sovereign freedom, one had to admit the possibility that the apparent perception of a person sitting might not be caused by sensory perceptions of physical objects as usual, but might instead be a vision impressed on the mind directly by God without using any physical processes or objects. At any rate, what was so brilliant about suppositional reasoning was that it admitted that science had been in need of presuppositions, and yet it granted those presuppositions in the context of science's business as usual.

Furthermore, suppositional reasoning provides partial and yet substantial common ground for all scientists, regardless whether an individual's worldview is Christianity, Islam, naturalism, or something else. Scientists holding different worldviews think of different philosophical and theological worries about science. They also think of different deep answers to those worries. For Albertus, the world was comprehensible to humans because the same good God made both the physical world and those humans, with their senses and minds suitable for comprehending the world around them. Confidence that nature was open to human investigation tremendously encouraged the growth of science in the great universities of Western Christianity. For medieval Christian scholars, Augustine's divine illumination was the deep answer behind science's success in finding much real truth. But other

scientists with other worldviews offered different deep answers. Nevertheless, all scientists could agree that given the common-sense presupposition
of business as usual between us and the world, science made sense.

Hence, suppositional reasoning became a device for demarcating a
human-sized and public science apart from philosophical differences and
theological debates. Albertus "proposed to distinguish between philosophy
[including natural philosophy] and theology on methodological grounds, and
to find out what philosophy [or science] alone, without any help from theology, could demonstrate about reality.... He acknowledged (with every other
medieval thinker) that God is ultimately the cause of everything, but he
argued that God customarily works through natural causes and that the natural philosopher's obligation was to take the latter to their limit" (Lindberg
1992:231).

Thomas Aquinas, the Angelic Doctor, was an enormously influential student of Albertus Magnus, and he accepted his teacher's view of science.
Aquinas's support alone would have been sufficient to ensure widespread acceptance in all medieval universities of Albertus's suppositional reasoning for
legitimating and demarcating science. Although primarily a theologian, the
Angelic Doctor also wrote extensive commentaries on several of Aristotle's
books, including the *Physics*.

Aquinas further refined Augustine's handmaiden formula for philosophy,
which was also developing into a formula for science. Aristotelian philosophy and Christian theology, though methodologically distinct, with reliance
on reason and revelation, were regarded as compatible roads to truth that
sometimes led to different truths but never to contradictory truths:

> This may seem like a simple reassertion of the Augustinian handmaiden
> formula, but in fact Thomas has subtly but significantly altered its content.
> The handmaiden named "philosophy" is still subordinate to the theological
> enterprise, and therefore still a handmaiden; but in Thomas's view, she has
> amply demonstrated her usefulness and her reliability, and he therefore
> offers her enlarged responsibilities and elevated status.... Philosophy and
> theology both have their spheres of competence, he argues, and each can be
> trusted within its proper sphere.... Despite the methodological differences
> between philosophy and theology, there are regions where they overlap.
> For example, the existence of the Creator is known both by reason and by
> revelation.... What rules govern the relations of theology and philosophy
> in such cases? The fundamental principle is that there can be no true conflict
> between theology and philosophy, since both revelation and our rational
> capacities are God-given. Any conflict must, therefore, be apparent rather
> than real – the result of bad philosophy or bad theology. The remedy in such
> cases is to reconsider both the philosophical and the theological argument.
> (Lindberg 1992:232–233)

Contrary to that formula, other thinkers have sought to avoid conflicts between science and religion by saying that they have different kinds or dimensions of truth. But intellectual heirs of the Angelic Doctor would say that science and religion have different methods for finding truth, but not different kinds of truth, so reason and faith together pursue a unity of truth. As regards science more specifically, "that God could have created any world he wished led fourteenth-century natural philosophers to perceive that the only way to discover which one he did create is to go out and look – that is, to develop an empirical natural philosophy, which helped to usher in modern science" (Lindberg 1992:242–243). So theological understanding of creation's contingency did motivate scientific methods of empirical observation, though, as Lindberg explains, that development was complex and slow.

John Duns Scotus, the Subtle Doctor, also supported Albertus's suppositional reasoning. He developed a distinctive view of science that sounds like greater emphasis on empirical evidence, but such a reading of the Subtle Doctor would be superficial if that were seen as the entirety of his thrust. His conception of what counted as truth and his sophisticated handling of science's presuppositions were also pivotal. His thinking encouraged the growing bracketing of science apart from philosophy and theology that had been precipitated earlier by Albertus, Aquinas, and Grosseteste.

Duns Scotus placed great emphasis on empirical evidence: "Scotus' keen interest in the history of thought brings him face to face with the scepticism of the Academicians on the one hand, and with the illuminationism of the Augustinians on the other. Both of these schools agree in denying that man is able to obtain true knowledge with his natural faculties only. While the former despair altogether of arriving at any truth, the latter try to safeguard true knowledge by an appeal to some sort of illumination from on high. Scotus' teaching on evidence as a natural and absolutely secure criterion of truth represents a reaction against both of these extreme positions" (Vier 1951:ix; also see Curley 1996:2–3).

Duns Scotus held that certainty is attainable in three cases: "In general he held that certain knowledge was possible of three types of 'knowable things': first, self-evident first principles such as the laws of identity, contradiction, and excluded middle or the statement that the (finite) whole is greater than the part; secondly, sensory experience, which he held could not be deceived, though judgements about sensed things might be false, as when a stick appeared bent when seen with its lower end dipping into water; and thirdly, consciousness of personal actions and states of mind" (Crombie 1962:169).

Would it be right to question the reliability of knowledge merely on the grounds that it concerns contingent physical objects, or to deny the scientific character of such knowledge? In contrast to Henry of Ghent, Duns Scotus does not think so: "'The perfection of scientific knowledge,' he says, 'consists in its being certain and evident.' He does not hesitate in dispensing with objective necessity as long as knowledge is based on objective evidence.... Scotus defends the reliability and sufficiency of sense perceptions as a source of evident certitude.... Scotus, therefore, has full confidence that under the influence of objective evidence the powers of the human mind are perfectly sufficient to arrive at a certain, and in many cases, infallible knowledge of contingent things and truths" (Vier 1951:118–119).

Fifth and finally, medieval philosopher-scientists adopted a conception of scientific truth that was more broad, fitting, and attainable than Aristotle's stringent *scientia*. Ockham "made a distinction between the science of real entities (*scientia realis*), which was concerned with what was known by experience to exist and in which names stood for things existing in nature, and the science of logical entities (*scientia rationalis*), which was concerned with logical constructions and in which names stood merely for concepts" (Crombie 1962:172). Thus, the natural sciences had quit trying to be just like geometry. Likewise, Aquinas acknowledged unqualified or paradigmatic *scientia* in logic and geometry, whereas the natural sciences still could achieve only a partial or secondary *scientia*.

Aristotle and Albertus had accepted propositions as real knowledge even if they had been only generalizations or probabilistic, and Aquinas's further approval of such less-than-certain knowledge, which he called "for-the-most-part truths," made such knowledge fully respectable. Aquinas "even allows that generalizations and probabilistic propositions can be the object of *scientia* despite the fact that they are not, strictly speaking, universal and necessary. He holds that we can have demonstrations of for-the-most-part truths that begin from premises that are also for-the-most-part truths" (Scott MacDonald, in Kretzmann and Stump 1993:177).

Thus Aquinas had an extended and resilient *scientia* that provided a workable version of truth for the physical sciences. That medieval version of *scientia*, unlike the *scientia* of antiquity, is commensurate with the ordinary usages of "truth" by scientists today.

Those five great ideas account for much of the medieval reinforcement of scientific method that vitalized science's pursuit of truth and claims of truth. Of course, much more could be said. For example, Grosseteste and others developed quantitative, mathematical methods, whereas Aristotle's science was mostly qualitative. Likewise, the adoption of Arabic numerals in the universities around 1250 greatly expedited calculations. Also, more attention might be given to the growing realization, articulated so eloquently by Roger Bacon, that science not only was of philosophical interest but also offered tremendous practical benefits. But the most important observation is that the main deficiencies of Aristotelian science were remedied in the thirteenth century. Experiments with manipulated objects were seen to provide relevant data with which to test hypotheses. Also, a workable integration of presuppositions, evidence, and logic emerged that endowed scientific method with accessible truth. Medieval philosopher-scientists also demarcated science apart from philosophy and theology, thereby granting science substantial intellectual and even institutional independence.

The thirteenth century began with a scientific method that lacked experimental methods and lacked an approach to truth that applied naturally to physical things. It concluded with an essentially complete scientific method with a workable notion of truth. Because of Robert Grosseteste at Oxford, Albertus Magnus at Paris, and other medieval scholars, it was the golden age of scientific method. Never before or since that century have the philosophy and method of science been advanced so greatly. The long struggle of sixteen centuries, from Aristotle to Aquinas, had succeeded at last in producing an articulated and workable scientific method with a viable conception of truth. Science had come of age. Since then, both the defenses of science and the attacks on science have largely had the character of embellishments to a basically fixed superstructure of old ideas and ongoing debates.

Modern Scholars

Skipping forward a final time in this brief history of truth, science's method and concept of truth have been developed further in modern times (from 1500 to the present). Developments since about 1960, which have been the primary determinants of the current scene, will be taken up in detail in the next chapter.

The development of increasingly powerful scientific instruments has been a prominent feature of scientific method during the modern era. An influential early example was the observatory of Tycho Brahe (1546–1601), with an unprecedented accuracy of four minutes of arc, about the limit possible without a telescope. Galileo Galilei (1564–1642) constructed an early telescope and invented the first thermometer. He carefully estimated measurement errors and took them into account when fitting models to his data. Blaise Pascal (1623–1662) invented the barometer and an early calculating machine.

Mathematical tools were also advanced. John of St. Thomas (John Poinsot, 1589–1644) developed a material logic applicable to truths about material objects, in contrast to the more familiar formal logic pertaining only to abstract ideas, so that the natural sciences, as well as geometry, could have an adequate logic. Pascal, Pierre de Fermat (1601–1665), Jacob Bernoulli (1654–1705), Thomas Bayes (1701–1761), and others developed probability theory and elementary statistics. Isaac Newton (1642–1727) and Gottfried Leibniz (1646–1716) invented calculus. Thomas Reid (1710–1796) invented the first non-Euclidean geometry in 1764. That discovery, that Euclid's axioms are not uniquely self-evident and true, further eroded the ancient reputation of geometry as the paradigmatic science. Although syllogistic logic was axiomatized by Aristotle, and geometry by Euclid, about twenty-three centuries ago, arithmetic was first axiomatized by Giuseppe Peano (1858–1932) a mere one century ago.

The application of science to the furtherance of mankind's estate was popularized by Francis Bacon (1561–1626) with his enduring slogan, "Knowledge is power." His attempt to win financial support for science from the English crown failed in his own lifetime, but bore fruit shortly thereafter.

Newton continued Aquinas's broad perspective on truth in science, in contrast to Aristotle's narrow vision. Newton believed that science could make valid assertions about unobservable entities and properties. For example, from the hardness of observable objects, one could infer the hardness of their constituent particles that were too small to be observed. Likewise, on the basis of observing inertia inherent in all bodies used in one's experiments, one could properly infer that inertia was a property of all bodies, including those not actually observed. He also believed that science should generally trust induction: "In experimental philosophy we are to look upon propositions inferred by general induction from phenomena as accurately or very nearly true, notwithstanding any contrary hypotheses that may be imagined, till such time as other phenomena occur, by which they may either be made more accurate, or liable to exceptions" (Cajori 1947:400). Also, Newton insisted, contrary to Leibniz, that science could claim legitimate knowledge even in the absence of deep explanation. Thus, the observed inverse-square law applying to gravitational attraction counted as real knowledge, even without any deep understanding or explanation of the nature or cause of gravity.

Newton's view of scientific method, which has influenced modern science so strongly, corresponded with that of Grosseteste: "Of his 'Rules of Reasoning in Philosophy' the first, second, and fourth were, respectively, the well-established principles of economy [parsimony], uniformity, and experimental verification and falsification, and the third was a derivative of these three. And when he came to describe his method in full, he described precisely the double procedure that had been worked out since Grosseteste in the thirteenth century," namely, induction of generalities from numerous observations, and deduction of specific predictions from generalities (Crombie 1962:317). "We reach the conclusion that despite the enormous increase in power that the new mathematics brought in the seventeenth century, the logical structure and problems of experimental science had remained basically the same since the beginning of its modern history some four centuries earlier" (Crombie 1962:318).

In 1562, the French scholar Henri Etienne (1531–1598) first printed, in Latin translation, the *Outlines of Pyrrhonism*, by the ancient skeptic Sextus Empiricus (fl. A.D. 150). "It was the rediscovery of Sextus and of Greek scepticism which shaped the course of philosophy for the next three hundred years" (Annas and Barnes 1985:5). That the preceding millennium had struggled rather little with skepticism may have been due to the perception that Augustine's refutation sufficed. Anyway, one does find René Descartes (1596–1650), John Locke (1632–1704), David Hume (1711–1776), Immanuel Kant (1724–1804), and other modern thinkers struggling mightily with Sextus's challenges. Even a contemporary philosopher has introduced his book on science in these terms: "Immodestly stated, my purpose is to succeed where Descartes failed: to submit our knowledge of the external world to an ordeal by scepticism and then, with the help of the little that survives, to explain how scientific rationality is still possible" (Watkins 1984:xi).

"In his *Outlines of Pyrrhonism* Sextus defends the conclusions of Pyrrhonian scepticism, that our faculties are such that we ought to suspend judgement on all matters of reality and content ourselves with appearances" (Woolhouse 1988:4). The skeptics' opponents, to use their own term, were the "dogmatists" who believed that truth was attainable. Sextus observed that two criteria for discovering truth were offered: reason by the rationalists, and the senses by the empiricists. He argued that neither reason nor sense perception could guarantee truth. To a considerable extent, the philosophies of Descartes and Leibniz can be understood as attempts to make reason work despite Sextus's skeptical criticisms. Likewise, the philosophies of Francis Bacon and John Locke attempt to make sense perception and empirical data work despite the ordeal by skepticism. So, despite their profound differences, rationalism and empiricism had in common the same implicit opponent, skepticism. Chatalian (1991) has argued persuasively that the implicit Greek opponents, Pyrrho of Elis (c. 360–270 B.C.) and Sextus Empiricus, were often superficially studied and poorly understood. Nevertheless, rationalism and empiricism sought to guard truth from skepticism's attacks.

René Descartes exemplified rationalism, which emphasized philosophical reasoning as the surest source of truth, rather than uncertain observations and risky inductions. Descartes agreed with Francis Bacon that science had both general principles and individual observations, but his progression was the reverse. The empiricist Bacon sought to collect empirical data and then progress inductively to general relations, whereas the rationalist Descartes sought to begin with general philosophical principles and then deduce the details of expected data. To obtain the needed stockpile of indubitable general principles, Descartes's method was to reject the unverified assumptions of ancient authorities and begin with universal doubt, starting afresh with that which is most certain.

His chosen starting point for indubitable truth was his famous *"Cogito ergo sum,"* "I think, therefore I exist." He then moved on to establish the existence of God, whose goodness assured humans that their sense perceptions were not utterly deceptive, so they could conclude that the physical world existed (Cottingham 1986, 1992). There were many layers of arguments before getting to the physical world's existence.

George Berkeley (1685–1753) was an empiricist. The battle cry of empiricists was "back to experience." In essence, "an empiricist will seek to relate the contents of our minds, our knowledge and beliefs, and their acquisition, to sense-based experience and observation. He will hold that experience is the touchstone of truth and meaning, and that we cannot know, or even sensibly speak of, things which go beyond our experience" (Woolhouse 1988:2). Berkeley was also an idealist, believing that only minds and ideas existed, not the physical world.

Berkeley applauded Newton's careful distinction between mathematical axioms and empirical applications, in essence, between ideas and things. But Berkeley was concerned that such a distinction would invite a dreaded skepticism: "Once a distinction is made between our perceptions of material things and those things themselves, 'then are we involved all in *scepticism*'. For it follows from this distinction that we see only the appearances of things, images of them in our minds, not the things themselves, 'so that, for aught we know, all we see, hear, and feel, may be only phantom and vain chimera, and not at all agree with the real things'" (Woolhouse 1988:110). In other words, once one distinguished appearance and reality, or ideas and things, one might never be able to build a secure bridge from appearances to reality. What was the solution? "Faced with the evidently troublesome distinction between things and ideas, Berkeley in effect collapses it; he concludes that *ideas are things*. As he explains, 'Those immediate objects of perception, which according to [some] . . . are only appearances of things, I take to be the real things themselves'" (Woolhouse 1988:113). Ideas and minds were all of reality; there were no such things as physical objects. Accordingly, science's proper goal was to account for the mind's experiences and perceptions, rather than an external physical reality.

Berkeley took the empiricist position to a new extreme, deeming minds and their ideas and experiences to constitute all of reality. Yet in his own estimation, he had countered skepticism: "We are not for having any man turn *sceptic*, and disbelieve his senses; on the contrary we give them all the stress and assurance imaginable; nor are there any principles more opposite to scepticism, than those we have laid down" (Woolhouse 1988:113–114). But his rejection of the physical world, so adamantly commended to us by common sense, was taken by many others to be "scepticism run riot" (Woolhouse 1988:113). So Berkeley slew skepticism and saved truth, but at the cost of dispensing with physical reality, which was too high a price for most scientists.

Incidentally, whereas Berkeley said that minds are real, but not physical things, some other philosophers have taken the opposite tack, saying that only physical things are real: Mental or conscious processes are just illusory epiphenomena caused by neural and chemical processes.

David Hume could be considered an empiricist or else a skeptic. "Among all the philosophers who wrote before the twentieth century none is more important for the philosophy of science than David Hume. This is because Hume is widely recognized to have been the chief philosophical inspiration of the most important twentieth-century school in the philosophy of science – the so-called logical positivists" (Alexander Rosenberg, in Norton 1993:64). Hume admired Francis Bacon and greatly admired Newton, "the greatest and rarest genius that ever rose for the ornament and instruction of the species" (Woolhouse 1988:135). Hume took himself to be discovering a science of man, or principles of human understanding more specifically, that was akin to Newton's science of mechanics in its method and rigor.

The standard reading of Hume is that he carried the empiricism of Locke and Berkeley to its logical and skeptical conclusion. But Norton (1993) has offered a more careful reading. Consider Hume's burning question: What grounds do scientific beliefs have, and accordingly what confidence do they merit? His analysis began with two fundamental moves.

First, Hume insisted that the objective was *human* understanding, so he examined human nature to assess our mental capacities and limitations. "There is no question of importance, whose decision is not compriz'd in the science of man; and there is none, which can be decided with any certainty, before we become acquainted with that science" (John Biro, in Norton 1993:34).

Second, Hume rigorously adopted an empiricist theory of meaning, requiring statements to be grounded in experience, that is, in sense perceptions and ideas based on them. For Hume, the question about a scientific belief was not whether or not it was true, because that question was impossible to answer, but rather was about how humans form that belief: "As to those *impressions*, which arise from the *senses*, their ultimate cause is, in my opinion, perfectly inexplicable by human reason, and 'twill always be impossible to decide with certainty, whether they arise immediately from the object, or are produc'd by the creative power of the mind, or are deriv'd from the author of our being. Nor is such a question any way material to our present purpose. We may draw inferences from the coherence of our perceptions, whether they be true or false; whether they represent nature justly, or be mere illusions of the senses" (David F. Norton, in Norton 1993:6–7).

It is difficult to induce contemporary scientists, who think that rocks and trees are real and knowable, to grasp the earnestness of Hume's empiricism. Hume's empiricist science concerned mental perceptions, not physical things. His concern was with "our perceptions, qua perceptions, with perceptions as, simply, the *elements or objects of the mind* and not as *representations* of external existences" (David F. Norton, in Norton 1993:8). For example, he was concerned with our mental perceptions and ideas of trees, not with trees as external physical objects. Accordingly, to report that "I see a tree" was, for Hume, a philosophical blunder, because this "I see" posits a mental perception, while this "tree" posits a corresponding physical object. He called that blunder the "double existence" (or "representational realism") theory – "the theory that while we experience only impressions and ideas, there is also another set of existences, namely objects" (Alexander Rosenberg, in Norton 1993:69). Of course, earlier thinkers, like Aristotle, had a more flattering name for that theory, the correspondence theory of truth. Anyway, for Hume, the corrected report would read something like "I am being appeared to treely," which skillfully avoids the double existence of perceptions and objects and instead confines itself to the single existence of perceptions.

So although Hume's avowed hero was Newton, their philosophies of science were strikingly different, because Newton's science concerned truth about a knowable physical world. Hume and Newton could agree on the truism that science was done by scientists – by humans. But Hume's "humans" were post-skeptical philosophers, whereas Newton's "humans" were common-sensical scientists. Likewise, Hume's "observations" were strictly mental perceptions, whereas Newton's "observations" were sensory responses corresponding reliably to external physical objects. Hume says, "I am being appeared to treely," but Newton says "I see a tree."

Given those two fundamental moves, Hume's analysis then derived three sweeping results: Cause and effect were meaningless concepts, inductive logic was bankrupt, and all scientific explanations were but obscure and uncertain speculations. Finally, those and related results led to the shocking conclusion that science could not establish any reliable link between internal mental beliefs and an external physical world. Nor could humans even be certain that they existed in the normal sense of a continuing, unified self (until death). Consequently, science's legitimate ambitions and claims had to be reduced to the mere cataloging of observed regular sequences of events in nature.

Just how skeptical was Hume? The once-standard reading was that Hume was a radical skeptic. But closer reading reveals deep ambivalence toward common sense, not a pure or simple skepticism. The accurate answer, even if somewhat disappointing, is that the extent of Hume's skepticism continually shifted, depending on what argument he was currently advancing and which audience he was addressing:

On the one hand, he is anxious to dispute the claims of transcendental metaphysicians and theologians that they possess rationally grounded beliefs. With his eye on such opponents, he argues that it is quite absurd to go in search of remote causes for the Universe when we cannot even give a satisfactory reason for believing that a stone will fall or that the sun will rise tomorrow.... On the other hand, Hume is equally anxious to destroy fanaticism and superstition.... He sometimes suggests, therefore, that a belief is rational provided only that it can be traced back to a constant conjunction; hence the rational justification for believing that the sun will rise tomorrow, as opposed to the irrationality of superstitious beliefs.... It is ridiculous, he says, to declare as only a probability that the sun will rise tomorrow or that all men are mortal. [John Passmore, in Gillispie 1970(6):558–559]

Thomas Reid, quite in contrast to his fellow Scot David Hume, grounded philosophy in an initial appeal to common sense:

Philosophy...has no other root but the principles of Common Sense; it grows out of them, and draws its nourishment from them. Severed from this root, its honours wither, its sap is dried up, it dies and rots....It is a bold philosophy that rejects, without ceremony, principles which irresistibly govern the belief and the conduct of all mankind in the common concerns of life: and to which the philosopher himself must yield, after he imagines he hath confuted them. Such principles [of common sense] are older, and of more authority, than Philosophy: she rests upon them as her basis, not they upon her. (Hamilton 1872:101–102)

Reid avoided the hopeless attempt to make natural science just like geometry by accepting both the deductions of geometry and the reliability of observation. "That there is such a city as Rome, I am as certain as of any proposition in Euclid; but the evidence is not demonstrative, but of that kind which philosophers call probable. Yet, in common language, it would sound oddly to say, it is probable there is such a city as Rome, because it would imply some degree of doubt or uncertainty" (Hamilton 1872:482).

Representing common sense as eyes and philosophy as a telescope, Reid offered the analogy that a telescope can help a man see farther if he has eyes, but will show nothing to a man without eyes (Hamilton 1872:130; Lehrer 1989:50). Accordingly, to the partial skeptic, Reid commended a dose of common sense as the best remedy; but to the total skeptic, Reid had nothing to say. Reid could give a man a telescope, but not eyes. Reid's common-sense philosophy is reviewed further in Chapter 4.

Immanuel Kant devised a new variant of rationalism intended to divert Hume's skepticism and to support a thoroughly subjective, human-sized version of scientific truth. His influential *Critique of Pure Reason* (1781, revised 1787) was followed by a popularization, the *Prolegomena to any Future Metaphysics That Shall Come Forth as Scientific* (1783). He was not happy with his predecessors. Kant rejected Descartes's starting point, "*Cogito ergo sum*," saying that it should instead read "*Cogito ergo ego sum*" or "I think, therefore *I* exist" to make explicit the subject "I" that is carried from the premise to the conclusion, rendering Descartes's argument circular (Carus 1902:169–170). Kant also reacted against Hume, who "ran his ship ashore . . . landing on scepticism, there to let it lie and rot" (Carus 1902:9). He also rejected Berkeley's idealism that would reduce the world to a mere "illusory appearance" (Carus 1902:199).

Nor did Reid fare any better in Kant's estimation, it being "positively painful" to see how badly Reid failed to understand and address Hume, "ever taking for granted that which he [Hume] doubted, and demonstrating with zeal and often with impudence that which he never thought of doubting" (Carus 1902:5). Kant chided Reid's appeal to common sense: "To appeal to common sense, when insight and science fail, and no sooner – this is one of the subtle discoveries of modern times, by means of which the most superficial ranter can safely enter the lists with the most thorough thinker, and hold his own. But as long as a particle of insight remains, no one would think of having recourse to this subterfuge. For what is it but an appeal to the opinion of the multitude, of whose applause the philosopher is ashamed, while the popular charlatan glories and confides in it?" (Carus 1902:6).

Against Descartes, Hume, Berkeley, and Reid and their failed metaphysics, Kant promises us a keen pilot that can steer our metaphysical ship safely. But Kant's thinking is remarkably complex and subtle.

Fortunately, however, the opening pages of his *Prolegomena* lead us quickly into the very heart of enduring themes in his philosophy of science. The centerpiece is his response to Hume's problem of causality. In the entire history of metaphysics, "nothing has ever happened which was more decisive to its fate than the attack made upon it by David Hume," specifically the attack upon "a single but important concept in Metaphysics, viz., that of Cause and Effect" (Carus 1902:3–4). Given that Hume "suffered the usual misfortune of metaphysicians, of not being understood" (Carus 1902:5), Kant is most anxious that we, the students of his *Prolegomena*, understand precisely which problem Hume tackled: "The question was not whether the concept of cause was right, useful, and even indispensable for our knowledge of nature, for this Hume had never doubted; but whether that concept could be thought by reason *a priori*, and consequently whether it possessed an inner truth, independent of all experience, implying a wider application than merely to the objects of experience. This was Hume's problem. It was a question concerning the *origin*, not concerning the *indispensable need* of the concept" (Carus 1902:5).

Before offering his own solution to the problem of causality, Kant made two preliminary points. First, Kant accepted Hume's proof that the concept of causality could not possibly be an *a priori* and necessary product of pure reasoning. Second, Kant greatly enlarged the significance of that problem of causality by generalizing it, showing that all metaphysical concepts underlying science suffered from the same questionable origins, including the concepts of space and time, possibility and actuality, necessity, and substance.

The problem with causality, or any other general law of nature, was that such laws made claims that went beyond any possible empirical support. Empirical evidence for a causal law could only be of the form "All instances of *A* observed in the past were followed by *B*," whereas the law asserted the far grander claim that "All instances of *A*, observed or not and past or future, are followed by *B*." But that extension was inductive, excessive, and uncertain, exceeding its evidence. Consequently, something else had to be added to secure such a law.

Accordingly, Kant's solution combined two resources: a general philosophical principle of causality known by *a priori* reasoning, and specific causal laws discovered by *a posteriori* empirical observation and induction. By that combination, "particular empirical laws or uniformities are subsumed under the *a priori* concept of causality in such a way that they thereby become necessary and acquire a more than merely inductive status" (Michael Friedman, in Guyer 1992:173). For example, "The rule of uniformity according to which illuminated bodies happen to become warm is at first merely empirical and inductive; if it is to count as a genuine law of nature, however, this same empirical uniformity must be subsumed under the *a priori* concept of causality, whereupon it then becomes necessary and strictly universal" (Michael Friedman, in Guyer 1992:173).

"Kant recognizes at least two distinct types of necessity (and thus apriority). The transcendental principles of the understanding are absolutely necessary and *a priori*: they are established entirely independent of all perception and experience. Empirical laws that somehow fall under these transcendental principles are then necessary and *a priori* in a derivative sense. They, unlike the transcendental principles themselves, indeed depend partially on inductively obtained regularities (and thus on perception), yet they are also in some sense grounded in or determined by the transcendental principles and thereby acquire a necessary and more than merely inductive status" (Michael Friedman, in Guyer 1992:174). Thus, a general principle of causality upgraded the evidence for particular causal laws. Incidentally, Aquinas had made some similar moves when legitimating a non-paradigmatic *scientia* suited to our human epistemic situation (Scott MacDonald, in Kretzmann and Stump 1993:174–180).

Additional transcendental principles besides causality received similar analysis, such as space and time. Kant taught that human cognition had two distinct faculties: the intellectual faculty of understanding or reason that represents things in themselves, and the faculty of sensibility that represents underlying reality as it appears to creatures so constituted as ourselves. Our perceptions of space and time would emerge from our faculty of sensibility. But other rational beings would have differently constituted faculties of sensibility and hence would not perceive reality in the same spatiotemporal manner: "Kant's radical division of the human cognitive faculties [of understanding and sensibility] has thus introduced a radically new element of *subjectivity* into the representations of space and time: space and time are '*subjective* and ideal and proceed from the nature of the mind by a constant law, as it were as a schema for coordinating with one another absolutely all external sensibles'" (Friedman 1992:31).

Space and time were appearances constructed in our human faculties, not real features of external objects and events: "Space and time are now based epistemologically on the nature of the mind rather than ontologically on the nature of things. . . . According to Kant, the human mind supplies the form of experience (time, space, and the categories of the understanding); but the content of experience is empirically given in sensation from a source outside the human self (the real material world)" [James W. Ellington, in Gillispie 1970(7):226, 234]. Kant claimed that "our faculty of cognition does not conform to the objects, but contrariwise, that the objects conform to cognition" and that objects "do not in themselves possess form, but our mind is so constituted that it cannot help attributing form and everything formal to the object of our experience" (Carus 1902:204). "Kant radically and irreversibly transformed the nature of Western thought. After he wrote, no one could ever again think of either science or morality as a matter of the passive reception of entirely external truth or reality. In reflection upon the methods of science, as well as in many particular areas of science itself, the recognition of our own input into the world we claim to know has become inescapable" (Paul Guyer, in Guyer 1992:3).

Such thinking was the beginning of constructivist or anti-realist views of science, that truth is constructed by us rather than discovered about nature. In contemporary philosophy of science, "An underlying battle rages between realists who believe that there is a world to be investigated which exists independently of human belief and language, and anti-realists. The latter wish to build our conceptions of truth and reality on the way human language is integrated with a particular way of acting. Truth is constructed and not discovered, they would maintain" (Trigg 1993:6).

However strange or sensible Kant's view of science might sound to a contemporary scientist, in his own estimation it was the perfect reply to Hume's skepticism: "Kant conceded that if the form and content of scientific laws wholly derive from sense experience, as Hume had urged, then there is no escape from Hume's [skeptical] conclusion. However, Kant was unwilling to grant Hume's premiss. Against Hume, he argued that although all empirical knowledge 'arises from' sense impressions, it is not the case that all such knowledge is 'given in' these impressions.... He held that sense impressions provide the raw material of empirical knowledge, but the knowing subject itself is responsible for the structural-relational organization of this raw material" (Losee 1993:112).

That escape from skepticism, however, had been purchased at a price. In Kant's system, "we have actual cognition of things as appearances and a problematic concept of the reality behind the appearance" [James W. Ellington, in Gillispie 1970(7):229]. "Thus Kant took the fateful first step of arguing that the possibility and indeed the certainty of the spatiotemporal framework of Newtonian physics could be secured only by recognizing it to be the form of our own experience, even though this meant that the certainty of the foundations of Newtonian science could be purchased only by confining them to objects as we experience them through the senses – 'appearances' or 'phenomena' – rather than those objects as they might be in themselves and known to be by a pure intellect – 'noumena'" (Paul Guyer, in Guyer 1992:10). Kant's philosophy of science offered knowledge of mental appearances, not knowledge of external realities.

Moving forward about a century after Kant to almost a century ago, the period around 1920 was pivotal for the philosophy and method of science. Although the current scene is one of vigorous debate among several sizable schools, for a few decades following 1920 a single school dominated, logical empiricism (also called logical positivism, just positivism, and the Vienna Circle). Some of the leading members, associates, visitors, and collaborators were A. J. Ayer, Rudolf Carnap, Albert Einstein, Herbert Feigl, Philip Frank, Kurt Gödel, Hans Hahn, C. G. Hempel, Ernest Nagel, Otto Neurath, W. V. Quine, Hans Reichenbach, Moritz Schlick, and Richard von Mises. Karl Popper, who would become the circle's most influential critic, often attended, but was not a member or associate. "Almost all work, foundational or applied, in English-language philosophy of science during the present century has either been produced within the tradition of logical empiricism or has been written in response to it. Indeed it is arguable that philosophy of science as an academic discipline is essentially a creation of logical empiricists and (derivatively) of the philosophical controversies that they sparked" (Richard Boyd, in Boyd, Gasper, and Trout 1991:3).

Looking backward, logical empiricism was a reaction to, or a synthesis of, earlier rationalist and empiricist traditions. Looking forward, it was the previous paradigm drawing serious criticisms from Popper, Kuhn, Lakatos, and Feyerabend. To understand why logical empiricism was so attractive and dominant after 1920, and why it was so criticized and dead by 1960, one may consult Carnap, Hahn, and Neurath (1929), Neurath and Cohen (1973), Schilpp (1974:1–181), Rescher (1985), Boyd et al. (1991:3–35), Trigg (1993), Misak (1995), Laudan (1996), Friedman (1999), and Salmon (1999).

As its apt name suggests, "logical empiricism" combines logic and empiricism. "Logical empiricism arose in the twentieth century as a result of efforts by scientifically inclined philosophers to articulate the insights of traditional empiricism, especially the views of Hume, . . . using newer developments in mathematical logic" (Richard Boyd, in Boyd et al. 1991:5). The central idea was to limit meaningful scientific statements to sensory-experience reports and logical inferences based on those reports. Considered separately, the rationalist tradition with its logic and the empiricist tradition with its sensory experience were deemed inadequate for science (in view of the problems listed earlier), but a clever integration of logic and experience was expected to work. Logical empiricism included three important tenets.

First, examining the "logical" part of logical empiricism, the truth or falsity of an analytic statement was determined solely by the meanings of its words and its logical form. By contrast, the truth or falsity of a synthetic statement was determined not only by its meaning but also by the way the world is. That analytic–synthetic distinction was important for Aquinas and was elaborated by Kant. As an example, "All bachelors are unmarried" is analytic, true by definition; whereas "Some bachelors are rich" is synthetic, depending for its truth or falsity on the way the world is.

That distinction assigned geometry to the analytic realm and the natural sciences to the synthetic realm, thereby ending the ancient effort to make the natural sciences just like geometry. Breathtaking advances in logic by Frege, Russell, Carnap, and others in the 1920s and 1930s unified all of mathematics in a single grand scheme with logic at its base. For instance, arithmetic was considered a part of logic and was treated as an entirely analytic and abstract subject, without any empirical or physical content. Ordinary logic, arithmetic, and Euclidean geometry were recognized to result from just one among infinitely many possible choices of axioms, with other choices leading to nonstandard logics and arithmetics and non-Euclidean geometries. Anyway, the domain of exclusively deductive thinking, exemplified by geometry in ancient and medieval times, had come to be understood to extend throughout all of logic and mathematics.

Second, examining next the "empiricism" part of logical empiricism, meaningful scientific statements were limited to sensory experience and logical inferences based on that experience. Logical empiricists were keen to have a demarcation criterion to separate science from non-science. Their proposal was that legitimate, meaningful scientific knowledge was grounded in the evidence of sensory experience. Evidence for or against the truth of a synthetic statement about the world would be provided solely by empirical observations. A statement about the world that failed to make any claims or predictions about observable phenomena was literally meaningless. Furthermore, if two theories predicted the same observations, then they were empirically equivalent, meaning that there was no basis for preferring one such theory over the other.

Third, presuppositions were not part of logical empiricism. Rather, the primary target of the demarcation criterion was metaphysics – philosophical presuppositions or theories about what exists. Indeed, "the fundamental motivation for logical empiricism" was "the elimination of metaphysics," including "doctrines about the fundamental nature of substances," "theological matters," and "our relation to external objects" (Richard Boyd, in Boyd et al. 1991:6). The problem with metaphysical presuppositions was that they were not analytic truths demonstrable by logic, and neither were they synthetic truths demonstrable by observational data, so for a logical empiricist, such ideas were just meaningless nonsense. Accordingly, science and philosophy parted ways. "The Circle rejected the need for a specifically philosophical epistemology that bestowed justification on knowledge claims from beyond science itself" (Thomas Uebel, in Audi 1999:956).

Clearly, the motivation of logical empiricism was to create a purified, hard-nosed, no-nonsense version of science based on solid data and avoiding philosophical speculation. Yet serious problems emerged that eroded its credibility by 1960.

One problem was that science parted ways not only with philosophy but also with common sense. Even the primitive theory, for instance, that a person's perception of a cat results from the eyes seeing an actual physical cat *is* a metaphysical theory about what exists. Common sense has metaphysical content, so the elimination of metaphysics would also unsettle common sense. "Given such a view" as logical empiricism, "difficult epistemological gaps arise between available evidence and the commonsense conclusions we want to reach about the world around us," including "enormous difficulty explaining how what we know about sensations could confirm for us assertions about an objective physical world" (Richard A. Fumerton, in Audi 1999:515). That loss of common sense and the physical world would be quite a high price for scientists to pay.

Another problem was that theories cannot be tested individually by means of their observational consequences, because theories always function together with auxiliary theories. For example, theories about planetary motions implicate additional theories about space and time. So if a theory's predictions fail when observational data are collected, it can be unclear whether the theory itself is at fault or one of its auxiliary theories is wrong. More generally, observations are theory-laden in a messy way that unsettles the logical empiricist's expectation that observations could adjudicate between rival theories.

Still another problem was that logical empiricism gave insufficient thought to scientists' humanity. Logical empiricism's stringent science used logic and data in a rather mechanical fashion, guaranteed to be scientific and to guard truth. It had no human face. Subsequently, however, other scholars were quite willing to say that science is done by scientists, by fallible humans. And their human face for science was inimical to logical empiricism's tidy story. It was a dirty face.

Those and other deficiencies of logical empiricism provided inviting targets for the next generation of philosophers of science, especially Popper, Kuhn, Lakatos, and Feyerabend. Because their criticisms will be reviewed in the following chapter, it suffices here merely to list several of their concerns: (1) Empirical data can neither falsify nor confirm a theory, so truth is elusive. (2) Theory choices are underdetermined by data, observations are theory-laden, and multiple criteria for theory choice lead to ambiguity. (3) Paradigms are incommensurable, so science cannot progress toward truth. (4) After admitting scientists' humanity and social situatedness, what makes a belief scientific is simply that it is what scientists say, and there is nothing special about science anyway.

The foregoing review of twenty-three centuries in about as many pages must suffice as a feast of divergent opinions and verdicts on science! As this brief history of truth draws to a close, it must be emphasized how atypical the past century has been compared with the preceding two millennia. Previously, all scientists also knew their philosophy well. To use a term fittingly applied to Einstein by Schilpp (1951), they were philosopher-scientists. Einstein rightly insisted that "Science without Epistemology is – in so far as it is thinkable at all – primitive and muddled" (Rosenthal-Schneider 1980:27). By contrast, the past century has been the one and only century in science's rich history to have produced mostly scientists rather than philosopher-scientists. Consequently, many contemporary scientists cannot defend science's credibility from various intellectual attacks, and they cannot optimize the methods and productivity of their own research programs. Weakness in scientific method is costly, wasting research dollars and compromising competitive advantages.

Finally, turning from the past to the future, who will harness the world's most advanced scientific thinking to develop successful theories and innovative products as the third millennium begins? This book's message is that a disproportionately large share of future developments will come from those scientists who have mastered their specialties like everyone else but who also have mastered the basics of science's philosophy and method. Method precedes results; method affects results. Method matters. It is to be hoped that the scientific community will soon return to the practice, which served previous millennia well, of producing philosopher-scientists.

SUMMARY

Unquestionably, science is one of the liberal arts and must be understood and taught as such. The essence of that vision is science in its historical, philosophical, cultural, social, economic, and ethical context, that is, a humanities-rich version of science.

Science's four traditional claims are rationality, truth, objectivity, and realism. Reason holds the double office of regulating belief and guiding action. Rational methods of inquiry use reason and evidence correctly to achieve substantial and specified success in finding truth, and rational actions use rational and true beliefs to guide good actions. Statements are the bearers of truth, and true statements correspond with reality. Objective beliefs concern external physical objects. Such beliefs can be tested and verified so that consensus will emerge among knowledgeable persons, and they do not depend on controversial presuppositions or privileged worldviews. Realism is correspondence of human thoughts with an external and independent reality, including physical objects.

To promote an understanding of science's method and claims in historical perspective, this brief history of truth has examined the standards and evidence expected for truth claims during the past twenty-three centuries, from Aristotle to the present. The basic question remains: What human-sized and public method can provide scientists with real truth about the physical world? Aristotle got much of scientific method right, but he disregarded experimental methods and had a somewhat confused expectation that a mature version of the natural sciences should be much like geometry in its method and certainty. Those deficiencies were remedied in the fledgling medieval universities in the 1200s. From 1500 to the present, tremendous advances have been made, especially regarding deductive and inductive logic.

The modern period has yielded a tremendous diversity of views on science. The rationalist tradition has emphasized reason and logic, whereas the empiricist tradition has emphasized sensory experience and empirical evidence. The skeptical tradition has revealed, but not settled, the matter of science's presuppositions. At this time in history, the way ahead for science's general principles will require a deep integration of these three elements: presuppositions, evidence, and logic.

For twenty-two centuries, from Aristotle until a century ago, it was the universal practice of the scientific community to produce philosopher-scientists, scholars who understood both philosophy and science. But most contemporary scientists receive meager training in the history and philosophy of science, the principles of scientific method, and even logic. It would be beneficial for the scientific community to return to the venerable tradition, which served previous generations well, of producing philosopher-scientists. This book's message is that a disproportionately large share of future developments will come from those scientists who have mastered their specialties like everyone else, but who also have mastered the basics of science's philosophy and method.

CHAPTER THREE

SCIENCE WARS

Does science have a rational method of inquiry that provides humans with objective truth about physical reality? Certainly a reply of yes represents the traditional claims of science, as delineated in Chapter 2. And certainly that would be the reply of most contemporary scientists, as well as the general public. Anyone who confidently believes the scientific stories that water really is H_2O and that table salt really is NaCl gives every appearance of being in the camp that replies yes to this question.

Nevertheless, at present there is a controversy raging over science's claims of rationality and truth, a controversy of such intensity that it often goes by the name of the "science wars." These intellectual wars have been so noteworthy that they have even made the front pages of the world's leading newspapers. This chapter will examine the controversy.

An enormous literature, some of it for and some against science's claims of rationality and objective truth, reveals a bewildering array of positions, motivations, attitudes, temperaments, rhetorics, and intended audiences. Accordingly, the first task is to locate a reasonable and constructive attitude for scientists to take toward this debate. Then we shall examine the four principal problems that convince some philosophers that science's claimed rationality is in big trouble. Finally, reactions from scientists will be explored, and some preliminary suggestions will be offered for shifting the debate to more fertile ground.

AUDITORS AND ATTITUDES

Owners of businesses have long been accustomed to having external auditors check their financial assets, liabilities, and ratings. But increasingly, scientists are having to get used to facing an accounting, although in their case the account is intellectual rather than financial. Many philosophers, historians, sociologists, and others have become self-appointed external auditors of science. Their ratings of science's claims of rationality and truth are becoming increasingly influential, strongly affecting public perceptions of science.

Of course, both businesses and scientific institutions also have internal auditors. In science, these take the form of peer review for publications and grants, and, more generally, the day-to-day interactions among colleagues. Naturally, the internal audit of science's intellectual assets by scientists themselves has been generally quite positive, showing confidence in science's methods and results. One might expect that the internal audit by scientists themselves and the external audit by philosophers and others would reach similar verdicts about science's intellectual soundness.

Before considering or comparing these verdicts from different sources, however, a more basic issue is simply whether or not science should be subjected to an external audit. Is there some legitimate propriety in having philosophers and other non-scientists check science's claims? Or should scientists have the prerogative of setting their own standards for their truth claims without any interference from anyone else?

Precisely because science is one of the liberal arts and because such fundamental intellectual notions as rationality and truth pervade the liberal arts, certainly it is within the purview of philosophy and history and other fields to have a voice in the weighing of science's intellectual claims. With keen insight and compelling logic, Frank (1957) showed that for any scientific conclusion, a claim of truth (expressed either as a certainty or as a probability) has both scientific and philosophical dimensions. Especially the general principles of scientific method, as contrasted with specialized techniques, have strong connections with many disciplines across the humanities.

But precisely because an external audit of science is legitimate, there is great responsibility on those auditors to give an accurate account. They must neither underestimate nor overestimate science's assets and liabilities.

Given that external auditing of science's rationality is inevitable and legitimate, what attitude should scientists take toward this audit? For starters, scientists who are confident that science's internal audit is positive and justified should welcome any competent external audit, because perforce it should complement and confirm the internal audit. The traditional independence of external auditors makes their verdicts particularly credible to the general public. On the other hand, when scientists encounter external audits greatly at variance with their own internal audits, they have the right to respectfully question the competency and accuracy of such unfavorable audits.

Much of the probing of science's method and rationality by philosophers, historians, and sociologists is simply an earnest attempt to determine exactly what science's actual methods imply for science's legitimate claims. It would be decidedly unrealistic, however, not to recognize that some of these

Figure 3.1. Science fights the dragon of postmodern and skeptical views of science. (This drawing by Andrew Birch is reproduced with his kind permission.)

probings constitute a militant call for scientists to promise less and for the public to expect less – much less!

Philosophers with relativistic, skeptical, and postmodern leanings routinely reach verdicts on science that are much more negative than what even the most cautious scientists reach. A particularly severe critique of science has arisen in recent years from some sociologists. It is called "Science Studies." The intensity of science's struggle with severe critiques is captured in a cartoon that appeared in *Nature* (Macilwain 1997). Figure 3.1 shows this cartoon with its combative motif of St. George and the dragon.

The attitude taken in this book, however, is that a trial by fire, by being exposed to arguments against science's rationality, offers scientists their best chance to uproot complacency and thereby to really understand the principles of scientific method. Science's enemies may appear to be the dragons of skepticism, relativism, and postmodernism, and indeed, where such philosophies prevail in earnest, nothing recognizable as science's traditional claims can survive. But the dragons that reside in one's own house are more dangerous and damaging than those that live in someone else's house. So this chapter warmly welcomes the dragons of severe critiques for a visit, because they can attack and defeat the resident dragons, especially complacency and overspecialization.

Figure 3.1 is correct in depicting a representative severe critique, Science Studies, as a smaller and less threatening dragon than St. George ever encountered. Were this cartoon to show instead science's battle with its own dragon of complacency, that dragon would be huge and menacing, like the dragons that St. George ordinarily encountered!

Only at one other time in all of history has the topic of science's method and credibility been as severely probed as it is today. Seven centuries ago in fledgling medieval universities, science was the new kid on the block, and it struggled mightily to formulate a method that could demonstrate respectability relative to the accepted standards already prevailing in theology and philosophy. That severe trial motivated incisive thought and led to unprecedented advances in the method and philosophy of science, which in turn produced the great flowering of science itself that occurred a few centuries later. Among scientists in medieval universities, the dragon of complacency was banished, and brilliant thinking flourished. But today, in a rather cynical and postmodern culture, science is the old kid on the block, struggling to preserve its credibility in a world in which rationality and objective truth are challenged in every sphere. It is hoped that the scientific community will meet the challenges and produce a truly reflective and exciting account of science's method and rationality.

The opening words of Macilwain (1997) can set the tone for this chapter: "Temperatures remain high in US faculties as what started as a relatively esoteric debate on the nature of scientific knowledge continues to reverberate. Whatever their differences, both sides agree that important issues are at stake – including the role of universities themselves. Universities across the United States are slowly awakening to a rumbling debate about the nature of scientific practice and knowledge, about who is qualified to pronounce on either – and about the motivation of those who do so."

Indeed, the debate whether or not science has a credible rationality is an enormously significant matter. Accordingly, the scientific community owes its critics a respectful hearing and a thoughtful reply. Modest preparation can equip scientists to speak to the issues at the informal, basic level of ordinary exchanges about science's credibility. Beyond that, it is desirable for at least a few scientists to know the issues well enough to be able to dialogue at a scholarly level.

This chapter will review the attacks on science's rationality, but will leave it for later chapters to sort out the ideas. It is hoped that this chapter will set forth the opposition's arguments with sufficient detail and clarity that scientists will find this account of the science wars to be genuinely disturbing and unsettling, not merely silly or aggravating. Science needs to feel the hot breath of the dragon that is breathing down its neck!

FOUR DEADLY WOES

The war between science and anti-science is tangled and emotional: "As is the case with most wars, it is often difficult to differentiate the aggressor from the victim. On the one hand, scientific realists claim that science is under siege by an unholy alliance of feminists, multiculturalists, literary and cultural theorists, and sociologists of science. They contend that rationality and science are threatened by this motley group of postmodernists who want relativism to replace realism and politics to supplant reason. On the other hand, postmodernists claim that they are only illustrating the human, all too human, characteristic of all knowledge. It is scientists who are guilty of denying the political and cultural situatedness of their own knowledge form" (Ward 1996:ix).

The attacks on science are of several different kinds. Some attacks are essentially political or emotional, such as displeasure at science's domination (especially historically) by white males. A radical anti-science element claims that "Science is sexist, classist and racist to its core . . . and at worst simply meaningless" (Herbert 1995). Even Einstein's familiar equation $e = mc^2$, saying that energy equals mass times the square of the speed of light, has been denounced as a mass delusion or even a deliberate deception concocted by power-hungry scientists. The more radical critics of science would erase that equation, as depicted in Figure 3.2.

Figure 3.2. Radical critics would dismiss science as a sexist and racist concoction of power-hungry scientists and on that basis would even seek to erase Einstein's famous equation. (This drawing by Carl R. Whittaker is reproduced with his kind permission.)

However, Holton (1993), Gross and Levitt (1994), Gross, Levitt, and Lewis (1996), Ross (1996), Koertge (1998), Segerstråle (2000), and Trachtman and Perrucci (2000) have already said all that can be said about political and emotional attacks on science, so that matter will not receive further attention here. Rather, the focus here will be on intellectual and philosophical attacks. Those attacks deny science's claim of objective truth about physical reality. For example, criticism of $e = mc^2$ because that equation's author was a white male is not of interest here, but allegations that the scientific method is fundamentally arational or unreliable will be of interest. The concern here is with serious intellectual attacks, not entrenched emotional reactions.

An important historical development for science occurred around 1920, when the philosophy of science began to be differentiated from philosophy, becoming an academic discipline in its own right, complete with its own journals: *Philosophy of Science* began publication in the United States in 1934, and *The British Journal for the Philosophy of Science* opened in 1950. That development has been largely positive, stimulating philosophical reflection on science. But it has also meant, for better or for worse, that many influential commentators on science are not themselves accomplished scientists. It is fair to say that prior to 1920, virtually all commentators on science were themselves distinguished scientists (with two notable exceptions, Francis Bacon and David Hume). Not only philosophers but also historians and sociologists have become influential interpreters and auditors of science.

During the past several decades, four philosophers of science have been especially prominent: Sir Karl Popper, Imre Lakatos, Thomas Kuhn, and Paul Feyerabend. They are the four inquisitors of Newton-Smith (1981), the four irrationalists of Stove (1982), and the four villains of Theocharis and Psimopoulos (1987). Among philosophers, numerous philosophers of science are well known; but among scientists, Popper and Kuhn (Figure 3.3) probably are better known than all the others combined (Horgan 1991, 1992).

The earliest of these four philosophers is Popper, whose reassessment of science's claims began with the publication of *Logik der Forschung* in 1934, which appeared in English as *The Logic of Scientific Discovery* in 1959 (second edition 1968). An especially influential contribution has been *The Structure of Scientific Revolutions,* by Kuhn (1962, second edition 1970). Several years ago, it had already sold nearly a million copies in 16 languages, and it is commonly considered "the most influential treatise ever written on how science does (or does not) proceed" (Horgan 1991; also see Fuller 2000:1).

The incisive thinking and penetrating analyses of Popper, Kuhn, and other philosophers have had many positive effects. Especially valuable are their effective criticisms of the logical empiricism that had preceded their generation,

Figure 3.3. Two important philosophers of science, Sir Karl Popper (left) and Thomas Kuhn (right). (The photograph of Karl Popper by David Levenson is reproduced with kind permission of Black Star, and the photograph of Thomas Kuhn with kind permission of Stanley Rowin Photography.)

Popper's insistence on falsifiability, and Kuhn's recognition of the human and historical elements in science. Nevertheless, it is also the case that the writings of those philosophers have mounted a sustained and influential attack on science's rationality, even though scientists often fail to recognize that. Ever since their writings challenged science's traditional claims of rational realism and objective truth, "whether science can be said to be rational, and if so in what sense, have been vexed questions" (Banner 1990:2).

Make no mistake: Much hangs in the balance in this grand debate. For example, Feyerabend has attacked science's rationality and concluded that scientists have "more money, more authority, more sex appeal than they deserve," so "It is time to cut them down in size" (Broad 1979). If his criticisms are valid, scientists need to realize that and give up their traditional claims of rationality and truth; but if not, scientists need to defend their claims.

To benefit from studying the philosophy of science, however, scientists' primary posture should not be defensive, but rather inquisitive. An instructive example is given by Newton-Smith (1981:xi), who says that in his book, "Popper, Kuhn, Feyerabend and Lakatos are severely criticized," and yet he also says that "Their writings have given form to the most important contemporary questions about science." Likewise, "those who know only their own side of a case know very little of that" (Susan Haack, in Gross et al. 1996:57). Intellectual growth often is limited more by lack of good questions than by lack of accurate answers. So scientists can benefit greatly from penetrating and disturbing questions that are intended to unsettle science, even if the final result for most of them is to affirm science's traditional claims with new insight and greater conviction.

The expectation here is that an indiscriminate and superficial reading of Popper and Kuhn will do scientists more harm than good, but a discerning and critical reading will do much more good than harm. Anyway, the first necessity is to appreciate some vexed questions. Popper, Kuhn, and other philosophers have challenged science with four deadly woes: elusive truth, underdetermined theory, incommensurable paradigms, and redesigned goals.

Elusive Truth

What criterion can demarcate science from non-science? In 1919 that question triggered Popper's interest in the philosophy of science (Popper 1974:33). He clearly distinguished that question, of whether or not a theory was scientific, from the different question of whether or not a theory was true. That interest was occasioned by various claims that Einstein's physics, Marx's history, Freud's psychology, and Adler's psychology were all scientific theories, whereas Popper suspected that only the first of those claims was legitimate. Exactly what was the difference?

The received answer, from Francis Bacon several centuries earlier and from logical empiricists more recently, was that science was distinguished from pseudoscience and philosophy (especially metaphysics) by its empirical method, proceeding from observations and experiments to theories by means of inductive generalizations. But that answer did not satisfy Popper, because admirers of Marx, Freud, and Adler also claimed an incessant stream of confirmatory observations to support their theories. Whatever happened was always and readily explained by their theories. What did that confirm? "No more than that a case could be interpreted in the light of the theory" (Popper 1974:35).

By sharp contrast, Einstein's theory of relativity could not sit easily with any and all outcomes. Rather, it made specific and bold predictions that put the theory at risk of disconfirmation if observations should turn out to be contrary to expectation. Einstein's theory claimed that gravity attracts light just as it attracts physical objects. Accordingly, starlight passing near the sun would be bent measurably, making a star's location appear to shift outward from the sun. Several years later, in 1919, a total eclipse of the sun afforded an opportunity to test that theory. An expedition led by Sir Arthur Eddington made the observations, clearly showing the apparent shift in stars' positions and thus confirming Einstein's theory. What impressed Popper most of all was the risk that relativity theory took, because an observation of no shift would have proved the theory false.

So, comparing supposedly "scientific" theories that are ready to explain anything with genuinely scientific theories that predict specific outcomes and thereby risk disconfirmation, Popper latched on to falsifiability as the essential criterion that demarcates science from non-science. "Irrefutability is not a virtue of a theory (as people often think) but a vice.... [The] criterion of the scientific status of a theory is its falsifiability" (Popper 1974:36–37).

Notice that science was distinguished by falsification, not verification. Popper insisted that conjectures or theories could be proved false, but no theory could ever be proved true. Why? Because he respected deductive logic, but agreed with David Hume that inductive logic was a failure. "Hume, I felt, was perfectly right in pointing out that induction cannot be logically justified. He held that there can be no valid logical arguments allowing us to establish '*that those instances, of which we have had no experience, resemble those, of which we have had experience*'....In other words, an attempt to justify the practice of induction by an appeal to experience must lead to an *infinite regress*" (Popper 1974:42). A single contrary observation allows deduction to falsify a general theory. Indeed, "*A* implies *B*; not *B*; therefore not *A*" is the valid argument *modus tollens*, denying the consequent. By contrast, no quantity of observations can possibly allow induction to verify a general theory, because further observations might bring surprises.

Hence, the best that science could do was to offer numerous conjectures, refute the worst with contrary data, and accept the survivors in a tentative manner. Conjecture followed by refutation was the scientific method, by Popper's account. But that implies that although "we search for truth...we can never be sure we have found it" (Popper 1974:56): Science could produce no truth, none whatsoever. So Popper offered his demarcation criterion of falsifiability to separate science from non-science, but at the cost of separating science from truth.

Underdetermined Theory

A prominent feature of the logical empiricism that dominated the philosophy of science preceding Popper and Kuhn was a sharp boundary between observation and theory. According to that view, true and scientific statements were based on empirical observations and their deductive logical consequences – hence the name, logical empiricism. By contrast, Popper insisted that observations were deeply theory-laden: "But sense-data, untheoretical items of observation, simply do not exist.... We can never free observation from the theoretical elements of interpretation.... We *cannot* justify our knowledge of the external world; *all* our knowledge, even our observational knowledge, is theoretical, corrigible, and fallible" (Karl Popper, in Lakatos and Musgrave 1968:163–164). Why? That claim has many facets, but here it must suffice to mention three principal arguments.

First, in order to make any observations at all, scientists must be driven by a theoretical framework that raises specific questions and generates specific interests. Popper (1974:46) explained the point nicely: "But in fact the belief that we can start with pure observations alone, without anything in the nature of a theory, is absurd. . . . I tried to bring home the same point to a group of physics students in Vienna by beginning a lecture with the following instructions: 'Take a pencil and paper; carefully observe, and write down what you have observed!' They asked, of course, *what* I wanted them to observe. Clearly the instruction, 'Observe!' is absurd. . . . Observation is always selective. It needs a chosen object, a definite task, an interest, a point of view, a problem."

Second, what may appear to be a simple observation statement, put to work to advance one hypothesis or to deny another hypothesis, actually has meaning and force only within an involved context of theory (Matthews 2000:248–251). For example, a pH meter may give a reading of 6.42, but that observation presumes a host of chemical and electronic theories involved in the design and operation of that instrument. For another example, observations of planetary positions may indicate that the earth goes around the sun, rather than the reverse, but those observations are understood in a complex framework that includes theories of space and time as well as presuppositions about the uniformity of nature and the general reliability of sense perception.

Third, theory choice involves numerous criteria that entail subtle trade-offs and subjective judgments. For example, scientists want theories to fit the observational data accurately and also want theories to be simple or parsimonious. But if one theory fits the observations more accurately but another theory is more parsimonious, which theory accords better with the observations? Clearly, not only the observations themselves are guiding theory choice, but also some deep theories about scientific criteria and method.

This problem that observations are theory-laden is related to a similar problem, the underdetermination of theory by data. For any given set of observations, it is always possible to construct infinitely many different and incompatible theories that will fit the data equally well (Quine 1975; Hoefer and Rosenberg 1994; Kukla 1996; Leplin 1997:152–163). Such theories are said to be empirically equivalent. Furthermore, besides being empirically equivalent, one may also specify that these competing theories be equally parsimonious and equal as regards any additional reasonable principles of evidence, in which case these theories are evidentially equivalent. The problem of underdetermination of theory by data is that there are always incompatible rival theories that are empirically equivalent or even evidentially equivalent.

Consequently, no amount of data is ever adequate to determine that one theory is better than its numerous equal alternatives.

What do such problems mean for science? "But if observations are theory-laden, this means that observations are simply theories, and then how can one theory falsify (never mind verify) another theory? Curiously, the full implications of this little complication were not fully grasped by Popper, but by Imre Lakatos: not only are scientific theories not verifiable, they are not falsifiable either" (Theocharis and Psimopoulos 1987).

So the first woe cited earlier was that science could not verify truths. Now this second woe is that science cannot falsify errors either. Science cannot declare any theory either true or false! "So back to square one: if verifiability and falsifiability are not the criteria, then what makes a proposition scientific?" (Theocharis and Psimopoulos 1987). These are huge problems, and yet there follows a third woe.

Incommensurable Paradigms

"Thomas S. Kuhn unleashed 'paradigm' on the world," reads the subtitle of an interview with him in *Scientific American* (Horgan 1991). It reports that "Kuhn ... traces his view of science to a single 'Eureka!' moment in 1947. ... Searching for a simple case history that could illuminate the roots of Newtonian mechanics, Kuhn opened Aristotle's *Physics* and was astonished at how 'wrong' it was. How could someone so brilliant on other topics be so misguided in physics? Kuhn was pondering this mystery, staring out of the window of his dormitory room ('I can still see the vines and the shade two thirds of the way down'), when suddenly Aristotle 'made sense.' Kuhn realized that Aristotle's views of such basic concepts as motion and matter were totally unlike Newton's. ... Understood on its own terms, Aristotle's physics 'wasn't just bad Newton,' Kuhn says; it was just different. ... He wrestled with the ideas awakened in him by Aristotle for 15 years. ... 'I sweated blood and blood and blood,' he says, 'and finally I had a breakthrough.' The breakthrough was the concept of paradigm."

Just what is a paradigm? The meaning of Kuhn's key concept is disturbingly elusive. Margaret Masterman (in Lakatos and Musgrave 1970:59–89) counted 21 different meanings, and later in that book, Kuhn himself admitted that the concept was "badly confused" (p. 234). In response to criticisms, Kuhn clarified two main meanings: A paradigm is an exemplar of a past scientific success, or is the broad common ground and disciplinary matrix that unites particular groups of scientists at particular times (Banner 1990:9).

The latter, broad sense is most relevant here. A paradigm is a "strong network of commitments – conceptual, theoretical, instrumental, and methodological" (Kuhn 1970:42). Scientific ideas do not have clear meanings and evidential support in isolation, but rather within the broad matrix of a paradigm. For example: "The earth orbits around the sun" is meaningless apart from concepts of space and time, theories of motion and gravity, observations with the unaided eye and with various instruments, and a scientific methodology for comparing theories and weighing evidence.

The history of science, in Kuhn's view, has alternating episodes of normal science, which refine and apply an accepted paradigm, and episodes of revolutionary science, which switch to a new paradigm because anomalies proliferate and unsettle the old paradigm. A favorite example is Newton's mechanics giving way to Einstein's relativity when experiments of many kinds piled up facts that falsified the former but fit the latter theory.

But why does Kuhn speak of revolutions in science when others have been content to speak of progress? Because different paradigms, before and after a paradigm shift, are incommensurable. What does that uncommon word mean? Literally, it means lacking a common measure. No common measure or criterion can be applied to competing paradigms to make a rational, objective choice between them. The problem is that when there is a paradigm shift, not only do scientific beliefs about nature change, but also the standards and criteria of scientific judgment change. "In learning a paradigm the scientist acquires theory, methods, and standards together, usually in an inextricable mixture. Therefore, when paradigms change, there are usually significant shifts in the criteria determining the legitimacy both of problems and of proposed solutions" (Kuhn 1970:109). Also, in a paradigm shift, the very meanings of key terms shift, so scientists are not talking about the same thing before and after the shift, even if some words are the same. For example, "Kuhn realized that Aristotle's view of such basic concepts as motion and matter were totally unlike Newton's" (Horgan 1991).

Kuhn uses three images to emphasize that no rational choice is possible among incommensurable paradigms. Scientific revolutions are like political upheavals that radically change the very constitution and rules of government, rather than working small changes within a fixed and continuing government. Scientific revolutions are also like religious conversions, a transfer of allegiance that is not forced by solid logic and objective evidence. And scientific revolutions are like aimless, purposeless evolution that has change but no goal. Paradigms just change, but not toward truth.

Well, if successive paradigms are incommensurable, what does that imply for science's claims? What is the bottom line? In his interview in *Scientific American*, Kuhn remarked, "with no trace of a smile," that science is "arational" (Horgan 1991). What does this uncommon word mean? To say that science is arational is to say that science is among those things, like cabbage, that have neither the property of being rational nor the property of being irrational. Consequently, saying that science is arational is a much stronger attack on science's rationality than saying that science is irrational.

After rationality and objectivity are gone, what happens to realism? The interview by Horgan (1991) says that Kuhn's "most profound argument" is that "scientists can never fully understand the 'real world,'" with the real world sequestered here in scare quotes. "There is, I think, no theory-independent way to reconstruct phrases like 'really there'; the notion of a match between the ontology of a theory and its 'real' counterpart in nature now seems to me illusive in principle" (Kuhn 1970:206). Banner (1990:40, 12) recognized the finality in those words, "illusive in principle," and perceptively discerned that although "Kuhn gives a central place to his study of the history of science," in fact "the crucial arguments are found to be philosophical, not historical." Incommensurable paradigms imply arational choices.

And after rationality, objectivity, and realism are gone, what happens to truth? Science has no truth. Indeed, what should be said is "Not that scientists discover truth about nature, nor that they approach ever closer to the truth," because "we cannot recognize progress towards that goal" of truth (Thomas Kuhn, in Lakatos and Musgrave 1970:20). Likewise, the back cover of the current edition (1970) of Kuhn's book quotes, unashamedly and approvingly, the review in *Science* by Wade (1977) saying that "Kuhn does not permit truth to be a criterion of scientific theories."

Obviously scandalized, Theocharis and Psimopoulos (1987) observed that "according to Kuhn, the business of science is not about truth and reality; rather, it is about transient vogues – ephemeral and disposable paradigms. In fact three pages from the end of his book *The Structure of Scientific Revolutions*, Kuhn himself drew attention to the fact that up to that point he had not once used the term 'truth'. And when he used it, it was to dismiss it: 'We may have to relinquish the notion that changes of paradigm carry scientists...closer and closer to the truth.'" With rationality, truth, objectivity, and realism gone, there follows a fourth and final woe for science.

Redesigned Goals

For more than two millennia since Aristotle, and preeminently in the fledgling universities in Oxford and Paris during the 1200s, philosopher-scientists labored to develop, refine, and establish scientific method. The intention was for scientific method to embody and support science's four traditional claims of rationality, truth, objectivity, and realism. But various arguments developed during the past century, such as the three problems discussed in the preceding subsections, have led many philosophers to relinquish science's traditional goals and hence to redesign its goals.

"According to the common-sense [and traditional] view, of course, the assent of the [scientific] community is dictated by certain agreed standards, enabling us to say that the preferred theory is the better one. But Kuhn turns this upside down. It is not a higher standard which determines the community's assent, but the community's assent which dictates what is to count as the highest standard.... Against the realist he asserts that the explanation of scientific progress 'must, in the final analysis, be psychological or sociological.'... Elsewhere he tells us that his 'position is intrinsically sociological'" (Banner 1990:12, 27). Theocharis and Psimopoulos (1987) summarize Kuhn's view: "a proposition is scientific if it is sanctioned by the scientific establishment." What makes a statement scientific is that scientists say it; nothing more. Sociology replaces method in that account of what is scientific.

A potential embarrassment for that account, however, is the stubborn, common-sensical rejoinder that scientific progress proves that the scientific method finds much truth – that progress and truth are intertwined. Accordingly, a heroic philosophical industry aims to avert this rejoinder, separating progress from truth. Laudan (1977:125, 24) says that

> I am deliberately driving a wedge between several issues that have hitherto been closely intertwined. Specifically, it has normally been held that any assessment of either rationality or scientific progress is inevitably bound up with the question of the *truth* of scientific theories. Rationality, it is usually argued, amounts to accepting those statements about the world which we have good reason for believing to be true. Progress, in its turn, is usually seen as a successive attainment of the truth by a process of approximation and self-correction. I want to turn the usual view on its head by making rationality parasitic upon progressiveness. *To make rational choices is, on this view, to make choices which are progressive....* By thus linking rationality to progressiveness, I am suggesting that we can have a theory of rationality *without presupposing anything about the veracity or verisimilitude of the theories* we judge to be rational or irrational.... [One] need not, and scientists generally do not, consider matters of truth and falsity when determining whether a theory does or does not solve a particular empirical problem.

On that view, though science and technology have, for instance, put men on the moon and designed effective new drugs, that counts for nothing – absolutely nothing – in the way of evidence for science's traditional claim of getting at the truth. Technological progress is affirmed, but scientific truth is denied. For another example, within the scope of a single sentence, Yager (1991) remarks that "Modern science does not give us truth; it offers a way for us to interpret events of nature and to cope with the world." For scientists accustomed to regarding science's ability to interpret and cope with nature as fruits of science's truth, it is extremely difficult to fully grasp how quickly and easily some other intellectuals can completely sever progress and truth.

A subtle move in some attacks on science's traditional claims is to separate things that, to scientists, seem naturally to go together. That is the first step in a divide-and-conquer strategy. For example, "rationality must be clearly disconnected from truth" (Stenmark 1995:221), and "truth is an inadmissible criterion of rationality" (Theodore R. Schatzki, in Natter et al. 1995:156). Likewise, Laudan (1977), as summarized by Leplin (1984:196), says that supposing truth to be a goal of science is confused and irrational, because truth is unreachable.

A particularly radical reinterpretation of science comes from Paul Feyerabend, "the worst enemy of science," according to Broad (1979), Theocharis and Psimopoulos (1987), and Horgan (1993). Like Lakatos, Feyerabend was also a student under Popper. In an interview with Feyerabend in *Science*, "Equal weight, he says, should be given to competing avenues of knowledge such as astrology, acupuncture, and witchcraft. . . . 'Respect for all traditions,' he writes, 'will gradually erode the narrow and self-serving "rationalism" of those [scientists] who are now using tax money to destroy the traditions of the taxpayers, to ruin their minds, to rape their environment, and quite generally to turn living human beings into well-trained slaves.' . . . Feyerabend is dead set against what has been called 'scientism' – the faith in the existence of a unique [scientific] 'method' whose application leads to exclusive 'truths' about the world" (Broad 1979).

Similarly, a more recent interview with Feyerabend in *Scientific American* says that "For decades, . . . Feyerabend . . . has waged war against what he calls 'the tyranny of truth.' . . . According to Feyerabend, there are no objective standards by which to establish truth. 'Anything goes,' he says. . . . 'Leading intellectuals with their zeal for objectivity . . . are criminals, not the liberators of mankind.' . . . Jutting out his chin, he intones mockingly, 'I am searching for the truth. Oh boy, what a great person.' . . . Feyerabend contends that the very notion of 'this one-day fly, a human being, this little bit of nothing' discovering the secret of existence is 'crazy.' . . . The unknowability of reality is one theme of . . . Feyerabend" (Horgan 1993).

Ironically, these four woes bring the status of science full circle. Popper started with the problem of demarcating science from non-science in order to grant credibility to science and to withhold credibility from non-science. Science was something special. But a mere generation later, his student Feyerabend followed his teacher's ideas to their logical conclusion and thereby ended with the verdict that science is neither different from, nor superior to, any other way of knowing. Science started out superior to astrology; it ended equivalent. Science started with a method for finding truth; it ended with a sociology for explaining persuasion and consensus.

Finally, so what? As science's four claims give way to science's four woes, how do philosophers react? Well, occasionally it would seem that the philosophers' mandate is to disagree with each other endlessly, and collectively to take all possible positions. Anyway, in the present case, the philosophers have not been disappointing! Various ones have exhausted the possibilities by construing science's woes in a negative, positive, or indifferent manner.

Many philosophers construe these attacks on science negatively, considering them unjustified criticisms with tragic implications. Rather, they maintain science's traditional claims of rational realism and objective truth. Among these philosophers are Trigg (1980, 1993), Newton-Smith (1981, 1999), Scheffler (1982), Newell (1986), O'Hear (1989), Banner (1990:7–64), Howson and Urbach (1993), Leplin (1997), and Koertge (1998).

But other philosophers react positively, welcoming the demise of science's four traditional claims at the hands of the four recent woes. For example, Brown (1987:230) says that "I have offered here one detailed argument for the now familiar thesis that there is no fundamental methodological difference between philosophy and science.... [But] it has become progressively clearer that the sciences cannot provide certainty and have no a priori foundation.... [Admittedly,] earlier thinkers believed that both science and philosophy provide certain knowledge of necessary truths. We must conclude that neither do.... [The] human intellect ... seems unable to grasp a final truth." So chastened science has no truth, and now philosophy can enjoy the same! Indeed, the impossibility of finding any truth, already happily established in science, can now spread rapidly to all intellectual life.

The third and final possibility is to react indifferently, saying that whether or not science has rational realism and objective truth makes no difference. Kuhn exemplifies this posture. After having told us that science is arational, that science finds no truth, that reality is eternally illusive in principle, and that science's supposed claims are to be explained away in psychological and sociological terms, a calm Kuhn tells us that "I no longer feel that anything is lost, least of all the ability to explain scientific progress, by taking this position" (Thomas Kuhn, in Lakatos and Musgrave 1970:26). For Kuhn, the rationality, truth, objectivity, and realism that scientists are accustomed to just doesn't matter.

REACTIONS FROM SCIENTISTS

Having just reviewed the diverse reactions of philosophers to the supposed demise of science's traditional claims, what about the reactions of scientists?

Do they think that these philosophical criticisms are valid, forcing honest scientists to adjust and downgrade their claims? Do scientists think that it matters whether or not science has a rational realism and objective truth?

To get truly thoughtful reactions from scientists to the philosophers and other self-appointed external auditors of science, ideally there would be opportunity for a sustained exchange that would get past the superficial rhetoric and really get into the substantive issues. But such opportunities are rare, because scientists interact mostly within their own community, as do philosophers. Hence, the closest that one ordinarily would get to such a solid exchange probably would be to read a book from each side, such as Gross et al. (1996) or Koertge (1998) defending science, and Galison and Stump (1996) or Ross (1996) attacking that defense. Fortunately, there have been occasional instances of sustained exchange among scientists and philosophers, as well as some historians and sociologists.

An Early Exchange

The first notable exchange began with the provocative commentary by Theocharis and Psimopoulos (1987) in *Nature*. It stimulated a lively correspondence, from which *Nature* published 18 letters, until the editor closed the correspondence and gave the authors an opportunity to reply (*Nature* 1987, 330:308, 689–690; 1988, 331:129–130, 204, 384, 558; reply 1988, 333:389).

The gist of the article by Theocharis and Psimopoulos (1987), in view of the four woes of the preceding section, was that "By denying truth and reality science is reduced to a pointless, if entertaining, game; a meaningless, if exacting, exercise; and a destinationless, if enjoyable, journey." Three dangerous repercussions were feared: cutbacks in science budgets because of lowered expectations, broader social and cultural decline because of the languishing interest in truth, and impairment of scientific progress if beguiled scientists believed that no objective truth exists to be found. In view of the severe attack and its potential deplorable repercussions, those authors were shocked that scientists had done so little to respond and, worse yet, sometimes had even accepted and promulgated anti-scientific ideas. For instance, they had little patience with a popular undergraduate text for courses in the philosophy of science that breezily announced that "philosophy or methodology of science is of no help to scientists" and that told its readers that "We start off confused and end up confused on a higher level."

Those authors commended to scientists the three tasks of refuting erroneous and harmful views of science, putting forth adequate concepts of objectivity, truth, rationality, and scientific method, and putting good scientific work into productive practice. Only then would the true value of science and philosophy be convincingly demonstrated.

The dominant tenor of the 18 letters to the editor was that of numerous scientists rushing in to defend Popper and Kuhn from what they perceived as an unreasonable or even malicious attack by Theocharis and Psimopoulos. One letter even recommends that Theocharis and Psimopoulos (and perhaps also the journal *Nature*) offer Sir Karl Popper a public apology. Some letters rejected the claim that objective truth was important for science. One letter claimed that the "most basic truth is that there can be no objective truth." Another reader replaced truth with prediction: "This process of making ever better prediction *is* scientific progress, and it circumvents entirely the problem of defining scientific truth."

One of the strongest letters was from sociologist Harry Collins, who, several years later, co-authored a controversial book to be discussed in the next subsection. Apparently considering himself to have ascended to high moral ground indeed, he suggested that "The only thing that makes clear good sense in Theocharis and Psimopoulos is the claim that the privileged image of science has been diminished by the philosophical, historical and sociological work of past decades. One hopes this is the case. Grasping for special privilege above and beyond the world we make for ourselves – the new fundamentalism that Theocharis and Psimopoulos press upon us – indicates bankruptcy of spirit luckily not yet widespread in the scientific community." So pursuit of truth had been transmuted into bankruptcy of spirit!

Similarly, Science Studies experts David Bloor and Steven Shapin were dismissive, remarking that "Philosophers have always been accused of corrupting youth and subverting the state," and apparently "times have not changed" regarding the ill-treatment given to those who "pose uncomfortable questions." And a philosopher roundly chastised the authors, saying that "matters are not likely to be improved by attacking people [philosophers!] who in all probability have no influence."

Some other reactions, however, were favorable. One responder wrote sympathetically that "Philosophical complacency . . . will not do; contrary to what both sceptics and conservatives often seem to believe, philosophical questions do matter." Another responded that "There are very good reasons why twentieth century philosophy of science, under the malign influence of Popper through to Feyerabend, is profoundly hostile to science itself. . . . It is indeed unfortunate that many scientists, through ignorance, quote these philosophers approvingly. The most effective victories are those in which the losers unwittingly assist their opponents."

Having reviewed the various positive and negative reactions to Theocharis and Psimopoulos (1987), we may allow those beleaguered authors the last word in this subsection. They offered a poignant and intriguing remark about the uniqueness of the contemporary scene: "Natural philosophy has had enemies throughout its 2,600 or so years of recorded history. But the present era is unique in that it is the first civilized society in which an effective anti-science movement flourishes contemporaneously with the unprecedently magnificent technological and medical applications of modern science. This is a curious paradox which cries out for clarification." Their own conjecture was that "The explanation of this sad situation is, of course, the large number of errors regarding the nature, scope, method, powers and limitations of science."

Recent Exchanges

A more recent exchange has been precipitated by the publication of a controversial and popular book, *The Golem: What Everyone Should Know About Science*, by Collins and Pinch (1993). It was reviewed and criticized by N. David Mermin in *Physics Today* [1996, 49(3):11, 13; 49(4):11, 13], with further exchanges [1996, 49(7):11, 13, 15; 1997, 50(1):11, 13, 15, 92, 94–95]; also see *Physics Today* [1997, 50(10):11, 13; 1998, 51(4):15, 102, 104]. It was also reviewed in several science journals (W. Gratzer 1993, *Nature* 364:22–23; G. Vines 1993, *New Scientist* 138:38–39; J. Turney 1993, *New Scientist* 138: 41–42; U. Segerstråle 1994, *Science* 263:837–838), two philosophy journals (D. L. Hull 1995, *Philosophy of Science* 62:487–488; T. Nickles 1995, *The British Journal for the Philosophy of Science* 46:261–266), two history journals (J. Golinski 1994, *The British Journal for the History of Science* 27:487–488; B. Marsden 1995, *Isis* 86:357–358), and a sociology journal (M. Lynch 1995, *Contemporary Sociology* 24:114–115).

Incidentally, *The Golem* appeared in a second edition in 1998, but the following discussion addresses the first edition of 1993 that was the object of extensive review. Not one word in the following quotations was changed in the second edition. Meanwhile, Collins and Pinch (1998) have extended their Golem metaphor in another book about technology.

Those authors introduced *The Golem* thus: "Science seems to be either all good or all bad [in various persons' estimations].... Both these ideas of science are wrong and dangerous.... What, then, is science? Science is a golem. A golem is a creature of Jewish mythology...a humanoid made by man from clay and water.... It is powerful.... But it is clumsy and dangerous...a bumbling giant.... [This] creature of clay was animated by having the Hebrew 'EMETH', meaning truth, inscribed on its forehead – it is truth that drives it on. But this does not mean it understands the truth – far from it" (Collins and Pinch 1993:1–2). The book's strategy was to explain science "with as little reflection on scientific method as we can muster," instead relying on several episodes of ordinary, fallible, untidy science to show "what actually happens" (p. 2). "[Our] conclusion is that human 'error' goes right to the heart of science, because the heart is made of human activity.... Scientists should promise less; they might then be better able to keep their promises" (p. 142).

The book suggests that among themselves, scientists disagree because experimental data cannot really settle matters (for data underdetermine theory). Nevertheless, scientists present a unified story to the public, "transmuting the clumsy antics of ... [science] into a neat and tidy ... myth," which is all right for the authors' newly enlightened public, because "There is nothing wrong with this; the only sin is not knowing that it is always thus" (p. 151).

Mermin strongly rejected Collins and Pinch's conclusion that "Science works the way it does not because of any absolute constraint from Nature, but because we make our science the way that we do." The collective import of many such declarations was that science does not discover objective truths about physical reality, but rather just constructs consensus among scientists. Mermin's reaction to that thesis was nuanced: "The pertinent issue in assessing the claims of *The Golem* is not whether scientific truth is determined from nature or by social construction, but whether Collins and Pinch strike a satisfactory balance between these two aspects of the process." Mermin charged that "their book furnishes an instructive demonstration of what can go wrong if you focus too strongly on the social perspective."

Obviously, Mermin said, scientists are humans involved in a social structure that is a real and integral aspect of our science, but "Agreement is reached not just because scientists are so very good at agreeing to agree." Mermin suggested that a crucial feature of science that Collins and Pinch overlooked in their denigrating account was the role of interlocking evidence: "an enormous multiplicity of strands of evidence, many of them weak and ambiguous, can make a coherent logical bond whose strength is enormous." Incidentally, in 1840, William Whewell had called such accordance of several lines of inductive evidence "consilience" – a term recently revived by Wilson (1998).

Collins and Pinch replied to Mermin that their book "is about as inoffensive a social analysis of science as one could wish for," so they were surprised that it "seems to have engendered some responses more appropriate to the appearance of a heretical tract." In their view, "Many scientists, especially those who have become spokespersons, are evidently scientific fundamentalists. They think of science as the royal route to all knowledge; they think of it as able to deliver the kind of certainty that priests once delivered; they think of it as a complete world view or quasi religion.... Where we sociologists think of ourselves as trying to gain social and historical insight into science, the fundamentalists see heresy.... [Mermin's] reaction has the whiff of the offended priest."

Displaying their penchant for high moral ground, Collins and Pinch went on to say that "In *The Golem*, we argue that a science seen as a body of expertise is strong enough to manage without myths, self-delusions and overly defensive attitudes. Such a science can join in the debates about its history and sociology without fear and can face its public with pride. Only scientific fundamentalists need hide behind creation myths, witch-hunts, censorship and suppression. In doing so, however, they will destroy the democratic foundation of the science they say they love."

Then Mermin replied, perplexed that Collins and Pinch had transmogrified a sensible debate about science's status into inflammatory language about religion's conflicts: "Because I share Collins and Pinch's distaste for the pronouncements of 'scientific fundamentalists,' I was disappointed that they managed to sniff out an offended priest in my own analysis of their story.... I don't understand why my claim that this [interlocking evidence] is an important part of how science really works should elicit talk about religious inquisitions, McCarthy hearings, self-delusions and overly defensive attitudes." Mermin's last sentence in the exchange was a wholesome call for more content and less rhetoric: "We should all pay more attention to explaining clearly where we stand, and steer away from extravagant and sometimes inflammatory flourishes."

The letters to the editor were generally sympathetic to Mermin's defense of science's rationality. One letter remarked that "No modern-day consensus on the nature of science will be reached until we agree that what we are talking about is neither sociology nor science, but philosophy.... The successful use of science to predict future events correctly is the best argument that scientific procedures and protocols are a valid guide to objective reality." Another letter declared quite simply that "There really are results and facts."

The book reviews of *The Golem* are revealing. The reviews by philosophers pointed out that since the 1970s, Collins and Pinch had been advocating relativism, with Collins, at the University of Bath, originating the specific form known as Bath relativism. Also, in the *Physics Today* exchanges, Collins and Pinch themselves said that "we are engaged in a form of skepticism." The review by the philosopher Nickles was concerned that Collins and Pinch's sophisticated relativism was likely to induce in general readers a deplorably sophomoric relativism or skepticism.

Rising above the details, seeing the forest instead of the trees, what take-home message do scientists get from reading *The Golem*? In one sentence, the review by Turney captured the essence of *The Golem* and its seven tales of scientific perplexity: "They are, of course, moral tales, and the moral is that on close examination science is a much messier, ambiguous and problematic enterprise than is commonly made out." The crux expressed in Segerstråle's review follows: "What is the gist of the game being played in books such as this? The reasoning seems to be that if it can be shown that scientists for various philosophical or other reasons 'cannot' come to agreement on the basis of facts or experimental results, then it 'must' be social factors that are actually guiding the production of scientific knowledge.... [At] stake is the potential victory of sociology over experimental science!... It seems oddly obstinate for sociologists to tell readers that 'Nature poses much less of a constraint than we normally imagine.'... The authors in effect exclude Nature as a factor in scientific knowledge, in favor of Culture." Indeed, Collins (1981) had insisted that "the natural world has a small or nonexistent role in the construction of scientific knowledge."

The review by the philosopher Hull acknowledged that Collins and Pinch's claim that "theory and experiments *alone* do not settle scientific controversies" was a commonplace observation among philosophers, so "The interesting question is what sorts of *additional* factors contribute to such resolutions." The review by the historian Marsden took a constructive attitude: "If, however, the book sparks the curiosity of readers for fuller answers to these questions [about science], it will have served a very valuable purpose."

Another recent exchange to be considered here began with a commentary by Gottfried and Wilson (1997) in *Nature*. Subsequently, Colin Macilwain and David Dickerson continued the discussion in *Nature* (1997, 387:331–334); readers provided four letters, and Gottfried and Wilson replied (1997, 387:543–546), and then two more letters appeared (1997, 388:13; 389:538).

The main concern in Gottfried and Wilson's 1997 commentary was with attacks on science from sociologists, in contrast to Theocharis and Psimopoulos's 1987 commentary in the same journal a decade earlier that had focused on attacks from philosophers. A school of sociology, so-called Science Studies, had vigorously attacked science's traditional claims. Variants of that movement went under several names, such as the "strong program" or the Edinburgh school of sociology, but here the brief name "constructivism" suffices.

Gottfried and Wilson, in only three incisive sentences, got quickly to the very heart of the debate over science's status: "Scientists eventually settle on one theory on the basis of imperfect data, whereas logicians have shown that a finite body of data cannot uniquely determine a single theory. Among scientists this rarely causes insomnia, but it has tormented many a philosopher.... In 'cultural construction,' the Edinburgh school claims to have found the missing epistemological link."

Whereas the traditional account of scientific method explained how science achieved truth, with consensus as a by-product, the constructivist account offered consensus as the product, with no attendant claim of truth, ordinarily. Gottfried and Wilson replied that constructivists "dismiss or ignore a large body of concrete evidence" that contradicts their contention that science is merely a communal consensus, merely an agreeing to agree.

Seven lines of evidence were cited to show that science had a strong grip on reality: (1) steadily improving predictions, often unambiguous, precise, diverse, and even surprising, (2) increasingly accurate and extensive data, (3) increasingly specific and comprehensive theories, (4) interlocking evidence of diverse sorts, (5) progress over time in describing and explaining nature, (6) reproducible experiments, and (7) science-based technology that works. Of those seven witnesses to science's success, the first, "predictive power," was "the strongest evidence that the natural sciences have an objective grip on reality."

How did science's critics respond to that heap of evidence? What may surprise scientists is that the constructivist response was not to dispute that evidence, but rather to divert its force away from supporting any claims of realism or truth. Scientists must understand that in a constructivist's mind, two foregone conclusions loom large: First and foremost, philosophy had already proved that data underdetermine theory choice, so claims of truth were pretentious. Second, sociology had revealed the actual processes by which scientists come to agree, with cultural construction the central feature, rather than constraint from nature, so again claims of truth were pretentious. For instance, responding to the proffered evidence of technology that works, a letter from Andrew Pickering replied that "A machine works because it works, not because of what anybody thinks about it," such as thinking that the machine's design involved some scientific theories that were true. More generally, Pickering concluded that "In short, the continuing criticism of science studies in your pages remains wide of its mark."

Although constructivism or Science Studies may irritate scientists, that attack on science contains, implicitly, an instructive element. Science's critics, perhaps more clearly that most of science's defenders, are realizing that (1) science's big claim is the claim of truth, not merely some surrogate such as predictive success, and (2) a convincing claim of truth requires not only evidence of scientific successes but also philosophical proof of the rationality of the scientific method. More pointedly, (3) as long as the received philosophical view of scientific method features data that underdetermine theory choice, claims of truth are unsupportable, and hence explanations of consensus must invoke sociology instead of philosophy.

In short, science's detractors are asking science's defenders for a philosophical account of its rationality, not a scientific list of its successes. This is a noble request. The bottom line is that after a philosophical justification of science has been given, listing science's successes can seal the conviction of truth; but before a philosophical defense is given, and especially when a philosophical distress is already evident, various successes can be acknowledged without prompting any implication of truth.

Before closing this section on scientists' reactions, two other items merit mention. First, a particularly notorious episode in the science wars, mentioned by Gottfried and Wilson, was the so-called Sokal affair. To spoof postmodern and constructivist views of science, Sokal (1996) published an article in *Social Text*, only later to expose it as a hoax, as explained by Sokal (in Koertge 1998:9–22), Paul A. Boghossian (in Koertge 1998:23–31), Sokal and Bricmont (1998), and Steve Fuller (in Segerstråle 2000: 201–209). "'It took me a lot of writing and rewriting and rewriting before the article reached the desired level of unclarity,' he chuckles," in an interesting interview in *Scientific American* (Mukerjee 1998). That hoax provoked front-page articles in the *New York Times*, the *International Herald Tribune*, the London *Observer*, and *Le Monde*.

Second, the January 1997 issue of *Scientific American* contained an article on science and anti-science. The tone of the article, written by scientists for scientists, was somewhat complacent. The first sentence reassured readers that among various anti-science movements, "not all pose as much of a threat as has been claimed." Apparently, readers need not feel scared by what they were about to see. Nevertheless, the tone was also reasonably serious: "Science has long had an uneasy relationship with other aspects of culture. . . . Until recently, the scientific community was so powerful that it could afford to ignore its critics – but no longer."

The AAAS Posture

For several decades, particularly since the books by Popper and Kuhn appeared in 1959 and 1962, science's traditional claims of rational realism and objective truth have been under significant reappraisal and even sustained attack. As reviewed earlier, some individual scientists have reacted, as have some philosophers and others. But what about institutional replies, such as from the AAAS? Does the AAAS rebut these new ideas about science, or accept them, or just ignore them, or what?

The AAAS (1989, 1990) has offered a recent, mainstream assessment of science. Four objectives or perspectives set the tone. First, the AAAS pursues "balanced views" that are neither "uncritically positive [nor] antagonistic" (AAAS 1989:135). Regarding knowledge, "A liberal education in the sciences requires not only an understanding of what science knows, but also of what science does not know" (AAAS 1990:xiii). And regarding scientific and technological applications, both "the opportunities and risks inherent in the scientific enterprise" are acknowledged (AAAS 1990:xi), as well as the complexities of science both causing and solving various environmental and social problems (AAAS 1989:13). Second, the AAAS seeks to glean the best from science's long history, rather than being either nostalgic or trendy. "Over the course of human history, people have developed many interconnected and validated ideas about the physical, biological, psychological, and social worlds" (AAAS 1989:25). "As a liberal art, science is rooted firmly in intellectual tradition. It has a past, a present, and a future.... The past illuminates current scientific practice" (AAAS 1990:24). Third, the AAAS evaluates science and its ideas not as an isolated enterprise, but rather with intimate connections with "the wider world of ideas" (AAAS 1990:24). Accordingly, what science can claim legitimately depends not only on science itself but also on this wider world of ideas. Fourth and finally, science is not authoritarian, but rather each scientist must consider the evidence and draw his or her own conclusions (AAAS 1989:28–30). Indeed, the AAAS desires "independent thought" rather than indoctrination or uncritical support (AAAS 1989:135).

Their most general statements about science reveal most incisively just what the AAAS takes science to be. For instance, "science is the art of interrogating nature" (AAAS 1990:17). That simple but profound remark claims that science is objective in the fundamental sense of being about an object with its own independent existence and properties. Also, truth is sought: "When faced with a claim that something is true, scientists respond by asking what evidence supports it" (AAAS 1989:28). The basic elements in scientific method are observation and evidence, controlled experiments, and logical thought (AAAS 1989:25–28). Such remarks presume and express science's traditional claims of rationality, truth, objectivity, and realism.

But many modern ideas are also acknowledged. For instance, scientific thinking demands testability and falsifiability, as Popper insisted. "In fact, the process of formulating and testing hypotheses is one of the core activities of scientists. To be useful, a hypothesis should suggest what evidence would support it and what evidence would refute it. A hypothesis that cannot in principle be put to the test of evidence may be interesting, but it is not scientifically useful" (AAAS 1989:27; also see AAAS 1990:xiii).

Science advances by periods of slow "evolution" or normal science and by episodes of rapid "revolution," as Kuhn observed (AAAS 1990:24). "In the short run, new ideas that do not mesh well with mainstream ideas may encounter vigorous criticism, and scientists investigating such ideas may have difficulty obtaining support for their research.... In the long run, however, theories are judged by their results: When someone comes up with a new or improved version that explains more phenomena or answers more important questions than the previous version, the new one eventually takes its place" (AAAS 1989:28).

Scientific research is guided by paradigms that are "metaphorical or analogical abstractions" that "dictate research questions and methodology," as Kuhn emphasized (AAAS 1990:21, 24). "Because paradigms or theories are products of the human mind, they are constrained by attitudes, beliefs, and historical conditions" (AAAS 1990:21).

Data underdetermine theory: "No matter how well one theory explains a set of observations, it is possible that another theory may fit just as well or better, or may fit a still wider range of observations" (AAAS 1989:26). And data are theory-laden because theory guides the choice, organization, and interpretation of the data (AAAS 1989:27; 1990:17–18).

Science has a decidedly human face, quite unlike its earlier images offered by Francis Bacon in the 1600s or the logical empiricists in the early 1900s. Indeed, "human aspects of inquiry . . . are involved in every step of the scientific process from the initial questioning of nature through final interpretation" (AAAS 1990:18). "Science as an enterprise has individual, social, and institutional dimensions" (AAAS 1989:28). This humanity brings risks and biases: "Scientists' nationality, sex, ethnic origin, age, political convictions, and so on may incline them to look for or emphasize one or another kind of evidence or interpretation" (AAAS 1989:28). But on balance, scientists do attempt to identify and reduce biases: "One safeguard against undetected bias in an area of study is to have many different investigators or groups of investigators working in it" (AAAS 1989:28).

How much success does science enjoy in getting at the truth? The AAAS verdict is nuanced: "Scientific knowledge is not absolute; rather, it is tentative, approximate, and subject to revision" (AAAS 1990:20), and "scientists reject the notion of attaining absolute truth and accept some uncertainty as part of nature" (AAAS 1989:26). "Current theories are taken to be 'true,' the way the world is believed to be, according to the scientific thinking of the day" (AAAS 1990:21). Note that "true," here sequestered in scare quotes, is equated to nothing more real or enduring than "the scientific thinking of the day." Furthermore, science's checkered history "underscores the tentativeness of scientific knowledge" (AAAS 1990:24).

Nevertheless, "most scientific knowledge is durable," and "even if there is no way to secure complete and absolute truth, increasingly accurate approximations can be made" (AAAS 1989:26). However, deeming durability to be something good or admirable presumes that durability is serving as some sort of truth surrogate, because otherwise the durability or persistence of a false idea is bad. Hence, switching from "true" to "durable" is not a successful escape from the issue of truth. One of the most positive remarks is that "the growing ability of scientists to make accurate predictions about natural phenomena provides convincing evidence that we really are gaining in our understanding of how the world works" (AAAS 1989:26).

All in all, the AAAS verdict is a nuanced mix of positives and negatives: "Continuity and stability are as characteristic of science as change is, and confidence is as prevalent as tentativeness. . . . Moreover, although there may be at any one time a broad consensus on the bulk of scientific knowledge, the agreement does not extend to all scientific issues, let alone to all science-related social issues" (AAAS 1989:26, 30).

One must also observe, however, that the pages of AAAS (1989) catalogue hundreds of facts about the universe, the earth, cells, germs, heredity, human reproduction and health, culture and society, agriculture, manufacturing, communications, and other matters. Undoubtedly, the vast majority of these facts are presented with every appearance of truth and certainty and without even a trace of revisability or tentativeness. For instance, science has declared that the earth moves around the sun (and around our galaxy), and the former theory that the earth is the unmoving center of the universe is not expected to make a stunning comeback because of some new data or theory!

Admittedly, it is awkward that some AAAS declarations sound as though they reflect the concept that all scientific knowledge is tentative, whereas other statements apparently present settled certainties. Perhaps the AAAS verdict on science and truth is a bit unsettled, or perhaps some isolated statements lend themselves to an unbalanced or unfair reading relative to the overall message. Anyway, it may be suggested that, given a charitable reading, the AAAS position papers say that some scientific knowledge is true and certain, some is probable, and some is tentative or even speculative and that scientists usually have good reasons that support legitimate consensus about which level of certainty is justified for a given knowledge claim.

Although the AAAS acknowledges revolutionary changes in paradigms, they explicitly deny that successive paradigms are incommensurable or fail to move closer to the truth: "Albert Einstein's theories of relativity – revolutionary in their own right – did not overthrow the world of Newton, but modified some of its most fundamental concepts" (AAAS 1989:113; also see p. 26). Newton's mechanics and Einstein's relativity lead to different predictions about motions that can be observed, so they are commensurable, and relativity's predictions have never failed, so it is the better theory (AAAS 1989:114). And yet, looking to the future, physicists pursue "a more [nearly] complete theory still, one that will link general relativity to the quantum theory of atomic behavior" (AAAS 1989:114).

Likewise, the AAAS repeatedly discusses the logic of falsifiability. But unlike Popper, they also repeatedly discuss the logic of confirmation, including sophisticated remarks about the criteria for theory choice (AAAS 1989: 27–28, 113–115, 135). More generally, the AAAS acknowledges science's human face. But this causes no despair about humans rationally investigating an objective reality with considerable success.

Besides publishing position papers that bid fair as mainstream assessments by scientists of science, the AAAS is also called upon occasionally in an official capacity to represent the scientific community. For instance, when the Congress of the United States wanted a current assessment of science's rationality and objectivity, a 1993 symposium was co-convened by George Brown (who chaired the committee that controls 20% of the United States' entire science research budget) and the AAAS for the purpose of providing "a philosophical backdrop for carrying out our responsibilities as policymakers"

(Brown 1993:iii). The result was a report to Congress entitled *The Objectivity Crisis: Rethinking the Role of Science in Society* (Brown 1993).

Citing Kuhn (1970) and the interview with him in *Scientific American* (Horgan 1991), that report includes a call to scientists to accept the new picture of science as myth: "Some scientists are still scandalized by the historical insight that science is not a process of discovering an objective mirror of nature, but of elaborating subjective paradigms subject to empirical constraints. Other scientists, especially in established specializations, are largely unaware that basic assumptions shape scientific progress. Nevertheless, it is important to understand the nature, function, and necessity of scientific paradigms and other myths.... [We] must depend on *a priori* faith in our various myths and sub-myths to exploit our limited capacity for reason" (Ronald D. Brunner, in Brown 1993:6). Scientists wanting additional sobering reading can turn to three further reports to Congress in 1995 on scientific integrity and public trust (ISBN 0-16-052519-5, 0-16-052761-9, and 0-16-052655-8).

The AAAS also publishes the journal *Science*, which occasionally carries articles about the philosophical foundations of science and even the recent science wars. Unlike position papers that express AAAS views, such articles represent individual persons' views. Nevertheless, they reveal something about what the AAAS in general or the *Science* editors more specifically find noteworthy. Two items merit brief mention.

When Wade (1977) reviewed Kuhn's *The Structure of Scientific Revolutions* (1970), he began by saying that historians and philosophers of science have become more influential than scientists themselves in shaping notions of science's method and process. So, in the terms used here, science's external auditors are influential. Wade recounted Kuhn's ideas about incommensurable paradigms and elusive truth. He cited diverse reactions from scientists, sociologists, historians, and philosophers. But, curiously, he communicated no sense that Kuhn's huge theses tasked the scientific community with any imperative to accept them if true or else reject them if false.

More recently, the historian Forman (1997) has reviewed Gross et al., *The Flight from Science and Reason* (1996). This is one of the main books that defends science's rationality, particularly against postmodern and constructivist attacks. Forman takes the book's aim as "placing science back on its pre-postmodern pedestal." But Forman, who clearly identifies with "our postmodern world," finds the arguments mostly unconvincing and often self-defeating. Also, he perceives its rhetoric as being distasteful, an "invective written by certain rabid rationalists and incidental soreheads attracted to this outlet for outrage."

Finally, the AAAS cherishes independent thought and despises indoctrination, so neither they nor any other sensible persons could expect all scientists to agree with all statements in the AAAS position papers. So it seems that the self-chosen destiny of the AAAS is to serve an unruly mob of independent thinkers! Nevertheless, it seems fair to expect that most scientists would agree with most of the AAAS positions summarized in this subsection. In any case, the AAAS position is a compelling point of departure, both for those who mostly agree and for those who mostly disagree.

With admirable candor, the AAAS (1990:26) recognizes that even their careful position papers "are set in a historical context and that all the issues addressed will and should continue to be debated." One curious feature of AAAS (1989, 1990) is that despite the numerous unmistakable allusions to Popper and Kuhn's influential ideas, those figures are neither named nor cited. Well, an idea's merit depends on its content, not its sponsor (and indeed the reverse sentiment is a logical fallacy, namely, the genetic fallacy). As the issues continue to be debated in future years, however, it will be interesting to see whether or not the AAAS decides to engage science's external auditors – particularly the philosophers, historians, and sociologists – more directly.

The Typical Accommodation

So far, this section has explored prominent scientists' reactions to the philosophers and to the science wars as reflected in individual and institutional responses. To complete the picture, however, something must be said about the typical reaction or accommodation of ordinary scientists as expressed in their routine work.

What is going on when an ordinary scientist is not engaging philosophers or science wars directly, but rather is just publishing a routine research paper, say in an astronomy or ecology journal, and that paper happens to cite the work of figures such as Popper and Kuhn? Indeed, such citations are moderately common, and most scientists have at least a passing familiarity with those influential intellectuals. What concepts are scientists drawing from such philosophers?

For the most part, despite occasional noteworthy exceptions, it must be said that the routine use of Popper and Kuhn's ideas by scientists is rather uninspired. Frankly, the appropriations tend to be selective, superficial, and self-serving.

For instance, the first of the 18 replies to Theocharis and Psimopoulos (1987) was quite revealing. It nicely illustrated the selective reading of the philosophers that is characteristic of scientists. The writer was a physiologist who also taught a course in the philosophy of science: "Popper began it all by his concern to distinguish good science from bad. He identified Einstein as good and Adler as bad by characterizing Einstein's predictions as *falsifiable* but not as false. Falsifiable predictions were simply firm, clear predictions; predictions which explicitly required that the contrary did not occur.... I share the view that the next stage of Popper's thought sees him following up certain ideas to unbalanced and therefore somewhat antirational conclusions, but this first, key perception [regarding falsifiability] is firmly on the side of objectivity and truth. This is the one bit of Popper that I teach." His sensible posture toward the philosophers was that "If their accounts are unbalanced, it is up to us [scientists] to balance, not dismiss them." So it was up to scientists to selectively pick and teach the good bits of the philosophers' writings.

A careful reading of Popper and Kuhn reveals many disturbing ideas, and even more so a reading that takes their ideas to their logical conclusions. But the selective, superficial reading by scientists is rather innocuous, essentially just confirming what scientists already think and do anyway. But writing a paper that quotes the philosophers imparts a certain appearance of sophistication and legitimacy, and that nice bonus comes at no cost, because a selective reading imposes no interference whatsoever with doing exactly what the scientist wanted to do anyway!

It is difficult to know what to say about such a reading. Such superficiality is astonishing and deplorable in otherwise scholarly scientific papers. But perhaps more significantly, the scientists' superficial reading merely reveals a wonderful innocence. Noticing the good and missing the bad is precisely the posture of innocence. Scientists are simply doing business with reality and truth, and they seem remarkably unable to really get the idea that someone else really is up to something else. Consequently, potentially disastrous philosophical claims, such as claims that science is arational or finds no truth, cause scientists rather little harm – at least in their routine scientific investigations – because those claims are not among the "one bit of Popper" or the one bit of Kuhn that scientists get.

The one bit of Popper that does show up frequently in scientists' research papers is that a proposed hypothesis must make testable predictions that render the hypothesis falsifiable. And, more pointedly, a scientist should give his or her own favored hypothesis a trial by fire, deliberately looking for potentially disconfirming instances, not just instances that are likely to be confirming. Doubtless, this is wholesome advice.

But is this some fancy, new insight? Hardly! It is as old as ancient *modus tollens* arguments (*A* implies *B*; not *B*; therefore not *A*) and the medieval Method of Falsification of Robert Grosseteste (Losee 1993:36–38). Looking for potentially contradictory evidence seems more in the province of simple honesty than fancy philosophy.

Furthermore, the standard outcome of a hypothesis test in scientists' papers is the declaration that a particular hypothesis is true. Often this takes the form of a statistical test, taken to mean, in essence, that there is only a tiny chance that the outcome was due merely to chance, so in a given instance the conclusion has a confidence level of, say, 95% or 99.9%. However, such scientists are wholly unaware that they are bad disciples of their presumed master, Popper. Unlike his unprincipled disciples, Popper himself talked only of falsifiability and disconfirmation, but never confirmation!

Similarly, the one bit of Kuhn that does show up frequently in scientists' writings is the dramatic idea of a paradigm shift. Needless to say, this idea is particularly popular among those scientists who take themselves to be the innovators who are precipitating some big paradigm shifts in their own disciplines! And if one scientist says of another that he or she caused a paradigm shift that advanced their discipline tremendously, then those are words of high praise indeed. Scientists love to propose paradigm shifts. Of course, the standard claim in scientists' papers is that their shiny new paradigms are a whole lot better than their predecessors, and even are true or at least approximately true.

But again, unwittingly such scientists are bad disciples of their presumed master, Kuhn. His own view was that successive paradigms are incommensurable. Consequently, it makes no sense to say that one paradigm is better than another, or to claim that there has been any progress toward the truth. Of course, Kuhn's own view takes all the fun and prestige out of coming up with a slick new paradigm! So it is not too surprising that scientists have generally failed to get that discouraging bit of Kuhn.

Well, what would the world of science look like if instead scientists became good disciples of philosophers of science such as Popper and Kuhn? For starters, what about that critical scientific document, the grant proposal? At present, among all of the documents generated around the globe, these grant proposals represent the most confident literature to be found in all the earth, full of bold promises of great things to come.

Science's claims of rationality, truth, objectivity, and realism are so indispensable and pervasive that we must exercise our imaginations to envision a science stripped of those claims. The following opening words from a scientist's fictitious grant proposal to the National Institutes of Health are offered to enliven our image of such a bereft science: "My scientific methods are arational and no more likely to find truth than a guess, and my scientific conclusions are subjective and make no claims about reality or about what other persons should also believe or do. Anyway, please send me $200,000 per year for the next four years to study dental fillings." Such is what a teachable, penitent disciple of Popper and Kuhn would write! But need it be said that this won't wash? Is this applicant going to get his or her $800,000 of taxpayers' money – your money? Do you want this scientist's findings applied to your teeth? Such meek science is unnatural and unattractive.

What happened? Why is the current relationship between scientists and prominent philosophers so messy, complex, and rather dingy?

The explanation is actually rather straightforward: different purposes. As regards any aspect of science's methodology, be that either general principles or specialized techniques, the primary purposes of theoretical and applied science are to wrest secrets from nature and to develop useful technologies. On the other hand, philosophy's main interests regarding science have been to demarcate science from non-science, to put science in historical and cultural context, to describe scientists' actual methods, to explain the reaching of consensus, and to evaluate science's claim of rationality. But conspicuously absent from the philosophers' goals has been the task of helping scientists become better scientists, that is, helping them to advance their own purposes. Therefore, because of their different purposes, when scientists have tried to put the philosophers' ideas to work for their scientific purposes, the result inevitably has been a mismatch.

This is not to say that it is impossible for philosophers to see in their work many ideas that could help scientists. Neither is it impossible for them to focus attention on such ideas in publications that scientists can easily encounter. Rather, this is only to say that such opportunities have not been widely implemented.

Finally, this subsection has characterized scientists' typical, day-to-day use of popular philosophers' ideas. What is the bottom line? The bad news is a recent history of bad disciples misappropriating bad philosophies for the self-serving purpose of purchasing an appearance of sophistication on the cheap. The good news is that nothing is keeping scientists from a future history of mastering good philosophy for the practical purposes of increasing productivity and gaining perspective.

TWO RULES OF ENGAGEMENT

When scientists themselves encounter philosophers and others in grand discussions of science's rationality – that is, when science's internal auditors encounter its external auditors – two rules of engagement are absolutely crucial. Of course, a longer list of rules could be given. But the two principal rules are so surpassingly important that it is better to propose two rules that might be remembered than to propose six or seven that could easily all be forgotten. They concern (1) clearly defined targets and (2) rhetoric and substance. These two rules of engagement hold the key to shifting the debate in the science wars from a history of progressive polarization to a future of meaningful communication.

Clearly Defined Targets

Consider any of the arguments, as reviewed earlier in this chapter, against science's rationality and truth. For instance, recall the arguments that data

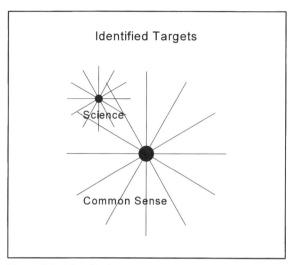

Figure 3.4. Identified targets for arguments against science's rationality. Arguments vary in their explosive force and intended targets. Some attack science only, whereas others attack both science and common sense. Because these two options call for very different analyses and responses, the targets of a given argument must be clearly identified.

underdetermine theory choice, that data are theory-laden and hence can neither falsify nor confirm a theory, and that successive paradigms are incommensurable and hence do not progress closer toward the truth.

Exactly what are the targets of such arguments? Is the target science only? Or are the targets science and common sense? These two options are depicted in Figure 3.4.

If the target is science only, then the argument presents a challenge that scientists really need to answer. But if the targets are science and common sense, then the argument is merely some variant of radical skepticism, and the scientific community is under no obligation to find it of any interest.

Science *begins* with the presumption that the physical world is comprehensible to us, as will be explored in detail in the following chapter on science's presuppositions (AAAS 1989:25; 1990:16). Therefore, any argument that assays to take down common sense and thereby to establish radical skepticism is simply outside science's purview. Science has many theoretical and practical tasks, but refuting skepticism is not one of them. A legitimate attack on science's rationality must target science alone, not science and common sense both. So whether an argument attacks science or else attacks science and common sense defines two wildly different cases. Why this is so can be illuminated by an analogy.

People often have the perception that many disease organisms are difficult to kill, because numerous terrible diseases still ravage millions of suffering persons, and scientists have no cure. But in fact, all of these viruses, bacteria, and other microbes are easy to kill – every last one of them. A strong dose of arsenic or cyanide could kill them all, not to mention the even easier expedient of merely heating them to 500˚ C. It is easy to kill any pathogen. The trick is not to kill the host at the same time! Medicine's challenge is to kill the pathogen *and* not kill the host. So a strong dose of arsenic fails to qualify as a medicine not because it kills too little, but because it kills too much.

Now, the same situation applies to science and common sense. Countless philosophical arguments and blunders kill common sense, which obviously kills science also. It is easy to kill common sense and science both. Indeed, a lackluster high-school sophomore can easily learn five skeptical objections in as many minutes – maybe our sense perceptions are unreliable, maybe some demon is deceiving us, maybe the physical world is only an illusion, maybe the future will be unlike the past, and so on. But for any discipline such as science that begins with a non-negotiable conviction that common sense delivers much truth because the world is comprehensible, a philosophical argument that kills science and common sense is unimpressive, not because it kills too little, but because it kills too much.

Consequently, the first and greatest burden placed on scientists when they read anti-scientific arguments is to ask this discerning question: Does this philosophical argument kill science alone, or does it kill science and common sense both? Any argument that, if understood clearly and applied consistently, would imply that we cannot really know trifling trinkets of common-sense knowledge is just plain ridiculous, whether or not the scientist has enough philosophical training and acumen to spot and refute the specific steps at which the argument goes awry.

Accordingly, as scientists encounter their external auditors, both individually and institutionally, they should put those auditors on notice that their science discipline begins with the presupposition that the world is comprehensible, that they want science's objectors to make it clear whether their arguments attack science only or both science and common sense, and that they have no interest in objections that amount to merely variants of radical skepticism. Scientists can politely insist that if a book's argument attacks both science and common sense, that fact is sufficiently critical that it should be stated on page 1 or 10, not delayed to page 95 or 395. And scientists can legitimately insist that an argument be coherent. If a book praises and presumes common sense in some passages, but rejects and attempts to undo it in other passages, that is incoherent at best or dishonest at worst.

In a word, the scientific community needs to make known to its external auditors that its rules of engagement require clearly defined targets. Whether an argument's target is science only or both science and common sense must be specified clearly in order to earn the interest and engagement of the scientific community.

Rhetoric and Substance

Theocharis and Psimopoulos (1987) mentioned an intriguing difference between two ancient philosophers. In the fifth century B.C. the Greek sophist Gorgias argued that "one can persuade anyone of anything, if one speaks well enough." But Aristotle replied that "One is more likely to win one's argument, if what one says is true." Well, every argument has some mixture of rhetoric and substance, and both affect persuasiveness. Nevertheless, the grand debate over science's rationality is so inherently complex that all participants should be urged to emphasize substance over rhetoric.

Downplaying rhetoric does not mean downplaying passion. Whether or not science is rational and can find truth is, after all, a hugely significant debate. It is only natural that both sides are passionate. What is distasteful, however, is to see an intelligent scientific debate hijacked into the combative language of ideological or religious wars, or to hear a careful argument rebutted with deliberately inflammatory and distracting rhetoric.

But simply advising more substance and less rhetoric is unlikely to work. The problem is that opinions about what is rhetoric and what is substance are strongly influenced by one's own position. It is all too easy, relative to one's own stance, to feel that the bad guys' arguments are just so much rhetoric, whereas the good guys' arguments are full of real substance! So, rather than recommending more substance directly, four other recommendations may be suggested that can indirectly but effectively promote the desired focus on substance.

First, it would help to identify a list of core topics that are essential for any assessment of science's rationality, either positive or negative. For instance, what is required to present a scientific argument with full disclosure? In other words, what are the inputs – all of the inputs – required to reach a scientific conclusion? This is a core topic because any sensible judgment about whether or not a given scientific argument is rational and its conclusion is true must begin with full disclosure of the argument's premises. Another core topic is the relationship between science and common sense. No satisfying or settled verdict on science can be reached without exploring this fundamental relationship. Still another core topic concerns the need for clearly analyzed conceptions and definitions of key terms such as "truth." A conceptual definition of truth must precede any attempt to analyze and assess the necessary and sufficient conditions for obtaining truth, or for failing to do so.

Second, science needs to wrestle its proper opponent. As Werner Callebaut (1993:xv) remarks, the "Philosophy of science as currently practiced is a reaction against a reaction." Too often a given position on science is shaped more by reaction to some other position than by reaction to nature. For instance, many commentators, having been scandalized by the excessive confidence of logical empiricism back in the 1930s, consider it their bounden duty to take science down a peg or two. But such a reaction to a re-action seems anachronistic and counterproductive in the current culture that is rather skeptical and postmodern, even in need of having science taken up a peg or two. Anyway, resolution of the four woes listed earlier in this chapter will require that one forgo reacting to reactions and instead react to nature.

Third, we must distinguish philosophical facts from interpretations, and at least attempt an interpretation of philosophical findings that will count for, instead of against, science's rationality. For example, philosophers have shown that data are theory-laden. That argument is compelling and correct. But then the scary spin put on that story is that therefore data and theory are so intertwined that data cannot guide theory choice, so truth is elusive. But how about at least attempting to explore a different line of interpretation? Perhaps data are theory-laden, and theory is data-laden, and perhaps that deep interaction between data and theory, with data having priority, as Aristotle insisted, is precisely what makes science work! For another example, many philosophical objections have been leveled at inductive logic, and that might seem a potentially disastrous challenge, because induction is an indispensable component of scientific method. Perhaps properly oriented, however, those objections can be seen as invitations to develop a more rigorous and more comprehensive understanding of induction.

Fourth and finally, we must distinguish between puzzles and problems. The philosopher Couvalis (1997:9–10) expresses this matter nicely:

> The history of philosophy is littered with arguments leading to absurd con-clusions, [arguments that] sound good but are obviously false – although it can take generations to figure out precisely what is wrong with the argu-ments. Unfortunately, rather than treating such arguments as entertaining puzzles, some philosophers have a lamentable tendency to take them seri-ously in the name of objectivity. The fact that some of their predecessors took other seemingly powerful philosophical arguments seriously and were shown to be wrong should have made much more of an impact in philos-ophy than it has. When dealing with apparently flawless arguments for a seemingly irrational or absurd claim, the common-sense rule of thumb is to start by asking: given the deplorable failure rate of apparently flawless philosophical arguments, is it likely that the conclusion of this argument is true? Typically, we will be able to note that the conclusion is highly unlikely to be true as it implies many things which are very likely to be false. This should be taken to indicate that something is wrong with the argument and

that it poses a mere puzzle rather than a serious problem. I shall be arguing that many of the arguments which purport to undermine the objectivity of scientific knowledge actually pose only puzzles.

The preceding four recommendations can promote an intellectual climate in which substance can trump rhetoric. These are the traits of contributions that can rightly earn the interest and engagement of the scientific community.

SUMMARY

Opinions about science's rationality and objective truth have always been strongly influenced by the claims of scientists themselves. But the assessments and judgments of philosophers, historians, sociologists, and others are becoming increasingly influential. So science's internal auditors are being checked by its external auditors. This setup is legitimate and beneficial. Even highly critical views are welcome, prompting scientists to really think things through.

Science's four traditional claims are rationality, truth, objectivity, and realism. But those claims are now under heavy attack, especially in terms of four intellectual problems: (1) Karl Popper claimed that empirical data could falsify a theory but never prove it, so science could never find truth. (2) Popper and others claimed that observations were theory-laden and that data underdetermined theory choice. From that, Imre Lakatos drew the implication that scientific theories could not be falsified either, so science could not declare any theory either true or false. (3) Thomas Kuhn said that paradigms were incommensurable, so science was arational and had no objectivity. (4) Kuhn also said that what made a statement scientific was merely that scientists said it. But then Paul Feyerabend went on to conclude that there was nothing special about science.

The first notable exchange among scientists, philosophers, and others regarding science's rationality began with the commentary by Theocharis and Psimopoulos (1987) in *Nature*. Those scientists felt that the critiques by Popper and Kuhn and other philosophers were unjustified and exaggerated, but nevertheless quite influential, so it was incumbent upon the scientific community to give a satisfying defense of science's rationality, objectivity, and truth. There have also been more recent exchanges. Position papers from the AAAS have provided a mainstream institutional response. Since about 1995, the debate has become so intense that people speak of the "science wars." But the typical reactions of scientists in their routine research papers, which sometimes cite Popper or Kuhn, tend to be rather unconcerned, selective, superficial, and self-serving.

To render the ongoing debate between the attackers and defenders of science's rationality more productive, two rules of engagement are recommended. First, for each argument against science's rationality, it must be made crystal clear whether the target is science only or both science and common sense. Those two targets call for very different analyses and responses. Second, several measures can be taken to promote more substance and less rhetoric. These include listing the core topics, wrestling with nature instead of with opinions, distinguishing between philosophical findings and their interpretations, and distinguishing between mere puzzles and serious problems.

CHAPTER FOUR

SCIENCE'S PRESUPPOSITIONS

Every conclusion of science requires presuppositions, just as necessarily as every conclusion of science requires evidence. Indeed, without appropriate presuppositions, evidence loses its evidential role, and that undoes science. Consequently, any reflective version of science must understand, disclose, and legitimate science's presuppositions.

The concept of "presupposition" will be defined more carefully later. Essentially, a presupposition is a belief that is required to reach a particular conclusion, and yet it cannot possibly be proved. A presupposition cannot be proved in the ordinary sense of marshaling definitive evidence because presuppositions precede and empower evidence. But that does not necessarily mean that presuppositions are arbitrary and shaky. Rather, presuppositions should be chosen carefully, disclosed, and then legitimated by appeal to common sense and sincerity.

Although presuppositions and evidence are equally essential, in ordinary scientific discourse the presuppositions are ignored, whereas the evidence is cited. Why? Basically that is because within the context of ordinary, common-sensical science, the presuppositions needed in science are sensible and unproblematic and are taken for granted. Nevertheless, "Our presuppositions are always with us, never more so than when we think we are doing without them" (O'Hear 1989:54).

Presuppositions and evidence are different in terms of implementation. Legitimating science's presuppositions can be done once and for all, whereas marshaling evidence must be done for each individual scientific inquiry. Envision reading several scientific papers about diabetes, cosmic rays, semiconductors, and turbulence over airplane wings. Each of those papers would contain its own distinctive and relevant set of evidence, different from those for the other studies. But all of those papers would involve the same presuppositions.

The method for discovering and legitimating science's presuppositions proceeds by philosophical reflection, which may be unfamiliar to most scientists and yet should be found to be reasonably simple and sensible. The goal is to identify presuppositions that are necessary and sufficient to support an ordinary realist interpretation of science, as well as unproblematic in the sense of presenting no obstacle to science being objective and public.

This chapter's topic of presuppositions bears primarily on this book's secondary objective of enhancing perspective. To the extent that answering the challenges from the science wars and defending science's rationality are important for science's long-term health, however, this chapter is crucial. It is primarily through its presuppositions (rather than its evidence or logic) that science interacts with various worldviews, resulting in positive or negative assessments of science's rationality and truth claims.

Furthermore, in order to construct a single, public version of science that will work equally well in (virtually) all worldviews, thereby preserving science's credibility and objectivity, great care must be exercised when selecting the roster of science's presuppositions. Too few presuppositions will leave science vulnerable to skepticism and thereby obliterate science's rationality. Too many presuppositions will wed science to one worldview, while divorcing it from others, and thereby obliterate science's objectivity.

By way of preview, this chapter argues that science requires more logic, more evidence, more instrumentation, more education, and more work than does common sense, but nothing more by way of presuppositions. Presuppositions cannot be proved by logic or established by evidence; rather, they can be disclosed by philosophy and accepted by faith.

HISTORICAL PERSPECTIVE ON PRESUPPOSITIONS

A historical review of presuppositions would be much easier to write if they were always called "presuppositions." But they go by many additional names, including assumptions, suppositions, starting points, *a priori* beliefs, axioms, premises, first principles, first philosophy, and first truths. Worse yet, they often affect and even dominate discussions without ever being mentioned or named explicitly. Even when presuppositions go unmentioned, they routinely exert their subtle and profound effects on other topics, especially attitudes toward empirical evidence and toward science's goals, credibility, and domain.

Hence, to understand the role of presuppositions in science, both historically and currently, we must do the work of detectives, noticing every little clue. Because presuppositions are the most subtle and problematic components of scientific method, some historical perspective will be invaluable.

Especially for scientists who have given science's presuppositions little thought, a brief historical survey will provide a window into the various options and their implications.

Aristotle (384–322 B.C.) said that "In every systematic inquiry (*methodos*) where there are first principles [*archai*], or causes, or elements, knowledge and science result from acquiring knowledge of these" (Irwin 1988:3). His *Metaphysics* was a sustained πρώτη φιλοσοφία or first philosophy: Underlying scientific results there were scientific methods, and underlying scientific methods there were scientific presuppositions. Aristotle's vision of an ideal or mature science was composed of first principles known to be necessarily true plus their deductive consequences that were also guaranteed to be true (Losee 1993:3–15). Hence, Aristotle's ideal science was analogous to Euclid's geometry, the former deriving numerous results from a few first principles, just as the latter derived numerous theorems from a few axioms. Naturally, Aristotle's writings on biology and other natural sciences reflected little of that ideal scheme, but rather relied on systematic observations, as does contemporary science. Consequently, subsequent generations of scientists have had to elaborate Aristotle's scientific vision to encompass the natural sciences.

The relative weight on presuppositions and evidence was problematic in the work of Aristotle. On the one hand, his admirable scientific realism committed him to serious emphasis on observation. Most fundamentally and strategically, Aristotle's scientific goal, expressed in his own terms, was to move from things "known by us" to things "known by nature," that is, from appearances to realities. Recall that he believed that "the facts about the world determine the truth of statements, but the converse is not true," and this "asymmetry in explanation . . . is taken to be a defining feature of truth about objective reality" (Irwin 1988:5). Thus data outweighed theory. On the other hand, Aristotle's science contained numerous beliefs that gripped scientists for centuries but were simply unmitigated and unsupported presuppositions, such as the belief that heavenly bodies had to move in perfect circles (Hübner 1983:51–52). Unfortunately, sometimes presuppositions outweighed the data.

Aristotle's *Posterior Analytics* has been called "one of the most brilliant, original, and influential works in the history of philosophy; it determined the course of philosophy of science – and to some extent of science itself – for two millennia" (Barnes 1975:xiii). The central question about scientific method posed in the *Posterior Analytics*, when boiled down to one sentence, is this: What logic and evidence are needed to demonstrate and organize scientific conclusions? In other words, what must go in so that scientific conclusions can come out?

Figure 4.1. Albertus Magnus, the patron saint of scientists. This 1352 fresco by Tommaso da Modena is in the Dominican monastery of San Niccolò, Treviso, Italy. (This photograph is reproduced with kind permission from Art Resource.)

That remains the most basic question that can be asked about scientific method. Aristotle's solution was highly technical and largely remote from contemporary scientific thinking, so it will not be explored here. But it will be salutary to keep in mind that incisive question that exercised the philosophy of science for two millennia. A satisfactory answer must list presuppositions among the inputs that justify science's outputs, as explained in the next section.

Albertus Magnus, or Albert the Great (c. 1200–1280), was a professor at Paris (Figure 4.1). He is remembered primarily as a theologian and as the teacher of Thomas Aquinas, but he also wrote extensive commentaries on Aristotle, including the *Physics*. "Albert was ... a remarkably acute firsthand observer of plant and animal life ... and he was perhaps the best field botanist of the entire Middle Ages. His intellectual energy was boundless, and his nontheological writings (less than half the total) included works on physics, astronomy, astrology, alchemy, mineralogy, physiology, psychology, medicine, natural history, logic, and mathematics" (Lindberg 1992:230).

Albertus Magnus handled science's presuppositions by an appeal to conditional necessity, a concept that ranks among the most important notions in the philosophy of science. He was building on the concept of suppositional reasoning that had been explained by Aristotle (Apostle 1969:40–41, 220–221; Charlton 1970:42–44, 127–128; Ferejohn 1991; Judson 1991). Something about nature, X, may be claimed to be true either unconditionally or conditionally, respectively symbolized as "X" and "X given Y." As an example of unconditional or absolute necessity, "A man is not taller than himself" is self-evidently true, standing on its own by virtue of the meanings of its words, without needing an appeal to any givens, suppositions, experiments, or evidence. By contrast, following Albertus's example, "You are sitting" has conditional necessity given "I see you sitting." That you are sitting is not self-evident or necessary from first principles, but it becomes necessarily true if I see that such is the case, at least on the supposition of business as usual between us and the physical world.

For the two options, unconditional necessity and conditional necessity, Aristotle used the terms "simple" and "hypothetical" (ὑπόθεσις). Albertus translated ὑπόθεσις into Latin as *conditio*, *suppositio*, and *positio*, and the usual Latin expressions are necessity *ex conditione* or *ex suppositione*.

Incidentally, the English term "conditional," derived from the Latin translation of the original Greek, is a fitting word in contemporary English; but a direct translation of the Greek is misleading, because "hypothetical" has come to have a very different meaning. Originally, a "hypo-thesis" was literally an "under-thesis," that is, a support for a thesis. So, in the conditional claim "X given Y," X is the thesis (θεσις), and Y is the under-thesis (ὑπόθεσις) or support for the thesis X. But now a "hypothesis" is an uncertain belief under consideration, possibly true and possibly false, that will require more research and evidence to determine whether or not it or some other competing hypothesis is actually true.

Albertus rendered the opening sentence of Book 2 of Chapter 9 of Aristotle's *Physics* as follows, with Aristotle's text in italic type and Albertus's amplification in roman type: "We ask therefore first *whether the necessity of physical things is a necessity simply or is a necessity 'ex suppositione'* and on the condition of some end that is presupposed. For example, a simple necessity is such that it is necessary that the heavy go down and the light go up, for it is not necessary that anything be presupposed for this for it to be necessary. Necessity 'ex conditione,' however, is that for whose necessity it is necessary to presuppose something, nor is it in itself necessary except 'ex suppositione'; and so it is necessary for you to sit if I see you sitting" (William A. Wallace, in Weisheipl 1980:116; also see p. 112).

In the subsequent formulations of Thomas Aquinas, Jean Buridan, and others, that notion of conditional necessity was gradually shifted in two significant ways. First, Aristotle and Albertus emphasized final causes, but later views of scientific explanation gave diminishing attention to teleology. Second, ontological concern with necessity gave way to epistemological concern with certainty. Often the motive for demonstrating that something was necessary had been to establish that it was certain. But necessity was a much richer concept than certainty, implicating a much larger (and potentially more controversial) story about how the world works. For example, right now, I may wave my arm, and be certain that I have just done so; but whether or not that action was necessary (given the previous states of the universe or whatever) implicates daunting debates over determinism (in either theistic or atheistic worldviews). Because establishing certainty by establishing necessity implicated an excess burden, scientific interest in certainty became progressively detached from inquiry into necessity.

So in its contemporary setting, the "suppositional necessity" of Albertus became what can be termed the "conditional certainty" of modern science. Updating Albertus's example, the statement that "You are sitting" has conditional certainty given the empirical evidence that "I see you sitting." As the saying goes, "Seeing is believing," or "What you see is what you get." The observation or under-thesis that "I see you sitting" counts as evidence for the claim or thesis that "You are sitting" precisely and only because common-sense presuppositions are accepted to the effect that "Seeing is believing."

So why can't biology be as certain as geometry? Well, in Albertus's view, biology *could* be as certain as geometry: "Surely Albert entertained no doubts that . . . one could have certain and apodictic demonstrations even when treating of animals, provided the proper norms of *ex suppositione* reasoning were observed" (William A. Wallace, in Weisheipl 1980:127–128). For example, that (normal adult) horses eat grass and that (Euclidean) triangles have interior angles totaling 180 degrees have equal degrees of certainty, even though they have different grounds of certainty.

Albertus proposed that science take ordinary human capacities to their limit without excessive dependence on philosophy and theology. That growing independence of science from theology was grounded in the concept of conditional necessity. Science started with common-sense presuppositions about the reliability of sense perception and the existence and uniformity of nature.

Thomas Aquinas endorsed his teacher's views of science's presuppositions and science's demarcation from theology and philosophy. Like Albertus, Aquinas wrote an extensive commentary on Aristotle's *Physics*, with his exposition on suppositional necessity in Lecture 15 of Book 2 (Blackwell, Spath, and Thirlkel 1963:124–128).

Other medieval masters endorsed Albertus's solution regarding science's presuppositions and demarcation. For John Duns Scotus, "evidence is considered a sufficient foundation of true and certain knowledge, regardless of whether the latter is of a necessary or contingent nature. By shifting the stress from objective necessity to objective evidence, Scotus has opened the road to a new concept of scientific knowledge which is both more realistic and more comprehensive than the aprioristic *scientia* of Aristotle and his followers" (Vier 1951:165; also see Bettoni 1961:123–129). In other words, conditional certainty was as certain as absolute certainty provided that the conditions were unproblematic; biology *could* be as certain as geometry. Likewise, William of Ockham held that the natural sciences could achieve conditional truths, which was different from mathematics achieving necessary truths, but those conditions would be unproblematic presupposing the common course of nature [Ernest A. Moody, in Gillispie 1970(10):171–175]. Jean Buridan further legitimated science's autonomy. Also, as "a brilliant logician in an age of brilliant logicians" (King 1985:xi), Buridan further developed the theory of supposition in Treatise IV of his *Summulae de dialectica*. When Copernicus was a student at Cracow, Buridan's works in physics were required reading.

The greatest tribute to the medieval grasp of science's presuppositions was that presuppositions handled rightly had endowed evidence with real power. "While the authority of Aristotle had often been challenged on the ground that his positions contradicted Christian doctrine, it had come, in Buridan's time, to be challenged on grounds of inadequacy as a scientific account of observed facts" [Ernest A. Moody, in Gillispie 1970(2):605].

Francis Bacon (1561–1626) was an influential proponent of science (Quinton 1980; Urbach 1987; Pérez-Ramos 1988; Martin 1992; Peltonen 1996). His *Novum Organum* claimed to offer a better scientific method than had his predecessors, especially Aristotle. It emphasized empirical evidence and despised philosophical presuppositions. The suppositional reasoning of Aristotle, Albertus, and Aquinas had evidence undergirded by presuppositions, whereas the new science of Bacon featured evidence untainted by presuppositions. Bacon charged that someone led captive by presuppositions would reach a conclusion before doing proper experiments.

Bacon's science had two key ideas: "the belief that science ought to proceed by means of presuppositionless observation and the idea that scientific research can be conducted by means of the systematic tabulation of data" (O'Hear 1989:16). How better to sweep away the previous "centuries of obscurantism" than by "cleansing the mind of all its presuppositions and prejudices, and reading the book of nature with fresh eyes?" (O'Hear 1989:16). Doubtless Francis Bacon's view of science, supposedly based on presuppositionless evidence, still typifies the way many scientists think about science.

René Descartes (1596–1650) is often regarded as the father of modern Western philosophy. He also invented analytic geometry and wrote on astronomy, physics, physiology, and psychology. In 1619 he had three dreams that greatly influenced his life. "He believed that he had been called by the Spirit of Truth to reconstruct human knowledge in such a way that it should embody the certainty heretofore possessed only by mathematics" (Losee 1993:74).

In the preface to *Meditations on First Philosophy*, Descartes promised that "each of us can become indubitably convinced first of his own existence, then of the existence of God, and finally of the essence of material things and the true nature of the human mind" (Cottingham, Stoothoff, and Murdoch 1988:viii; also see Cottingham 1986). Somewhat like Aristotle, Descartes sought to begin with powerful general principles and then deductively derive as much as possible. To obtain the needed stockpile of indubitable general principles, Descartes rejected the unverified assumptions of ancient authorities and began with universal doubt, starting afresh with that which was most certain.

His chosen starting point was the famous "*Cogito ergo sum*," that is, "I think, therefore I exist." That declaration appears in the second of the six meditations, titled "The nature of the human mind, and how it is better known than the body." In searching for the most certain of all possible starting points, Descartes started with the self. And given the self, with its mind and body, Descartes judged the mind and its thoughts to be better known and more certain than the body. Of course, if the goal was a plausible progression from most certain to least certain beliefs, many contemporary scientists would say that Descartes's argument ran backward, because it had layers of philosophical presuppositions and arguments before getting to the existence of the physical world or our own bodies.

Isaac Newton (1642–1727) had a vigorously empirical view of scientific method that was largely directed against Descartes's highly metaphysical approach. That science's presuppositions were unproblematic for Newton is shown by science's evidence being both needed and compelling: "If the workings of nature reflected the free agency of a divine will, then the only way to uncover them was by empirical investigation. No armchair science, premised on how God *must* have organized things, was permissible" (Brooke 1991:140). But George Berkeley reacted differently than Descartes and Newton, concluding that only minds and ideas existed, not the physical world.

David Hume (1711–1776) was a great skeptic. Certainly, skepticism undoes science's traditional claims, particularly of objective truth about physical reality, so skepticism has always been outside mainstream science. Nevertheless, for many scientists, the skirmish with skepticism has been the primary stimulus for their having any thoughts at all about science's presuppositions.

Hume thought that science's ambitions must be limited to describing our perceptions, avoiding philosophical speculations about some external physical world. Anyway, despite his philosophical convictions, Hume realized that common sense continually thrusts upon us vivid thoughts about real hands, houses, and such, and he conceded that common sense must rule in life's ordinary dealings. His writings reflect an awkward tension between common sense and philosophy that never gets resolved. The common-sense antidote to skeptical "amazement and confusion" was "simply to turn one's back on the arguments which produce it; dinner with friends and a game or two of backgammon are Hume's preferences. After this, the sceptical arguments cannot but seem 'cold, and strained'" (Woolhouse 1988:147). However, to say that common sense and science rest on animal instinct, despite rational philosophy's censure, cannot but seem a cold and strained defense of science. Hume was unwilling to grant the presuppositions that the physical world existed and that sense perceptions were generally reliable, so his science was shadowy and cold, forgoing any claims of truth about physical realities.

Thomas Reid (1710–1796) was the great protagonist of common sense as the only secure foundation for philosophy and science, in marked contrast to Hume. Of course, previous ancient and medieval thinkers had developed the scientific method within a common-sensical framework. But subsequent challenges, especially from Berkeley and Hume, had necessitated exposing science's common-sense roots with greater clarity and force. According to Reid,

> Hume's error was to suppose that it made sense to justify first principles of our faculties by appeal to [philosophical] reason. It does not.... To attempt to justify the first principles of our faculties by reasoning is to attempt to justify what is the most evident by appeal to less evident premises, those of philosophers.... Philosophy, properly understood, does not justify these principles of common sense but grows from them as a tree grows from its roots.... The attempt to justify a conclusion that is evident to begin with, such as that I see a cat, by appeal to premises that are philosophically controversial is doomed to absurdity. When the conclusion of an argument is more evident to begin with than it could be shown to be by a philosophical argument, the latter is useless as the justification of the conclusion.... No such [philosophical] argument has the evidential potency of innate [common-sense] principles of the mind. (Lehrer 1989:19, 294)

Reid's conception of science based on common sense had five main elements:

(1) The Symmetry Thesis. The dominant eighteenth-century science of the human mind, originating from Locke, Berkeley, and Hume, said that real knowledge could be achieved only for our sensations and relations among sensations, not for objects supposedly causing our sensations, so science must settle for appearances rather than realities. As a corrective, Reid adopted a symmetry thesis that gave the internal and external worlds equal priority and status, with both taken as starting points for philosophical reflection (Lehrer 1989:21).

(2) Harmonious Faculties. Hume and other skeptics granted philosophical reasoning priority over scientific observation. But Reid endorsed the basic reliability of all of our faculties, both sensory and mental, saying that

"He must be either a fool or want to make a fool of me, that would reason me out of my reason and senses" (Hamilton 1872:104). "Scepticism about the soundness of the sceptic's arguments is at least as justified as the scepticism which he urges upon us" (Dennis C. Holt, in Dalgarno and Matthews 1989:149).

(3) Parity among Presuppositions. Reid also claimed a parity between realist and skeptical presuppositions. Reid noted that we have two choices: to trust our faculties as common sense enjoins, or not to trust our faculties and become skeptics. A critic may complain that Reid's appeal to common sense is dogmatic or circular: "Because propositions of common sense are foundational, it is not possible to provide constructive, independent grounds for their acceptance. The propositions of common sense constitute the final court of appeal; they cannot themselves be justified, at least in the manner appropriate to derivative propositions. For that reason it must seem to the committed idealist or sceptic that the defender of common sense begs the question" (Dennis C. Holt, in Dalgarno and Matthews 1989:147).

But Reid's first reply was that such exactly was the nature of a foundational presupposition, that it could only be insisted upon. Furthermore, there existed a parity of presuppositions: The realist accepted common sense, and the skeptic rejected common sense. Either stance was adopted presuppositionally, without possibility of any independent justification or deeper proof. Hence, to say that "Realism is accepted by common-sense presuppositions" was to state a *truth* about realism, but that was not a *criticism* of realism, because skepticism's alternative was no less dependent on (different) presuppositions. Therefore, the contest between realism and skepticism had to turn on considerations other than the role of presuppositions.

(4) Reason's Double Office. Reid maintained that reason held the traditional "double office" of "regulating our belief and our conduct" [Selwyn A. Grave, in Edwards 1967(7):121]. Belief and action should match. If not, the diagnosis was not the logical problem of incoherence between one belief and another contrary belief, but rather the moral problem of insincerity or hypocrisy shown by mismatch between belief and action. As for the world of human actions, common sense was the only game in town. For example, a skeptic's mouth may say that we cannot be sure that a car is a real or hard object, but at a car's rapid approach, the skeptic's feet had better move!

Reid happily quoted Hume's own admission that a skeptic "finds himself absolutely and necessarily determined, to live and talk and act like all other people in the common affairs of life" (Hamilton 1872:485). Nevertheless, Hume went on to remark that "reason is incapable of dispelling these clouds" of skepticism. But if that was the case, it must be that Hume's version of reason held only the single office of regulating belief, rather than Reid's traditional reason that held the double office of regulating belief and action. In other words, only after first having adopted an impoverished notion of reason that pertained to belief but not to action was it possible for someone to regard as reasonable a skeptical philosophy that could not possibly be acted upon and lived out without jeopardizing the skeptic's survival.

(5) Asking Twice. What is the basis for science's presuppositions? Reid's reply depended on whether that question was asked once or twice. Asked once, Reid supported science's presuppositions by an appeal to common sense. But if asked twice, the deeper issue became why the world was so constituted as common sense supposed. For instance, why did the physical world exist, rather than nothing? And why were we so constituted that the world was comprehensible to us?

Clearly, those deeper questions could not be answered satisfactorily by a mere appeal to common sense, but rather required the greater resources of some worldview. Regarding that deeper appeal to a worldview, Reid had two things to say.

First, Reid said that his own worldview, Christianity, explained and supported science's common-sense presuppositions. That worldview said that God made the physical world and made our senses reliable. "In Reid's doctrine the existence of common sense has theistic presuppositions; its truths are 'the inspiration of the Almighty.' Reid did not maintain that belief in them depends upon belief in God; they are imposed upon us by the constitution of our nature, whatever our other beliefs. His implication is that we have to go behind common sense, if we are to explain its competence, to the fact that our nature has been constituted by God" [Selwyn A. Grave, in Edwards 1967(7):120–121].

Second, Reid claimed that virtually all other worldviews also respected the rudimentary common sense that provided science's presuppositions. Common sense was imposed on us "by the constitution of our nature," and that human nature was shared by all humans, regardless of whatever a person happened to believe or not believe about God. Accordingly, Reid's own appeal to Christian theology could be supplemented or generalized by an appeal to anyone's own favored worldview (with the exception of radical skepticism). For example, Keith Lehrer has found it easy work to adapt Reid's common-sense philosophy to his own worldview, evolutionary materialism. Reid had suggested that "What is necessary for the conduct of our animal life, the bountiful Author of Nature hath made manifest to all men." But Lehrer has substituted "the struggle for survival for the actions of God and what is necessary for the survival of species for what is necessary for the conduct of animal life" (Lehrer 1978; also see Keith Lehrer, in Daniels 1989:133).

Reid's strategy for supporting science's presuppositions had a wonderful clarity and balance. Worldview-independent, common-sense presuppositions preserved science's credibility without compromising its objectivity. At the same time, there was no confusion or pretense that mere common sense provided a deep or ultimate explanation of why the world was as it was. That job had to be done by some worldview. Fortunately, although worldviews differ on many other points, they do not challenge each other over rudiments of common sense such as "The earth exists" or "This is a book." By seeing common sense as a penultimate rather than an ultimate defense of science, Reid invited the humanities to complement science's picture of the world.

The bottom line on Reid's common-sense philosophy is expressed nicely in a mere one sentence by Duggan (1978): "Whether we accept Reid's account of empirical evidence will depend, as is the case with all interesting philosophical theories, very largely on our 'starting points' and our pre-theoretical views of what constitutes an absurdity." Given common-sense presuppositions, empirical observations were interpreted as informative experience of the physical world, and hence those observations or data could count as evidence when testing scientific theories.

Immanuel Kant (1724–1804) had a subtle and complex philosophy of science that is not easily summarized. "But *very* roughly speaking, he held that we see the world through colored glasses, or rather that our minds impress a certain pattern on the physical world," and consequently scientific propositions "express factual propositions, since they tell us something about our experiences, but they can be known by pure reason, since it was the mind that stamped them upon reality" (Kemeny 1959:16–17). Hence, although science required presuppositions, that need not concern us, because our minds have imposed those presuppositions on nature (Paul Guyer, in Guyer 1992:24). There has been little in the way of fundamentally new and different proposals after Kant.

In review, imagine the following thought experiment. Imagine that the contemporaries Berkeley, Hume, and Reid were brought together and were all patting a single horse. All three would report the same experience of a big furry animal, but their interpretations of that experience would differ. Berkeley would say that the physical horse did not exist, but only the mind's idea of a horse. Hume would say that science should concern our experience of the horse, but would not say that the horse did or did not exist. And Reid would say that philosophy and science should follow common sense with a confident and cheerful certainty that the physical horse did exist. Likewise, imagine stepping further back in the history of this debate and seeing the contemporaries Plato and Aristotle patting a single dog. For Plato, the dog would be but an illusory and fleeting shadow of its inaccessible but thoroughly real Form. But for Aristotle, the dog itself would be accessible to our sensory experience and would be completely real. Clearly grasp that that perennial debate was not about the sensory data as such, but rather was about the interpretation of the data. That debate was altogether about philosophical presuppositions and altogether not about scientific evidence.

Finally, position papers from the AAAS (1989, 1990) provide a contemporary, mainstream expression of science's presuppositions: "Science presumes that the things and events in the universe occur in consistent patterns that are comprehensible through careful, systematic study. Scientists believe that through the use of the intellect, and with the aid of instruments that extend the senses, people can discover patterns in all of nature. Science also assumes that the universe is, as its name implies, a vast single system in which the basic rules are everywhere the same" (AAAS 1989:25). "All intellectual endeavors share a common purpose – making sense of the bewildering diversity of experience. The natural sciences search for regularity in the natural world. The search is predicated on the assumption that the natural world is orderly and can be comprehended and explained" (AAAS 1990:16).

Furthermore, careful scientific argumentation should disclose all premises. Indeed, the AAAS (1989:139) lists several "signs of weak arguments" that are useful for checking both others' and one's own arguments. One sign of a "shoddy" argument is that "The premises of the argument are not made explicit." Likewise, "Inquiry requires identification of assumptions" (NRC 1996:23). Therefore, science's presuppositions should be explicitly and fully disclosed.

THE PEL MODEL OF FULL DISCLOSURE

A given scientific argument may be good or bad, and its conclusion may be true or false. But in any case, the first step in assessing a scientific conclusion is merely to disclose the argument fully. Then each and every piece of the argument can be inspected carefully and weighed intelligently, and every participant in the inquiry can enjoy clear communication with colleagues.

It is intellectually satisfying to be able, when need be, to present a scientific argument or conclusion with full disclosure. Also recall that the question "What goes in so that scientific conclusions can come out?" was asked by Aristotle and became the central question for the philosophy of science for two millennia. But unfortunately, precious few scientists are trained to be able to answer that, the most elemental question that could possibly be asked about scientific inquiry.

What does it take to present a scientific conclusion with full disclosure? The basic model of scientific method developed in this book, named by the acronym the PEL model, says that presuppositions (P), evidence (E), and logic (L) combine to support scientific conclusions. In essence, scientific method amounts to providing the presuppositions, evidence, and logic needed to support a given scientific conclusion.

Furthermore, these three components interact so deeply that they must be understood and defined together, not separately. Accordingly, this section discusses presuppositions, evidence, and logic together before the next section can explain presuppositions in greater detail. The situation is analogous to the three concepts of mothers, fathers, and children. It is easy to explain all three concepts together, but it would be impossible to give a nice explanation of mothers while saying nothing about fathers and children.

Remarkably, a very simple example suffices to reveal the general structure of scientific reasoning, no matter how complex. Consider the following experiment, which you may either just imagine or else actually perform, as you prefer. Either envision or get an opaque cup, an opaque lid for the cup, and a coin. Ask someone else to flip the coin, without your observing the outcome or the subsequent setup. If the flip gives heads, place the coin in the cup and cover the cup with the lid. If the flip gives tails, place the coin elsewhere and cover the cup with the lid. Now that the setup is completed, ask this question: "Is there a coin in the cup?"

The present assignment is to give a complete, fully disclosed argument with the conclusion that there is or is not a coin in the cup, as the case may be. This means that *all* premises needed to reach the conclusion must be stated explicitly, with nothing lacking or implicit. Before reading further in this section, you might like to write down your current answer to this problem for comparison with your response after studying this section.

To simplify the remaining discussion, the assumption is made that the actual state of affairs, to be discovered in due course through exemplary scientific experimentation and reasoning, is that "There is a coin in the cup." Those readers with this physical experiment before them may wish to make that so before proceeding with the assignment. Nevertheless, for purposes of the following story, we shall pretend that we do not yet know, and still need to discover, whether or not the cup contains a coin.

The question "Is there a coin in the cup?" can be expressed with scientific precision by stating its hypothesis set – the list of all possible answers. From the foregoing setup, particularly the coin flip, there are exactly two hypotheses:

H_1. There is a coin in the cup.
H_2. There is not a coin in the cup.

These two hypotheses are mutually exclusive, meaning that the truth of either implies the falsity of the other. They are also jointly exhaustive, meaning that they cover all of the possibilities. Because these hypotheses are both mutually exclusive and jointly exhaustive, exactly one hypothesis must be true.

How can we determine which hypothesis is true? The answer we seek is a contingent fact about the world. Thus, no armchair philosophizing can give the answer, because nothing in the principles of logic or philosophy can imply that the cup does or does not contain a coin. Rather, to get an answer, we must look at the world to discover the actual state of nature. We must perform an experiment.

Various satisfactory experiments could be proposed. We could shake the cup and listen for the telltale clicking of a coin. We could take an x-ray photograph of the cup. But the easiest experiment is to lift the lid and look inside. Here we presume a particular outcome, that we look and see a coin. That experimental outcome motivates the following argument and conclusion:

Premise. We see a coin in the cup.
Conclusion. There is a coin in the cup.

As a common-sense reply, this argument is superb, and its conclusion is certain that H_1 is true. Nevertheless, as a philosophical reply, this argument is incomplete and defective. Symbolize seeing the coin by "S," and the coin's existence in the cup by "E." Then this argument has the form "S; therefore E." It is a *non sequitur*, meaning that the conclusion does not follow from the premise. Something is missing, so let us try to complete this argument.

Another required premise is that "Seeing implies existence," or "S implies E," specifically for objects such as coins and cups. From the perspective of common sense, this is simply the presupposition that seeing is believing. In slightly greater philosophical detail, this premise incorporates several specific presuppositions, including that the physical world exists, our sense perceptions are generally reliable, human language is meaningful and adequate for discussing such matters, all humans share a common human nature with its various capabilities, and so on. The story of this premise can be told in versions as short or long as desired, with the short version given here and the long version developed later in this chapter.

With the addition of this second premise, the argument now runs as follows: "S; S implies E; therefore E." This is much better, following the valid argument form *modus ponens* (as will be explained in detail in the next chapter, on deductive logic). However, to achieve full disclosure, the logic used here must itself be disclosed by means of a third premise declaring that "*modus ponens* is a correct rule for deduction." Incidentally, to avoid a potential problem with infinite regress that philosophers have recognized for over a century (Jeffreys 1973:198–200), note that here *modus ponens* is not being implemented in a formal system of logic, but rather is merely being disclosed as a simple element in ordinary scientific reasoning.

Finally, a fourth premise is required. For a particular scientific argument for a given person, each of that person's beliefs is one of the following: the argument's conclusion itself, or a presupposition, or an item of evidence, or a rule of logic, or an inert item in the archive. Here "archive" is used as a technical philosophical term denoting all of a person's beliefs that are wholly irrelevant to a given inquiry. For example, given the current inquiry about a coin in a cup, my beliefs about the price of tea in China may be safely relegated to the archive. The archive serves the philosophical role, relative to a given inquiry, of providing for a complete partitioning of a person's beliefs. It also serves the necessary and practical role of dismissing irrelevant knowledge from consideration so that a finite analysis of the relevant material can yield a conclusion (whereas if one had to consider everything one knows before reaching a conclusion about the coin, no conclusion could ever be reached).

Of course, to reside legitimately in the archive, a belief must be genuinely irrelevant and inert. Sometimes progress in science results from showing that a belief accidentally relegated to the archive is in fact relevant and must be exhibited as a presupposition, item of evidence, or logic rule. Anyway, a final premise is required here, saying that "the archive dismisses only irrelevant beliefs." It contains nothing with the power to unsettle or overturn the current conclusion.

Rearranging the preceding four premises in a convenient order, they can now be collected in one place to exhibit the argument entirely, with full disclosure:

Premise 1 [Presupposition].   Seeing implies existence.
Premise 2 [Evidence].   We see a coin in the cup.
Premise 3 [Logic].   *Modus ponens* is a correct rule for deduction.
Premise 4 [Archive].   The archive dismisses only irrelevant beliefs.
Conclusion.   There is a coin in the cup.

On the basis of the foregoing experiment and argument, the conclusion follows that "there is a coin in the cup." This conclusion merits scientific claims of rationality, truth, objectivity, realism, and certainty.

This elementary argument exemplifies full disclosure according to the PEL model. It could be called the PELA model to recognize all four inputs, including the archive, but because the archive is essentially inert, I prefer the briefer acronym PEL that focuses on just the three active components. The formula of the PEL model is that presuppositions, evidence, and logic give the conclusion. This structure of a rational argument, flushed out by this simple coin example, pervades all scientific claims of knowledge about the world, regardless of how elementary or advanced. Figure 4.2 summarizes the components of the PEL model.

With this model, the basic nature of presuppositions can be understood clearly. A presupposition is a belief that is necessary in order for any of the hypotheses to be meaningful and true but that is non-differential regarding the credibilities of the individual hypotheses. Presuppositions emerge from

---

The PEL Model of Full Disclosure

Presuppositions + Evidence + Logic $\rightarrow$ Conclusions

Presuppositions are beliefs that are absolutely necessary in order for any of the hypotheses under consideration to be meaningful and true but that are completely non-differential regarding the credibilities of the individual hypotheses. Science requires several common-sense presuppositions, including that the physical world exists and that our sense perceptions are generally reliable. These presuppositions also serve to exclude wild ideas from inclusion among the sensible hypotheses under serious consideration.

Evidence is data that bear differentially on the credibilities of the hypotheses under consideration. Evidence must be admissible, being meaningful in view of the available presuppositions, and it must also be relevant, bearing differentially on the hypotheses.

Logic combines the presuppositional and evidential premises, using valid reasoning, to reach a conclusion. Science uses deductive and inductive logic.

A complete partitioning of a person's beliefs results from also recognizing an archive containing all beliefs that are irrelevant for a given inquiry, that is, beliefs that are not presuppositions or evidence or logic rules or conclusions. Irrelevant material must be ignored to avoid infinite and impossible mental processing. But the archive has no active role, and hence is not indicated in the acronym for the PEL model.

Figure 4.2. Scientific conclusions emerge from three inputs: presuppositions, evidence, and logic.

comparing the hypotheses to see what they all have in common, and, in turn, the hypotheses originate from the question being asked, which is the ultimate starting point of an inquiry. For example, in order to declare either H_1 or H_2 to be true, it must be the case that the physical world exists and that human sense perceptions are generally reliable. But these presuppositions are completely non-differential, making H_1 neither more nor less credible than H_2.

Presuppositions also serve a second role, that of limiting the hypothesis set to a finite roster of sensible hypotheses. Were common-sense presuppositions ignored, the foregoing hypothesis set with only two hypotheses, H_1 and H_2, might not be jointly exhaustive. Instead, it could be expanded to include countless wild possibilities such as H_3, that "We are butterflies dreaming that we are humans looking at a cup containing a coin." But no empirical evidence could possibly discriminate among those three hypotheses, so that expanded hypothesis set would prevent science from reaching any conclusion. Numerous wild hypotheses, due to abandoning common-sense presuppositions, can undo science.

Evidence has a dual nature, admissible and relevant (Shafer 1976; Achinstein 1983; Schum 1994). First, evidence is *admissible* relative to the available presuppositions. Hence, given common-sense presuppositions about the existence of the physical world and the general reliability of sense perceptions, it is admissible to cite the seeing of a coin, whereas without such presuppositions, such a claim would not be meaningful or admissible. Second, evidence is *relevant* relative to the stated hypotheses, bearing differentially on their credibilities. Hence, seeing a coin is relevant testimony because it bears powerfully on the hypotheses, making H_1 credible and H_2 incredible.

To avoid a possible embarrassment of riches, evidence can be further partitioned into two subsets: tendered evidence that is actually supplied, and reserved evidence that could be gathered or presented but is not because it would be superfluous. For example, before gathering any evidence whatsoever, the credibilities of hypotheses H_1 and H_2, that the cup does or does not contain a coin, can be represented by probabilities of 0.5 and 0.5. But after tendering the evidence that "We see a coin in the cup," those probabilities become 1 and 0. And after citing the additional evidence that "Shaking the cup causes a telltale clicking sound," those probabilities remain 1 and 0, as is still the case after also observing that "An x-ray photograph shows a coin inside the cup."

So after initial evidence has already established a definitive conclusion, additional evidence has no further effect on the hypotheses' credibilities. At this point, wisdom directs us to close the current inquiry and move on to other pressing questions that are not yet resolved. Likewise, sometimes a conclusion will have reached a high probability of truth that may be considered adequate, even though more effort and evidence potentially could further strengthen the conclusion. Because scientific evidence comes at a cost of time and money, a wise scientist will prioritize various projects and know how to gather only the evidence necessary to get maximal results.

Comparing briefly, presuppositions answer the question, How can we reach any conclusion to an inquiry? But evidence answers the question, How can we assert one particular conclusion rather than another? For example,

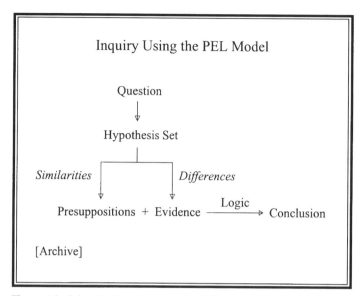

Figure 4.3. Scientific inquiry using the PEL model. Similarities among all of the hypotheses support presuppositions, whereas differences suggest potential evidence. Logic combines the presuppositions and evidence to reach the conclusion. Irrelevant knowledge is relegated to an inert archive.

presuppositions about the existence of the physical world and the reliability of our sense perceptions are needed to reach any conclusion about a coin in the cup, whereas the evidence of seeing a coin in the cup leads us to reach the particular conclusion that there is a coin in the cup.

Logic serves to combine the premises to reach the conclusion. For example, the foregoing argument has the form "S; S implies E; therefore E," which follows the valid rule *modus ponens*. Finally, the archive serves to avoid infinite mental processing, but does merit a check that its contents are truly irrelevant. Figure 4.3 summarizes inquiry using the PEL model, from starting question to final conclusion. Note that the PEL model closely interlinks the concepts of presupposition, evidence, and logic. For example, half of the concept of evidence involves admissibility, which is determined by the presuppositions. Consequently, if one's concept of presuppositions is fuzzy, inexorably the concept of evidence will also be fuzzy, which will be disastrous. When presuppositions are not rightly understood, they become inordinately influential, suppressing the proper influence of evidence.

AAAS statements about the basic components of scientific thinking correspond with the PEL model proposed here. Evidence and logic are the most evident components: "The process [of scientific thinking] depends both on making careful observations of phenomena and on inventing theories for making sense out of those observations" (AAAS 1989:26; also see pp. 27–28 and AAAS 1990:16). As mentioned earlier in this chapter, presuppositions are also discussed. Furthermore, these three elements are brought together as the basis for scientific conclusions in the statement that "the principles of logical reasoning … connect evidence and assumptions with conclusions" (AAAS 1989:27).

Finally, at most, a scientific argument may be correct; at the least, it should be fully disclosed. Full disclosure is the first and minimal requirement for clear scientific reasoning. Hence, when weighing scientific arguments and claims, it helps considerably to understand that when fully disclosed, *every* scientific conclusion emerges from exactly three inputs: presuppositions, evidence, and logic (and knowledge that is irrelevant for a given inquiry is relegated to an inert archive).

WHAT ARE PRESUPPOSITIONS?

Among scientists, even the very concept of a presupposition is often vague. Therefore, this section takes a hard look at the question, What are presuppositions? Its analysis unfolds in three stages, giving first an informal definition of presuppositions, then a formal definition, and finally a general definition.

What are presuppositions? As a first, informal reply, presuppositions (1) are starting points and (2) are not testable or provable. Besides these two basic properties, presuppositions also often have the incidental properties of being implicit and yet influential. So a presupposition is an unstated assumption or an arbitrary starting point.

Although this informal definition may seem sensible and adequate, actually it is rather confused and vague. Here the term "presupposition" has a specific technical meaning that emerges from the PEL model of full disclosure. A presupposition is a belief that is absolutely necessary in order for any of the hypotheses under consideration to be meaningful and true but that is completely non-differential regarding the credibilities of the individual hypotheses.

How do the informal and formal definitions compare? Does the formal definition merely refine and sharpen the informal definition? Or do they contradict each other?

The first claim in the informal definition was that a presupposition is a starting point. Is that correct? No! A scientific inquiry does not begin with presuppositions, but with a question, preferably stated precisely by a hypothesis set. Given this hypothesis set, presuppositions emerge mechanically as beliefs held in common by all of the hypotheses in the hypothesis set. Furthermore, presuppositions are not arbitrary. If a belief is held in common by all of the hypotheses, then it is a presupposition; if it is not so held, then it is not a presupposition – there is no arbitrary choice here whatsoever.

The second claim in the informal definition was that a presupposition is not testable or provable. Is that correct? At best, this is a diffuse claim that fails to identify the real action. Indeed, no "proof" can be given, for instance, that the physical world exists, because were its existence sincerely questioned, then no appeal to physical observations could provide admissible evidence. But this is not a particularly significant remark. The more constructive move is to identify an intended audience within which needed presuppositions are already held in common. That is, we cannot get presuppositions that have been proved, but we can get presuppositions that are unproblematic, at least for some specified audience.

The formal definition of presuppositions given here accords with the tra-ditional meaning in logic and philosophy: "The presuppositions of a proposi-tion are things that must be true for the proposition to be either true or false. Truth and falsity are referred to as the two truth values. So presuppositions are things that must be true for a proposition to have a truth value" (Davis 1986:259). Similarly, "the presuppositions of a question are things that must be true for the question to have an answer," that is, for any of the hypotheses of the hypothesis set to be true (Davis 1986:258). And more specifically in the domain of science, "Most scientists take for granted their metaphysical assumptions, but they are none the less necessary logically to the conclu-sions of science" (Caldin 1949:176). The following definitions summarize the formal or technical meaning of presuppositions:

> Proposition P is a presupposition of a statement S if and only if P must be true for S to have a truth-value, either true or false.

> Proposition P is a presupposition of a question Q expressed by the hypothesis set H_1 to H_N if and only if (1) P must be true for every hypothesis H_1 to H_N to be possibly true and (2) P makes no hypothesis more or less credible than another.

Moving along to the third and final stage of this section's analysis, pre-suppositions can be considered and defined from a general perspective that pertains to the overall scientific enterprise rather than to an individual sci-entific inquiry. To handle science's presuppositions with minimal effort, it helps to list those presuppositions, once and for all, that are always needed – rather than working them out again and again for every individual scien-tific inquiry. This strategy works because a basic set of presuppositions is needed throughout science. For example, if we applied the preceding formal definition of presuppositions to numerous scientific inquiries, repeatedly and consistently the presupposition would emerge that the physical world exists. Consider the following definition:

> A presupposition of science is a belief that is necessary for an ordinary realist implementation of science if that belief cannot possibly be proved by any evidence or reasoning whatsoever but rather must be accepted by common sense and faith.

An ordinary realist implementation assumes that science pursues and offers true and justified statements about an external physical reality. By mutating and downgrading science's goals, and especially by foregoing any claims of truth, we find that fewer presuppositions may be needed. However, anything less or other than a realist science whose business is truth simply is not science as any scientist would ordinarily conceive it. Nevertheless, science's presuppositions cannot be proved by any means whatsoever, meaning that science itself cannot prove its presuppositions, nor can it expect philosophy or any other discipline to prove them on its behalf. Indeed, what philosophy can prove is that science's basic presuppositions are unprovable. However, on balance, if something less than knockdown proof is considered, such as conformity with our most basic experiences and shared interpretations of life, then some presuppositions do have more philosophical warrant than others, as noted earlier in the thinking of Thomas Reid.

The general and formal aspects of presuppositions can be combined as follows. Every scientific inquiry assumes the general presuppositions of science and may have additional specific presuppositions. For example, consider the question "Are these leaves yellow because of nitrogen deficiency or magnesium deficiency?" This question and its ensuing inquiry need the general, basic presuppositions of science, including that the physical world exists. Such presuppositions are inevitable and reasonable. This question also has the specific presupposition that the cause of these yellow leaves is not nitrogen deficiency *and* magnesium deficiency, nor is it some other problem or combination of problems. Such a presupposition may be questionable and unreasonable. If reflection reveals an objectionably restrictive presupposition, the remedy ordinarily is to expand the hypothesis set to include additional plausible possibilities. To ask a better question is to improve science at its most strategic step.

It is useful to distinguish between presuppositions that appear in every scientific inquiry and other presuppositions that appear in only some scientific inquiries. Here these are called global and local presuppositions, respectively. This chapter's primary concern is with global presuppositions, which need to be settled once and for all to preserve science's rationality and credibility. However, it is also true that each individual scientific study should be checked to make sure that it is free from additional local presuppositions that are unrealistic, restrictive, or uncertain.

DISCLOSURE OF PRESUPPOSITIONS

The method used here for disclosing science's presuppositions proceeds in two steps, implemented in the following two subsections. First, a little exemplar of common-sense knowledge about the world, called a "reality check," is selected that is as certain and universally known as is anything that could be mentioned. Second, philosophical reflection on this exemplar flushes out its presuppositions and reveals that they also suffice for scientific thinking.

The reason for choosing this particular method for disclosing presuppositions is that it renders science's presuppositions as unimpeachable as our most certain knowledge. Science's presuppositions cannot be made unnecessary, but they can be made unproblematic. Also, this strategy fits historically with the thinking of many prominent scientists and mainstream philosophers.

A Common-Sense Reality Check

The text for this book's reality check, complete with its preamble, reads as follows:

> *Reality Check*
> It is rational, true, objective, realistic, and certain that "Moving cars are hazardous to pedestrians."

To serve as a suitable object for philosophical analysis, however, it is essential that this text stand as common ground, believed by author and reader alike. So do you believe this, that it is dangerous for pedestrians to step into the pathway of oncoming cars? I trust that this is the case. Indeed, readers who happen to have been star pupils in kindergarten will recognize this reality check as a sophisticated version of the command "Look both ways before crossing a road."

The choice, for or against accepting this reality check, is primordial in that it is a common-sense conviction logically prior to all subsequent fancier choices about the claims and methods of science. Common sense precedes science. Recall Reid's sentiment that the principles of common sense are older and of more authority than philosophy. Likewise, Wittgenstein insisted that rudimentary common-sense beliefs are oblivious to evidence because no evidence is more certain than such beliefs themselves: "my not having been on the moon is as sure a thing for me as any grounds I could give for it" (Ludwig Wittgenstein, in Anscombe and von Wright 1969:17e; Newell 1986:68).

Similarly, Blackburn (1973:2) remarked that "It should be evident already that … the belief that my typewriter will not suddenly become weightless, change colour, or talk to me, is not the result of scientific investigation. It is to be justified, if at all, by reasoning common to all men."

My appeal to common sense might be discredited or dismissed by a quick remark such as "Common sense isn't so common." Surely, this means that people sometimes spend more money than they earn, neglect the upkeep that could prevent costly repairs, steal other persons' property, and so on. But clearly, my chosen exemplar of common sense is not gathered from the glorious heights of common sense, with offerings such as "Spend less than you earn" or "A stitch in time saves nine." Rather, "Moving cars are hazardous to pedestrians" is an exemplar of rudimentary common sense. So my appeal to rudimentary common sense should not be misinterpreted or dismissed as the unrealistic assumption that everyone is a paragon of sense and virtue. Rather, it should be interpreted and taken seriously as the claim that all normal humans living on this one earth know some basics about physical reality. As an example of a person who holds common sense in my very modest version, it suffices for you to envision a wicked man who snatches a woman's purse and then looks both ways before crossing a street to run away.

In this academic book on scientific method, why fight for a meager scrap of common sense, that "Moving cars are hazardous to pedestrians"? What is the dreaded contrary of common sense? What is the threat?

Presumably, common sense's opponent is skepticism. But sincere skepticism is extremely rare. I, for one, have never met a single person who doubted, in any sense that could be taken as sincere, that "Moving cars are hazardous to pedestrians." Nor is it easy to imagine that such a person could survive apart from institutional care.

No, the opponent of common sense and science is not earnest skepticism, which is exceedingly rare. Rather, the real opponent of common sense and science is ambivalent skepticism, which is common, just as is any other kind of inconsistency or insincerity.

The skeptical tradition, from start to finish, has been characteristically ambivalent. The founding figure of ancient Greek skepticism, Pyrrho of Elis (c. 360–270 B.C.), claimed not to trust his senses and once essayed to walk over a cliff, as if it could not matter (Burks 1977:137; Chatalian 1991; Annas and Barnes 1994). That sounds like gratifying, serious skepticism! But he did that in the presence of disciples, who kept him from harm, and he lived to the ripe old age of ninety. Figure 4.4 depicts a similar humorous scenario featuring a careless scientist who unwittingly is about to repeat Pyrrho's intended experiment.

The attempted coherence of the Pyrrhonistic skeptics is quite charming. Against the dogmatic Academic skeptics such as Sextus Empiricus, who

**Dr. Leeb's experiments with virtual reality
were about to become reality reality.**

Figure 4.4. A scientist, one Dr. Leeb, on the brink of being an intellectual heir of
the ancient Greek skeptic, Pyrrho of Elis. Pyrrho started to walk over a cliff as if it
could not matter, as the future need not resemble the past, but fortunately he was
prevented by watchful students. (This drawing by Andrew Toos is reproduced with
kind permission from Cartoon Resource. © Andrew Toos / CartoonResource.com.)

claimed to show that knowledge was impossible, the Pyrrhonists claimed that
they did not even know that they could not know (Kelly 2000). They were
skeptical about whether or not they were skeptics! Anyway, despite their presumed doubts about everything, his fellow skeptics seemed proud and confident enough about reporting that Pyrrho traveled with Alexander the Great
to India. Much later, David Hume would continue that tradition of ambivalence, saying that the dictates of common sense must regulate ordinary daily
life, even though they are not philosophically respectable (Williams 1991).

A more recent example of that ambivalent tradition is Karl Popper. On
the one hand, some passages by Popper are reassuring to a common-sensical
reader. He writes of his "love" of common sense and says that "I am a great
admirer of common sense" (Popper 1974:43; 1979:viii). Likewise, in his auto-
biography, in Schilpp (1974:71), Popper says that common-sense knowledge,
such as "that the cat was on the mat; that Julius Caesar had been assassinated;
that grass was green," is all "incredibly uninteresting" for his work because
he focuses instead on genuinely "problematic knowledge" involving difficult
scientific discoveries.

On the other hand, he writes elsewhere that "We *cannot* justify our
knowledge of the external world; *all* our knowledge, even our observational

knowledge, is theoretical, corrigible, and fallible," which is clearly a blank check for skepticism (Karl Popper, in Lakatos and Musgrave 1968:164). More pointedly, he says that "The statement, 'Here is a glass of water' cannot be verified by any observational evidence" because of philosophical problems with induction and related matters that grip everyone, you and me included (Popper 1968:95). Now to say that you cannot know that "Here is a glass of water" is as plainly spoken a denial of common sense as to say that you cannot know my reality check that "Moving cars are hazardous to pedestrians." Clearly, common sense is under attack. But this attack is not consistent or sustained. Although Popper (1945:283) waxes eloquent about "the standards of intellectual honesty, a respect for truth, and ... modest intellectual virtues," regrettably the force of such fine rhetoric is undercut by his saying elsewhere that trifling truths like "Here is a glass of water" are beyond a human's reach.

The remedy for a disappointing, sophomoric, insincere skepticism is the sincere and cheerful acceptance of just one little scrap of common sense, such as that "Moving cars are hazardous to pedestrians." Reason must be restored to its double office of regulating both beliefs and actions. The sophomoric skeptic has a mouth whose words say that "Maybe cars are hazardous to pedestrians, and maybe not," but feet whose actions always say that "Moving cars are hazardous to pedestrians." The cure for this foot-and-mouth disease is sincerity.

Like the declaration "I love you," the reality check can be voiced with varying degrees of conviction. Because of the frequent problems with superficial or ambivalent skepticism, the degree of conviction intended here must be made clear. Exactly which plaudits attend this reality check? To voice the reality check with clear conviction and to connect it with science's ambitions announced at the start of this chapter, the reality check is proclaimed here with a preamble listing science's four basic claims: rationality, truth, objectivity, and realism. Furthermore, because this particular item of common-sense knowledge is so easy for all persons to learn and is absolutely exempt from controversy, it is proclaimed here with one additional plaudit: certainty. It is voiced cheerfully with absolute confidence and unlimited boldness. It is voiced with no ambivalence, no superficiality, and no insincerity.

Given a non-negotiable conviction that the reality check is true, what does this conviction imply for intellectual attacks on science's realism? It has a decisive implication, namely, any attack on science that also takes down common sense is simply incredible. Any attack that also targets common sense, including denying that we can know the reality check, fails because it does too much and thereby it loses credibility. Consequently, a legitimate attack on science's rationality must target science alone, not science and common sense both, as was explained in Chapter 3.

Philosophical Reflection

The foregoing remarks about science's presuppositions, particularly the AAAS statements, can be summarized as a single brief and comprehensive presupposition, as follows:

> **Science's Comprehensive Presupposition**
> The physical world is orderly and comprehensible.

This subsection unpacks this statement in order to disclose science's presuppositions in greater detail. Regardless of the level of detail, however, science's presuppositions do not individuate uniquely. So the following is *a* disclosure, not *the* disclosure. Nevertheless, it should be entirely adequate for this book's purpose of understanding scientific method.

Having selected a little exemplar of knowledge about the world that is shared by author and reader alike, that "Moving cars are hazardous to pedestrians," the next step is philosophical reflection to disclose the presuppositions that are necessary for this reality check to be meaningful and true.

> The logical premises of factuality are not known to us or believed by us *before* we start establishing facts, but are recognized on the contrary *by reflecting on the way we establish facts*. Our acceptance of facts which make sense of the clues offered by experience to our eyes and ears must be presupposed first, and the premises underlying this process of making sense must be deduced from this afterwards.... We do not believe in the existence of facts because of our anterior and securer belief in any explicit logical presuppositions of such a belief; but on the contrary, we believe in certain explicit presuppositions of factuality only because we have discovered that they are implied in our belief in the existence of facts. (Polanyi 1962:162)

At first blush, one might suppose that philosophical reflection on "Moving cars are hazardous to pedestrians" would flush out only meager, insignificant presuppositions. The falsity of that supposition should be evident momentarily. Rather, as McGinn (1989) argues, there is tremendous philosophical significance in little things that we would ordinarily claim to know. The presuppositions underlying the reality check can be organized in three broad groups: ontological, epistemological, and logical presuppositions.

(1) Ontological Presuppositions. The reality check presupposes distinctions, presupposes that physical reality has multiple things that are not all the same, such as cars that differ from pedestrians, or moving cars that differ from stationary cars. Because the universe is not merely one undifferentiated blob of being, there exists something to be comprehended.

It also presumes that reality has natural kinds, to use the philosophical term. This means that multiple objects can be of the same kind (at a given level of description), such as numerous cats each being a cat. And more generally, human artifacts can also be of a given kind, such as numerous cars each being a car. Ultimately, this means that the universe is parsimonious. It is not the case that the universe is maximally complex, with each object and each event being wholly different and unique. Such complexity would undo learning and comprehensibility.

One particularly important natural kind, from our perspective anyway, is the human being. The pedestrians or humans mentioned in the reality check share numerous properties, such as being soft and therefore vulnerable to strong impact from a large and hard object. It is not the case that car accidents are hazardous for some humans, whereas others are invincible.

The reality check also presumes that physical reality is predictable. Its implicit advice not to step in front of a rapidly moving car obviously agrees with past experience regarding car accidents. But equally, this advice is predictive, directed at preventing more accidents in the future.

(2) Epistemological Presuppositions. The reality check presumes that a human can know that an object is a rapidly approaching car and can act to move out of harm's way. This presupposes that our eyes, ears, and other sensory organs provide generally reliable information about the external world and that our brains can process and comprehend these sensory inputs. Furthermore, our brains can also direct our feet to move purposefully. Merely knowing without acting would not promote survival, so reason's double office of regulating belief and guiding action is evident.

Of course, not quite every human can implement the reality check. Some are merely too young to be allowed to be near streets on their own. Others, sad to say, never developed normally because of serious genetic or birth defects, or else have lost competence because of disease or senility. Apart from such exceptions, however, the capacities to know and to act to implement the reality check are universal for all human beings. It is not the case that certain races or nationalities or economic groups can handle the reality check, but not others.

Another epistemological presupposition is that human language is meaningful. The reality check is expressed by several words in English. Humans have abilities of language and communication.

(3) Logical Presuppositions. The reality check presumes coherence. To assert that "Moving cars are hazardous to pedestrians" legitimately, coherence demands that we not also assert the negation that "Moving cars are not hazardous to pedestrians." There is no credit for asserting the reality check if its opposite is also asserted.

The reality check also presupposes truth. Coherence requires merely that the reality check and its negation not both be asserted, but it does not indicate which one to assert. Truth is more choosy. The reality check expresses a true belief that corresponds with physical reality.

The reality check also presumes deductive and inductive logic. Logic is required to take a general principle and apply it to specific episodes of being near moving cars. Deduction is active in handling probability concepts, such as the idea of a thing being hazardous, meaning that harm is likely even if not certain. Induction is active in recognizing objects and in learning and using language.

The foregoing account of the reality check's presuppositions is not exhaustive. For example, the listed presupposition regarding predictive success is based in turn on a still deeper presupposition about the uniformity of nature, so another presupposition could be added here regarding the laws of nature. So the preceding remarks are offered merely as a moderately detailed account, illuminating but not exhaustive, of the reality check's presuppositions. If desired, the reader's own insight can be enlisted to detect additional presuppositions. But for most purposes, any further analysis would become technical and tedious.

The presuppositions flushed out by analyzing this one representative little scrap of common sense, the reality check, pervade common sense. That is, philosophical analysis of "This cup contains a coin" or "I have not been on the moon" or "Here is a glass of water" would evince the same presuppositions as this analysis of "Moving cars are hazardous to pedestrians." For any of these statements to be meaningful and true, the general makeup of ourselves and our world cannot follow just any conceivable story, but rather the world must be along the lines indicated by common-sense presuppositions.

Furthermore, the presuppositions pervading science cannot be less than those encountered in one little scrap of common sense.

> Although through our [scientific] theories, and the instrument-aided observations they lead to, we can go beyond and correct some of the pre-theoretical picture of the world we have by virtue of our being human, there is always going to be a sense in which all our knowledge and theory is based on elements in that [common-sense] picture.... More theoretical knowledge of the world is always going to have some connection, however remote, with the humdrum level if it is to count as science fact rather than science fiction.... We naturally and correctly expect observationally more remote theories to make some contact with the world of everyday experience, even if only at the level of registering meter-stick readings and traces on screens. This is because, in the end, theory has to answer to our pre-theoretical, everyday observations and experience. (O'Hear 1989:95–96)

On the other side, some have argued that the presuppositions of science must be more than those of common sense, or else at least partially different. They claim that these extra presuppositions provide science with needed

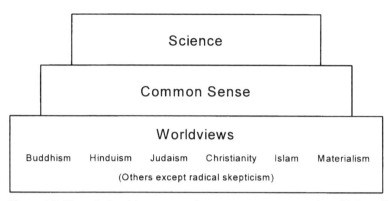

Figure 4.5. The relationships among science, common sense, and worldviews. Science does not provide its own foundation. Rather, science's penultimate foundation is common sense and its ultimate foundation is some worldview.

esoteric content that is wholly absent from common-sense knowledge and reasoning. For instance, science has overturned common sense with certain surprises: that the objects around us are mostly empty space, and that the rate at which time passes by is not constant but depends on an object's speed relative to the observer. However, those surprises are conclusions of science, not presuppositions. Indeed, those discoveries were eligible to become conclusions precisely because they were never presuppositions. Those surprises were established by empirical evidence that counted as evidence precisely because common-sense presuppositions were in effect. Science can overturn common-sense expectations and beliefs, but not common-sense presuppositions.

Likewise, some have argued that the presuppositions of science must be more than those of common sense, drawing also on some worldview to really explain why the world is as it is. However, the position taken here follows the mainstream scientific tradition of seven centuries, which began with Albertus Magnus, of distinguishing penultimate and ultimate accounts of science's presuppositions. The relationships among science and its penultimate foundation in common sense and its ultimate foundation in some worldview are depicted in Figure 4.5.

A penultimate account of its presuppositions must be included in science's own business, because these presuppositions are necessary to reach any conclusions. By contrast, pursuit of an ultimate account obliges science to enlist support from the humanities, so it seems wrongheaded to expect such an account from only science itself. Whereas there are many worldviews, there is only one common sense shared by all persons, including the ubiquitous belief that "Moving cars are hazardous to pedestrians." Therefore, invoking

science's presuppositions by a penultimate appeal to common sense pre-
serves science's objective and public character.

In summary, if you believe the reality check, that "Moving cars are hazardous to pedestrians," then you have already adopted all of the presuppositions needed for science to work. You have already delivered science from the
specter of skepticism. Compared with common sense, science requires more
experimentation, data, reasoning, and work, but absolutely nothing more
by way of presuppositions. Building science on a base of common sense is
a plausible and respected tradition. According to Einstein, "The whole of
science is nothing more than a refinement of everyday thinking" (Einstein
1954:290).

Philosophy's Big Bang

Why common sense? Why start a defense of scientific method by an appeal
to common sense? The most obvious reason for this path is the natural and
logical progression from that which is more certain to that which is less well
known. But there is another reason, more subtle but more profound.

But first, an analogy will be helpful. Physics has its Big Bang. In today's world, the four fundamental forces of nature appear quite distinct and
separate in ordinary circumstances. But physicists tell us that in the initial
moments following the Big Bang, those forces were mingled together in
a single superforce. Well, philosophy has a similar Big Bang. In ordinary
discourse, ontology, epistemology, and logic are reasonably distinct and recognizable topics within philosophy. But at the point where discourse begins,
those topics fuse together. The reason is that epistemology presumes ontology, because what we can know depends on what exists. But also ontology
presumes epistemology, because what we can become aware of depends on
our sensory and cognitive faculties. And logic is operating in any rational
discourse.

Consequently, it is not possible to initiate ontology, epistemology, and
logic separately or sequentially. Instead, their foundations present a problem
in simultaneous inference. At the very start of an inquiry into methods for
knowing about physical objects and events, ontology/epistemology/logic is a
single functional entity. The only way to access this complex entity initially is
through humble recourse to common sense, mixed with faith and sincerity.
So philosophy's Big Bang is common sense. Subsequently, fancier distinct
disciplines can emerge.

Interestingly, philosophy's Big Bang has been recognized more clearly in
Eastern than in Western philosophy: "[It] is of interest to note that in the
Indian tradition, logic, epistemology, and ontology have not been sharply
separated and discussed. On the contrary, thinkers in this tradition are of the

view that their interrelated discussion adds to the clarity of our understanding of the concerned concepts" (Chattopadhyaya 1991:xviii).

SENSIBLE QUESTIONS

Presuppositions exert their profound influence on scientific reasoning at many points, but most of all at the very first step in a scientific inquiry: formulating a question. Presuppositions affect the answers that scientists reach, but more fundamentally they affect the questions that scientists ask. Recall that the presuppositions of a question are the things that must be true in order for any hypothesis under consideration to be true, so that the question can have an answer within the inquiry's hypothesis set.

The question "Is there a coin in the cup?" was expressed precisely by this hypothesis set: H_1, that "There is a coin in the cup," and H_2, that "There is not a coin in the cup." Figure 4.6 depicts *The Thinker* contemplating a scientific question, such as this one. H_1 and H_2 are the two sensible alternative hypotheses within the common-sense perspective prevailing in science. Given common-sense presuppositions, these two hypotheses are mutually exclusive and jointly exhaustive. No other possibility requires consideration. Also, seeing the coin in the cup counts as definitive evidence, making the experiment's conclusion certain.

Comparing these two hypotheses, H_1 and H_2, to list their similarities and thereby to detect their presuppositions, both hypotheses presuppose that cups and coins exist. To assert either hypothesis, it is necessary to presuppose further that human sense perceptions are generally reliable, that our language suffices for discussing such matters, and so on. To make a long story short, this cup-and-coin story presupposes common sense.

But if common-sense presuppositions were not in force, then the preceding question or hypothesis set would not be sensible, interesting, or complete. Specifically, different versions of the alternative hypothesis H_2 could be proposed. Before moving on to these variants, however, the original H_2 can be described as a contextual negation of H_1.

A statement's contextual negation retains the statement's presuppositions and context, and negates only its principal contents. H_2 is a contextual negation of H_1 because H_2 retains the presuppositions in H_1 that the cup and coin exist and maintains the common-sense context with its additional presuppositions that sense perceptions are generally reliable and so on; and H_2 negates only the principal claim of H_1 by asserting that the coin is not, rather than is, in the cup. To emphasize the role of H_2 as a contextual negation, it is convenient here to also denote H_2 by H_C.

The other possibility, besides contextual negation, is logical negation. A statement's logical negation expresses the exact, total opposite. Logical

Figure 4.6. *The Thinker* tackles a scientific question with two hypotheses. (This photograph of the 1880 sculpture by Auguste Rodin is reproduced with kind permission from the Philadelphia Museum of Art.)

negation is customarily expressed by prefacing the original statement with the words "It is not the case that . . . " Thus, the hypothesis H_1, that "There is a coin in the cup," has the logical negation, here denoted H_L, that "It is not the case that there is a coin in the cup."

Because there are two possible negations, H_1 can be incorporated into two different questions. The original question was "H_1 or H_C?" The alternative question is "H_1 or H_L?" This section's objective is to discern which of these questions is sensible.

The difference between H_C and H_L is that H_L could be true for a far greater variety of reasons than H_C. Accordingly, the logical negation H_L can be regarded as a composite hypothesis that includes numerous specific subcases. The tame reason that could make H_L true is the old H_2, which can be written here with an expanded text as "It is not the case that there is a coin in the cup because the coin is outside the cup." Recalling that H_C is just another label for H_2, notice that H_L includes H_C as one of its subcases.

However, there are countless additional wild reasons that could make H_L true. For example, another subcase within H_L that can be denoted H_3 is that "It is not the case that there is a coin in the cup because the coin does not exist." Becoming wilder, H_4 says that "It is not the case that there is a coin in the cup because the entire physical world does not exist." And becoming wilder yet, H_5 says that "It is not the case that there is a coin in the cup because we are actually one-gram blue butterflies merely dreaming that we are humans observing a cup and coin," and H_6 says that "It is not the case that there is a coin in the cup because we are actually two-gram brown butterflies merely dreaming that we are humans observing a cup and coin." Leaving further possibilities to the reader's own rich imagination, H_L is a composite hypothesis containing as subcases the tame hypothesis H_2 plus infinitely many wild hypotheses $H_3, H_4, \ldots, H_\infty$.

Comparing H_1 and H_L, they are so radically dissimilar that no similarities remain to support common-sense presuppositions. For example, there is no presupposition common to both hypotheses that the cup and coin even exist. The original H_1 against H_C supported common-sense presuppositions, but H_1 against H_L does not. So "H_1 or H_L?" asks a question outside the common-sense world of science.

So the original question, "H_1 or H_2?", equivalently denoted by "H_1 or H_C?", pits H_1 against only one alternative, H_2. But the other question, "H_1 or H_L?" pits H_1 against infinitely many alternatives, H_2 *and* $H_3, H_4, \ldots, H_\infty$. Things get wild! These two questions, "H_1 or H_C?" and "H_1 or H_L?", beget different inquiries with three notable differences.

First, the huge question, "H_1 or H_L?", unlike the little question, "H_1 or H_C?", encompasses *all* conceivable hypotheses about the cup-and-coin experiment. Therefore, the huge question has the advantage that by exhausting absolutely all of the possibilities, it is certain to include the true hypothesis in its hypothesis set, regardless how tame or wild reality might actually be. By contrast, the little question considers merely two common-sensical hypotheses. Consequently, this little hypothesis set exhausts the possibilities only if science's common-sense presuppositions are in fact true of the world. If there is any skeptical worry about the truth of common sense, then the little hypothesis set risks not including the truth among the alternatives considered, which would beget an inquiry that could not possibly reach the truth. On the other hand, the huge question has no such risk, covering all of the possibilities regardless whether the truth lies inside or outside the realm of common sense.

Second, the two questions "H_1 or H_C?" and "H_1 or H_L?" have different implications for evidence. For the common-sense question, "H_1 or H_C?", seeing a coin counts as evidence for its existence. But for the skeptical question, "H_1 or H_L?", supposedly seeing a coin is inadmissible as evidence because there are no presuppositions saying that the coin exists or that our sight is reliable. So by shifting the question from "H_1 or H_C?" to "H_1 or H_L?", seeing a coin switches from evidence to non-evidence.

Third and finally, these two questions have different implications for the archive, composed of beliefs deemed irrelevant for the current inquiry. For the common-sense question, beliefs about the price of tea in China can be dismissed safely to the ignored archive. However, for the skeptical question, the physical world's existence is still up for grabs, so if startling news were to arrive about real tea in a real China, suddenly the world's existence would be affirmed at least in part, which would naturally increase the plausibility of cups and coins also being real. Therefore, the price of tea in China no longer resides in the background, but rather counts as evidence because it renders H_1 more plausible and H_L less plausible. So by shifting the question from "H_1 or H_C?" to "H_1 or H_L?", China's tea prices switch from non-evidence to evidence.

Combining these three points, the skeptical question, "H_1 or H_L?", has the advantage of being certain to contain a true hypothesis, but has the disadvantage of turning evidence topsy-turvy and thereby never reaching an answer. That disadvantage outweighs and nullifies the advantage. From a common-sensical or scientific perspective, determining whether or not a cup contains a coin is well within human capacities. Consequently, to ask whether or not a cup contains a coin in such a manner that no answer is possible is to ask a foolish question. "H_1 or H_C?" is an answerable and sensible question, but "H_1 or H_L?" is unanswerable and foolish.

This same conclusion can be expressed in different terms that some readers might find preferable or clearer. One could say that a scientist does ask the ambitious question "H_1 or H_L?", but uses two kinds of resources in reaching an answer. This question "H_1 or H_L?" can be expanded and rewritten as "H_1, H_2, H_3, . . . , or H_∞?" The wild hypotheses, H_3 through H_∞, are ruled out by common-sense worldview commitments, not by evidence (Kukla 2001). But the alternative hypothesis, H_2, is ruled out by the evidence of seeing a coin in the cup. These combined resources then leave a single hypothesis standing, H_1.

Whether one regards worldview commitments as proscribing H_3 to H_∞ from consideration or alternatively as rejecting them after consideration is essentially a matter of taste. I happen to prefer the former, which employs the rhetoric of sensible questions, but the latter is functionally equivalent. Either way, the salient conclusion is that common-sense commitments exclude conceivable but wild hypotheses, leaving fewer hypotheses that express a sensible question that can be answered with attainable evidence.

SCIENCE'S CREDIBILITY AND AUDIENCE

Typically, the defenders of science's rationality argue as if everyone should buy their arguments (and likewise the attackers of science expect everyone to be convinced). But in fact, science is not judged credible by everyone, as the vigorous debate reviewed in Chapter 3 makes plain. Indeed, given certain presuppositions, such as that the physical world is nonexistent or unknowable or illusory, no ordinary version of science can possibly be considered credible. Science is not immune to a person's presuppositions. Rather, commonsensical presuppositions render science credible, whereas some other presuppositions render it incredible. So not everyone will reach the same verdict on science, largely because of different initial presuppositions.

Consequently, it is futile for proponents of science to try to co-opt everyone into their program, and likewise it is equally wrongheaded for attackers of science to expect to co-opt everyone into their view. Such goals are philosophically shallow and morally questionable.

Instead, a productive dialogue that can engage everyone is to delineate those presuppositions that do or else do not render science credible. It is worthwhile for the academic community to work together to help each person see the logically consistent implications of his or her own presuppositions for science's credibility. The goal is to enlighten everyone, not to co-opt the opposition. Accordingly, the following two subsections explain why earnest skepticism cannot endorse science's credibility, whereas other worldviews can.

The Dismissal of Skepticism

Were this book claiming to address proponents of *every* worldview, including skepticism, then it would not be fair or correct to pretend that the preceding reality check is shared knowledge. Skepticism and the reality check are incompatible. If skepticism is true, then the reality check with its grand preamble is definitely unwarranted; but if the reality check is true, then skepticism is false. So something must go. My choice is to hold on to the reality check, accepting the consequence that this limits the range of worldviews engaged here.

This book's project is to presuppose common sense and then build scientific method, not to refute the skeptic and thereby establish common sense. The skeptic is unanswered here, not because of ill will on my part, but simply because I do not know what a skeptic wants from a non-skeptic or realist such as myself.

Fortunately, real skeptics are quite rare. My university work, including professional meetings of scientific and philosophical societies, causes me to meet many people on many continents. However, I have not yet personally met one real skeptic. Or, to be a little more accurate, I have met some people with skeptics' mouths, but have not encountered any skeptics' feet. Their mouths may counter the reality check, saying it is uncertain; but their feet obey it with all diligence, as their survival attests. "Skeptics are like dragons. You never actually meet one, but keep on running across heroes who have just fought with them, and won" (Palmer 1985:14).

Realists may be perplexed that skepticism tends to be such an extreme position as to reject even the simple reality check. Recall, for instance, that Karl Popper (1968:95) judged that even the simple common-sense belief that "Here is a glass of water" lies outside the bounds of human competence. But that extremism has a logical explanation. Imagine that you and I are enjoying lunch and beer at a pub. Suddenly, I am struck with remorse that I have never experienced being a skeptic, and forthwith give it my best attempt. After struggling manfully for an hour, I proudly exclaim "I've got it; I doubt that this salt shaker exists! Everything else still exists, but this salt shaker is gone – clean gone!" Understanding that I have lived for decades without the slightest inclination toward skepticism, doubtless your charity will move you to praise my fledgling skepticism. Nevertheless, you might be sorely tempted to say something like "Well, let me move this pepper shaker that you can see right next to the salt shaker that you cannot see. Now can you see them both?" The same embarrassment would attend any other modest version of skepticism, such as doubting my beer but not yours, or doubting one chair in the pub but not anything else. Only the radical doubt of everything leaves no easy refutation close at hand.

Skepticism is not dangerous because its doctrines might actually convince anyone to abandon caution when crossing a road or cooking a dinner. Common sense and animal fear prevent such foolishness. Rather, skepticism is dangerous because it disintegrates a personality, severing belief from action.

The Worldview Forum

A worldview is a person's beliefs about the basic makeup of the world and life. Depending on a person's intellectual maturity, a worldview may be more or less explicit, articulate, and coherent. But everyone has a worldview. It supplies answers to life's big questions, such as, What exists? What can we know? What is true and good? What is the purpose of human life? What happens after death? Many worldviews are rooted in a religion or a philosophical position, but some persons hold views not affiliated with any widespread movement. There are many minor worldviews, but relatively few major ones. The world's population, currently around 6 billion persons, is approximately 32% Christians, 19% Muslims, 19% atheists, 14% Hindus, 9% tribal or animist religions, 6% Buddhists, and 1% other, which includes 0.3% Jews (McManners 1993:648–649).

The most significant question that can be asked about worldviews is, Which one is true? However, that is not this book's question. Rather, this book is about scientific method, so its question is, How much does worldview pluralism affect science's claims and fortunes? It is simply a fact of life that historically there have been diverse views and that worldview pluralism is likely to continue for the foreseeable future. Is this a problem for science, or not? Does worldview diversity present insurmountable problems, motivating separate versions of science for each worldview or even rendering science invalid for adherents of some worldviews? Or can science, preferably in one single version, work for essentially everyone, including atheists, materialists, Buddhists, Jews, Christians, Muslims, and others?

The answer offered here is that a single version of science works fine for nearly every worldview, but not quite all. Science works in all worldviews that cheerfully assert the reality check, but it fails in all those that reject the reality check. Accordingly, the "worldview forum" is defined as the set of all worldviews that assert the reality check.

The Worldview Forum

This book's development and defense of science are intended to count within all worldviews willing to assert that it is rational, true, objective, realistic, and certain that "Moving cars are hazardous to pedestrians," but its reasoning does not count outside this worldview forum.

The great divide is the reality check. This book's worldview forum is depicted in Figure 4.7.

The great conflict between confident science and skeptical anti-science has seen a thousand skirmishes and battles, frequently waged with extremely technical philosophical arguments. But that war's outcome turns on but one event, the initial acceptance or rejection of common sense. Recall Reid's view that philosophy has no other root but the principles of common sense. For example, if we can know that "Moving cars are hazardous to pedestrians," or that "Here is a glass of water," or that "This cup contains a coin," then plausibly we can also know that "Table salt is composed of sodium and chlorine" or that "The earth orbits around the sun." But if the common-sense belief that "Moving cars are hazardous to pedestrians" is deemed too difficult for us to grasp, then all bets are off for science's prospects.

To make a long story short, if common sense is not respected, science's defenders face an impossible task; but if common sense is respected, science's debunkers face a losing battle. Common sense and science are bound to share

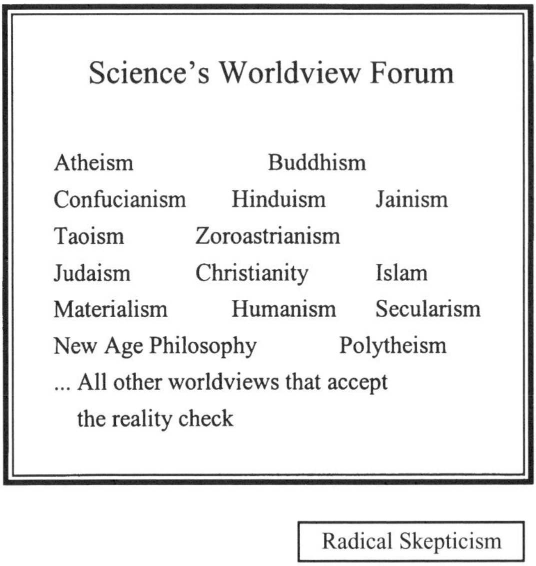

```
┌─────────────────────────────────────────────┐
│  ┌───────────────────────────────────────┐  │
│  │                                       │  │
│  │        Science's Worldview Forum      │  │
│  │                                       │  │
│  │                                       │  │
│  │   Atheism              Buddhism       │  │
│  │   Confucianism    Hinduism    Jainism │  │
│  │   Taoism      Zoroastrianism          │  │
│  │   Judaism     Christianity    Islam   │  │
│  │   Materialism     Humanism  Secularism│  │
│  │   New Age Philosophy      Polytheism  │  │
│  │   ... All other worldviews that accept│  │
│  │       the reality check               │  │
│  │                                       │  │
│  └───────────────────────────────────────┘  │
└─────────────────────────────────────────────┘
```

                        ┌──────────────────┐
                        │ Radical Skepticism│
                        └──────────────────┘

Figure 4.7. Science's worldview forum, which includes every worldview except radical skepticism. This broad forum is essential for science's objectivity.

similar fates. The war of science versus anti-science turns on one primordial choice, not a thousand technicalities.

Clearly, science could not be objective and public if science needed to depend on controversial philosophies or cultures. Fortunately, underneath these philosophical and cultural differences, there exists on this one earth a single human species with a shared human nature and common sense, and that commonality provides adequate resources for science's presuppositions. "Cultures may appear to differ, but they are all rooted in the same soil.... Human nature precedes culture and explains many of its features" (Roger Trigg, in Brown 1984:97).

SCIENCE'S REALISM AND FAITH

Recall the distinction between local and global presuppositions. A local presupposition is not proved in a given argument, but is or might be proved in another argument within some overall discourse or discipline, whereas a global presupposition is demonstrably unprovable in any argument whatsoever. This distinction requires further discussion here to illustrate how it plays out in ordinary scientific discourse.

Envision the following mundane scientific experiment. An agronomist grows soybeans with and without fertilizer in a careful replicated experiment, finding yields of 4,700 and 3,500 kg/ha, respectively. Hence, the reported conclusion is that fertilizer causes a yield increase of 1,200 kg/ha.

The PEL model can be applied to this agricultural experiment to flush out its presuppositions. Here this will not be done for the usual purpose of full disclosure, however, but rather for the current objective of flushing out one local presupposition and one global presupposition to illustrate their distinction.

One presupposition in this experimental report is the validity of its particular calculation: Subtraction of 3,500 from 4,700 gives the reported conclusion of 1,200. But obviously agronomists do not reiterate Peano's axioms for arithmetic and derive the proof that $4,700 - 3,500 = 1,200$. Rather, an agricultural journal just presupposes that branch of deductive logic called arithmetic. So the particular presupposition, that "$4,700 - 3,500 = 1,200$," is a local presupposition, because one could leave the agriculture library and go to the mathematics library to construct a proof that this subtraction is correct.

Another presupposition in this experimental report is that the physical world exists, particularly the agronomist, field, soybeans, and fertilizer. But as mentioned earlier in this chapter, no scientist or philosopher can prove that "The physical world exists" because it enters science's foundation as a presupposition. Hence, the presupposition that "The physical world does exist" is a global presupposition, proved neither in this agricultural report nor anywhere else. Neither are we hoping for some brighter philosopher or more innovative scientist to discover the long-awaited proof tomorrow. Rather, we already possess solid proof that no such proof is or ever will be possible.

Obviously, the philosophical reasoning that proves that "The physical world does exist" is unprovable likewise proves that "Sense perceptions are generally reliable" and "Inductive logic works" are forever unprovable. So science is pervaded by numerous global presuppositions. However, "global presuppositions" is a tedious phrase, so an equivalent brief word may be desired that means a belief held without possibility of proof. Such exists. It is the word "faith." Science rests on faith. Without this faith, technology's progress is still viable, but science's truth is undone.

Intellectuals accustomed to the pat formula that "Science has facts but religion has faith" may be shocked to see that science also has faith. Nevertheless, science *is* built on faith. If scientists rarely grasp or even sometimes contradict this, all that proves is that many scientists have a superficial understanding of their discipline's foundations. For scientists, seeing a coin in a cup counts as definitive evidence for the conclusion that "There is a coin in the cup" precisely, only, and decidedly because of scientific faith.

To say that science rests on common-sense faith may give some readers the uncomfortable feeling that science is therefore risky and shaky. Such an

Figure 4.8. While weighing the reality check's truth, a scientist ponders whether to trust Doubt or Faith. (This photograph of *The Sorrowing Soul between Doubt and Faith*, by Elihu Vedder, is reproduced with kind permission from the Membership Purchase Fund, The Herbert F. Johnson Museum of Art, Cornell University.)

impression is mistaken. The presuppositions underlying common sense and science cannot be proved, but neither can they be disproved, as Reid insisted regarding the parity among presuppositions. So there are exactly two faith choices: by faith to accept science's presuppositions, or by faith to reject science's presuppositions. The choice is not between faith and non-faith, but between one faith and another faith.

Science rests on faith, and faith accepts no substitutes. Science's common-sense faith makes sense within all worldviews in the worldview forum, but not in other worldviews outside it. The struggle between doubt and faith is depicted in Figure 4.8.

Finally, a few words may be said to clarify the relationship between this book's reality check and philosophical foundationalism. One might equate these two, thinking the reality check *is* my foundation, but that would be a careless reading. Foundationalism seeks a starting point that is indubitable and self-evident, or at least evident to the senses. Thereby, it seeks to function without presuppositions, without faith, and without any vulnerability to worldview differences. Clearly, the reality check has no such ambition or role. Instead, the reality check is a call to sincerity and community. It is also a call to humility, to an acknowledgment that a big world outside of ourselves and not of our own making controls what we little persons ought to believe about that world. The valid insight in foundationalism, however, is that a philosophical story does have to start somewhere, and my story begins with the reality check, not to instruct the reader about things previously unknown, but rather to prompt the reader to perceive the significance of things already known.

A REFLECTIVE OVERVIEW

For many scientists, the study of science's presuppositions moves them into relatively unfamiliar territory. The problem with approaching unfamiliar material is that it is easier to grasp technical details than the big picture. Accordingly, as this chapter draws toward a close, this section will provide a reflective overview.

Given that presuppositions are necessary, what do they invite scientists to contemplate? To a scientist who has not previously given presuppositions much thought, presuppositions may seem to invite scientists to check off one more thing when full disclosure is required – rather like an extra and unwelcome homework assignment. But actually, metaphysical presuppositions offer scientists a wonderful invitation to contemplate and cultivate a humanities-rich version of science.

This invitation is expressed especially well by Collingwood (1940). By way of introduction, his remarks concern the delineation of the subject matter of metaphysics given by its inventor, Aristotle. Collingwood uses the word "science" in its original meaning of a body of systematic and orderly thought about a specific subject, and hence he deems that Aristotle's term, directly translated as "first philosophy," is better translated into English as "First Science." Collingwood calls metaphysics both First Science, because it comes first in the logical order of subjects, and Last Science, because it comes last in the curriculum after the more obvious subjects have already been mastered.

> First Science is the science whose subject-matter is logically prior to that of every other, the science which is logically presupposed by all other sciences, although in the order of study it comes last. Sometimes he [Aristotle] calls it Wisdom, σοφία, with the implication that this is the thing for which φιλοσοφία, science, is the search; this again implying that in addition to their immediate function of studying each its own peculiar subject-matter the sciences have a further function as leading to a goal outside themselves, namely the discovery of what they logically presuppose.... As the Last Science it will be the ultimate goal of the scientist's pilgrimage through the realms of knowledge. The person who studies it will be doing what in all his previous work he was preparing himself to do. Hence if any particular

science is described as some particular form or phase of or search for a wisdom which within its own limits it never quite achieves, this First and Last Science must be described not as φιλοσοφία but as σοφία, the Wisdom for which every kind of φιλόσοφος is looking. (Collingwood 1940:5–6, 9–10)

In other words, a science that asks life's big questions is more human and more appealing than an impoverished science divorced from the humanities. Scientists wearied by long hours of exploring a minute topic, such as one part of one protein in one kind of bacterial flagellum, can suffer from a lack of meaning. They need an occasional refreshing glance at the big picture. Consequently, the invitation to possess a humanities-rich science can be welcome indeed. This invitation comes at many junctures in scientific thinking, but especially from science's presuppositions and faith. Healthy science includes a goal outside itself.

SUMMARY

By perceiving science's presuppositions, scientists can understand their discipline in greater depth and offer scientific arguments with full disclosure. An inquiry's presuppositions are those beliefs held in common by all of the hypotheses in the inquiry's hypothesis set. Analysis of a single scrap of common sense, such as the reality check that "Moving cars are hazardous to pedestrians," suffices to flush out the presuppositions that pervade common sense and science alike.

Aristotle clearly accepted science's ordinary, common-sense presuppositions, but his deductivist vision worked better for mathematical sciences than for natural sciences. Albertus Magnus resolved that deficiency by conditional or suppositional reasoning, granting the natural sciences definitive empirical evidence on the supposition of common-sense presuppositions. That device also granted science substantial independence from philosophy and theology, a view subsequently endorsed by Aquinas, Duns Scotus, Ockham, Buridan, and others. A tremendous diversity of views on science's presuppositions and claims emerged from the work of Francis Bacon, Descartes, Newton, Berkeley, Hume, Reid, and Kant.

Philosophical analysis of the reality check reveals ontological, epistemological, and logical presuppositions. But science's presuppositions do not individuate in any unique manner, and they can be listed in greater or lesser detail as one's current objectives dictate. Expressed as a single grand statement, science presupposes that the physical world is orderly and comprehensible. The most obvious components of this comprehensive presupposition are that the physical world exists and that our sense perceptions are generally reliable.

Science's presuppositions supply a basic and indispensable picture of our actual world, in contrast to a wild picture of all possible worlds that could be imagined. When asking a scientific question, these presuppositions focus the roster of viable hypotheses to a limited set that can be sorted out with ordinary, attainable evidence. But without such focus, an infinite hypothesis set would emerge that could not be sorted out with any finite, attainable quantity of data. Science works in the actual world commended to us by common sense and daily experience; science perishes in the imaginary world of all conceivable possibilities.

Philosophy proves that science's fundamental presuppositions are forever unprovable, so they constitute science's faith. But this is unproblematic within a worldview forum composed of all worldviews that accept the reality check.

CHAPTER FIVE

DEDUCTIVE LOGIC

This is the first of five chapters directed mainly at this book's primary goal of increasing productivity (the others being Chapters 6–9). Logic is the science of correct reasoning and proof, distinguishing good reasoning from bad. Logic sorts out the relationships that are fundamental in science, the relationships between hypotheses and evidence, between premises and conclusions.

In the context of logic, an "argument" is not a nasty dispute, but rather is a structured set of statements in which some statements, the premises, are offered to support or prove others, the conclusions. A deductive argument is valid if the truth of its premises entails the truth of its conclusions, and is invalid otherwise. Many deductive systems, including arithmetic and geometry, are developed on a foundation of predicate logic in the modern and unified vision of mathematics.

"The most immediate benefit derived from the study of logic is the skill needed to construct sound arguments of one's own and to evaluate the arguments of others" (Hurley 1994:ix). Every college offers courses in logic from its philosophy or mathematics faculty, and many fine introductory books on logic are available (such as Kearns 1988 and Hurley 1994). Unfortunately, most scientists never take such a course or read such a book.

Of course, given the simple premises that "All men are mortal" and "Socrates is a man," one trusts scientists to reach the valid conclusion that "Socrates is mortal," even without formal study of logic. But given the more difficult problems that continually arise in science, the rate of logical blunders can increase substantially in the absence of elementary training in logic. Fortunately, most such blunders involve a small number of common logical fallacies, so even a little training in logic can produce a remarkable improvement in reasoning skills.

The aim of this chapter differs from that of an ordinary text or course on logic. One short chapter cannot do the work of a whole book on logic, that is, teach logic. What it can do, however, is convey an insightful general impression of the nature and structure of deductive logic. The PEL model introduced in Chapter 4 identifies logic as one of the three essential inputs (along with presuppositions and evidence) required to support scientific conclusions. Consequently, the credibility of science depends on having a logic that is coherent and suitable for investigating the physical world. Incidentally, to convey modern logic accurately, this chapter will occasionally present some rather technical material, but readers are expected to master only the general structure, not to memorize technicalities.

This chapter first will distinguish the two basic kinds of logic: deductive logic, explained in this chapter, and inductive logic, explored in Chapter 7. The history of logic will be reviewed briefly, followed by elementary and formal accounts of logic and a brief look at arithmetic. Common invalid arguments will be analyzed to refine reasoning skills. Finally, formal logic will be connected with additional resources to produce material logic for justifying knowledge claims about the physical world. One branch of deductive logic, probability theory, is deferred to the next chapter.

DEDUCTION AND INDUCTION

The distinction between deduction and induction can be explained in terms of three interrelated differences. Of these three differences, the one that will be listed first can be regarded as the fundamental difference, with the others understood to be consequences or elaborations. Because of these differences, custom dictates distinct appellative terms for good deductive and inductive arguments. A deductive argument is valid if the truth of its premises guarantees the truth of its conclusions, and is invalid otherwise. An inductive argument is strong if its premises support the truth of its conclusions to a considerable degree, and is weak otherwise. The following deductive and inductive arguments, based on the work of Salmon (1984:14), can be used to make the discussion more concrete:

Valid Deductive Argument
Premise 1. Every mammal has a heart.
Premise 2. Every horse is a mammal.
Conclusion. Every horse has a heart.

Strong Inductive Argument
Premise 1. Every horse that has ever been observed has had a heart.
Conclusion. Every horse has a heart.

(1) The conclusion of a deductive argument is already contained, usually implicitly, in its premises, whereas the conclusion of an inductive argument

goes beyond the information present, even implicitly, in its premises. The technical terms for this difference are that deduction is non-ampliative, but induction is ampliative. For example, the conclusion of the foregoing deductive argument simply states explicitly, or reformulates, the information already given in its premises. All mammals have hearts according to the first premise, and that includes all horses according to the second premise, so the conclusion follows that every horse has a heart. On the other hand, the conclusion of the foregoing inductive argument contains more information than its premise. The premise refers to some group of horses that have been observed up to the present, whereas the conclusion refers to all horses, observed or not, and past or present or future.

Note that this difference, between ampliative and non-ampliative arguments, concerns the relationship between an argument's premises and conclusions, specifically whether or not the conclusions contain more information than the premises. This difference does not pertain to the conclusions as such, considered in isolation from the premises – indeed, the foregoing two arguments have exactly the same conclusion.

(2) Given the truth of all of its premises, the conclusion of a valid deductive argument is true with certainty, whereas even given the truth of all of its premises, the conclusion of an inductive argument is true with, at most, high probability, but not absolute certainty. This greater certainty of deduction is a direct consequence of its being non-ampliative: "The [deductive] conclusion must be true if the premises are true, *because* the conclusion says nothing that was not already stated by the premises" (Salmon 1984:15). The only way that the conclusion of a valid deductive argument can be false is for at least one of its premises to be false. On the other hand, the uncertainty of induction is a consequence of its being ampliative: "It is because the [inductive] conclusion says something not given in the premise that the conclusion might be false even though the premise is true. The additional content of the conclusion might be false, rendering the conclusion as a whole false" (Salmon 1984:15). For example, the foregoing inductive conclusion could be false if some other horse, not among those already observed and mentioned in this argument's premise, were being used for veterinary research and had a mechanical pump rather than a horse heart.

Deductive arguments are either valid or invalid on an all-or-nothing basis, because validity does not admit of degrees. But inductive arguments admit of degrees of strength. One inductive argument might have an average error rate of, say, 1 error in every 5 applications, whereas another might have only 1 error in every 1,000 applications.

The contrast between deduction's certainties and induction's probabilities can easily be overdrawn, however, as if to imply that induction is second-rate logic compared with deduction. If we represent certain truth by a probability of 1 and certain falsehood by 0, then an inductive conclusion can have any probability from 0 to 1, including values arbitrarily close to 1 representing certainty of truth (or 0 representing certainty of falsehood).

Pomerance (1982) gives an interesting example regarding the search for large prime numbers. Primes more than 100 digits long are used in encryption schemes for military communications. Were an encryption accidentally based on a composite number rather than a prime number, the national security of the United States could be put at risk. Unfortunately, no deductive proof of primality is known that could offer a feasible computation for such large numbers, using even the largest computers. But a feasible inductive procedure is known that can easily drive the error rate down to one in 10^{30} cases, which is considered a negligible uncertainty even when national security is in the balance. Furthermore, the strength of this inductive procedure's conclusion increases not linearly but exponentially with the amount of computation. So were the computer granted twice as much time for checking whether or not a big integer was prime, the original error rate of one in 10^{30} cases would not drop to one in 2×10^{30}, but rather to one in 10^{60}. Given sufficient computer time, induction can come arbitrarily close to certainty. Induction can deliver practical certainties, although it cannot deliver absolute certainties.

(3) Typically, deduction reasons from the general to the specific, whereas induction reasons in the opposite direction, from specific cases to general conclusions. That distinction was prominent in Aristotle's view of scientific method (Losee 1993:6–9) and remains prominent in today's dictionary definitions. For instance, the *Oxford English Dictionary* defines "deduction" in the context of logic as "inference by reasoning from generals to particulars," as "opposed to induction," and it defines "induction" as "The process of inferring a general law or principle from the observation of particular instances," as "opposed to deduction." It also makes evident the long history of these meanings, such as the remark about the word "induction" that it is "Directly representing L. *inductio* (Cicero), rendering Gr. ἐπαγωγή (Aristotle), in same sense." Definitions conveying the same distinction regarding the direction of reasoning can be found in practically any dictionary. Of course, there are exceptions (such as a tautology, which is a valid, though boring, deductive argument, with identical premise and conclusion). But typical reasoning justifies the dictionary definitions. Deduction reasons from a given model to expected data, whereas induction reasons from actual data to an inferred model, as depicted in Figure 5.1.

As encountered in typical scientific reasoning, the "generals" and "particulars" of deduction and induction have different natures and locations. The general principles exist in models or theories in a scientist's mind, whereas the particular instances pertain to physical objects or events that have been observed. Often the observations or data compose a limited sample from

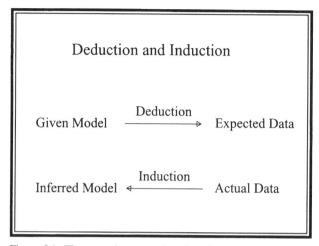

Figure 5.1. The opposite reasoning directions of deduction and induction. Deduction reasons from the mind to the world, whereas induction reasons from the world to the mind.

a larger population of interest. For example, given the general theory or model that supplemental vitamins promote chickens' growth, by deduction we expect to see faster growth in particular chickens given supplements than in those not given supplements. Likewise, given the particular data that 40 chickens given supplements grew considerably faster than another 40 chickens without supplements, by induction we infer the general conclusion that vitamins promote chickens' growth.

Statistical tests help scientists determine whether an observed result strongly supports the truth of a given hypothesis or model or likely was caused merely by chance sampling fluctuations. Incidentally, to make deductions or predictions easier when need be, sometimes the entities in a theoretical model are simplified or idealized, such as a "fair coin" with exactly equal chances of heads and tails (unlike an actual coin, which is always biased to at least a minute degree).

Deduction is not better than induction, nor the reverse. Rather, they pursue answers to different kinds of questions, with deduction reasoning from a mental model to expected data, and induction reasoning from actual data to a mental model. Both are indispensable for science.

HISTORICAL PERSPECTIVE ON DEDUCTION

Aristotle (384–322 b.c.) wrote extensively on logic (Rose 1968; Kneale and Kneale 1986:23–112; Parry and Hacker 1991; Patterson 1995). Although his works on some topics, including natural science, suffered much neglect until the early 1200s, his corpus on logic, the *Organon*, or tool (of reasoning), fared better. Aristotle's logic built on ideas from Socrates and Plato. Epicurean, Stoic, and Pythagorean philosophers also developed logic and mathematics. Besides Greece, there were impressive ancient traditions in logic in Babylon, Egypt, India, and China. Largely because of Augustine's early influence, the Aristotelian tradition came to dominate logic in the West, so that tradition is emphasized here.

In his *Prior Analytics*, Aristotle taught that every belief comes through either deduction (συλλογίσμος) or induction (ἐπαγωγή) (Book 11, Chapter 23) (McKeon 1941:102). Induction often supplies the premises used subsequently in deductive arguments (Pérez-Ramos 1988:201–215; Chakrabarti 1995:77). Aristotle's syllogistic logic was the first deductive system that we know of, predating Euclid's geometry and deducing numerous theorems from just several axioms and rules (Parry and Hacker 1991).

Aristotle proposed an inductive-deductive model of scientific method that features alternation of deductive and inductive steps (Losee 1993:6–9). This alternation, moving from mental model to physical world and back again, leads scientists to a mind-world correspondence – to truth. In this process, any discrepancy between model and world is to be resolved by adjusting the model to the world, because the actual data in the inductive step have priority over the expected data in the deductive step. Assuming that the data are not faulty or excessively inaccurate, actual data contrary to a model's expectations imply that something is wrong with the model. Adjusting one's model to the world is the basis of scientific realism.

Euclid (fl. c. 300 B.C.) was the great master of geometry. Many truths of geometry were known before Euclid. For example, earlier Babylonians and Egyptians knew that the sum of the interior angles of a triangle equals 180 degrees. But that was known by empirical observation of numerous triangles, followed by inductive generalization. Their version of geometry was a practical art related to surveying, in line with the name "geometry," literally meaning "earth measure."

But Euclid's *Elements of Geometry*, in one of the greatest paradigm shifts ever, instead demonstrated those geometrical truths by deduction from several axioms and rules. Euclid's geometry had five postulates concerning geometry, such as that a straight line can be extended in either direction, plus five axioms or "common notions" concerning correct thinking and mathematics in general, such as that the whole is greater than the parts (Todhunter 1933; Mueller 1981). Euclid's combination of geometrical postulates and logical axioms represented a nascent recognition that logic underlies geometry. Countless theorems can be deduced from those postulates and axioms, including that the sum of the interior angles of a triangle equals 180 degrees.

Subsequently, beginning with the non-Euclidean geometry of Thomas Reid (1710–1796), consistent alternatives to the standard axioms were discovered, rendering Euclid's work *a* geometry rather than *the* geometry (Daniels 1989). In Reid's alternative geometry, the sum of the interior angles of a triangle equals more than 180 degrees.

Despite some of its shortcomings, Euclid's geometry stands as one of the most magnificent and influential contributions to deductive reasoning. As mentioned in Chapter 2, geometry's towering edifice of deductive certainty left a profound mark on philosophy and science as an ultimate standard of truth and proof. A millennium and a half would pass before Albertus Magnus and other medieval scholars would work out a concept of truth equally applicable to physical objects, as contrasted with mathematical ideas.

Anicius Manlius Severinus Boethius (A.D. 480–524) translated, from Greek into Latin, many parts of Aristotle's logical works, Porphyry's *Introduction to Aristotle's Logic*, and parts of Euclid's *Elements*. His *On Arithmetic*, based on earlier work by Nicomachus of Gerasa, became the standard text on arithmetic for almost a millennium.

Peter Abelard (c. 1079–1142) wrote four books on logic. He and his students, John of Salisbury and Peter Lombard, greatly influenced medieval logic. The use of Arabic numerals was spread into Europe by Alexandre de Villedieu (fl. c. 1225), a French Franciscan, John of Halifax (or Sacrobosco, c. 1200–1256), an English schoolman, and Leonardo of Pisa (or Fibonacci, c. 1180–1250), an Italian mathematician. The modern mind can hardly imagine the tedium of multiplication or division using Roman numerals, or how few persons in medieval Europe could perform what we now regard as elementary calculations.

Albertus Magnus (c. 1200–1280) wrote 8,000 pages of commentary on Aristotle, including much logic. He also wrote a commentary on Euclid's *Elements* (Paul M. J. E. Tummers, in Weisheipl 1980:479–499).

Robert Grosseteste (c. 1168–1253) founded the mathematical-scientific tradition at Oxford (Crombie 1962; Losee 1993:36–38). He affirmed and refined Aristotle's inductive-deductive scientific method, which he termed the "Method of Resolution and Composition" for its inductive and deductive components, respectively. Also, his Method of Verification involved deriving the deductive consequences of a theory beyond the original facts on which the theory was based and then observing the actual outcome in a controlled experiment to check the theory's predictions. That method recognized the priority of data over theories, in accord with Aristotle. And Grosseteste's Method of Falsification eliminated bad theories or explanations by showing that they implied other things known to be false. To increase the chances of eliminating false theories, he recommended that conclusions reached by induction be submitted to the test of further observation or experimentation.

Putting all those methods together, the objective of Grosseteste's new science was to make theory bear on the world and the world bear on theory, thereby bringing theory into correspondence with the world. That scientific method sought to falsify and reject false theories, to confirm and accept true theories, and to discern which kinds of observational or experimental data would help the most in theory evaluation.

There is substantial similarity between Grosseteste's medieval science and modern science, as summarized in the astonishingly wise and admirably brief account of scientific method in Box, Hunter, and Hunter (1978:1–15). "Modern science owes most of its success to the use of these inductive and experimental procedures, constituting what is often called 'the experimental method'. The ... modern, systematic understanding of at least the qualitative aspects of this method was created by the philosophers of the West in the thirteenth century. It was they who transformed the Greek geometrical method into the experimental science of the modern world" (Crombie 1962:1). With that assessment, that a basically correct and complete scientific method emerged in the thirteenth century, I concur.

William of Ockham (c. 1285–1347) wrote a substantial logic text, the *Summa logicae*, illustrated in Figure 5.2. The principle of parsimony is often called Ockham's razor because of his influential emphasis on this principle.

Jean Buridan (c. 1295–1358) wrote the *Summulae de dialectica*, a then-modern revision and amplification of the earlier logic text by Peter of Spain (fl. first half of the thirteenth century), and two advanced texts, the *Consequentiae* and *Sophismata* (Pinborg 1976; King 1985). His theory of supposition, Treatise IV in his *Summulae*, is a theory of reference intended to connect the terms in a logical argument with objects in the physical world. Hence, he explored the philosophical basis for applying logic to science. "Buridan was a brilliant logician in an age of brilliant logicians" (King 1985:xi).

John of St. Thomas (John Poinsot, 1589–1644) wrote on the usual formal logic but also on material logic applicable to the material world. René Descartes (1596–1650) was the founder of analytic geometry. Blaise Pascal (1623–1662) contributed to projective geometry, arithmetic, combinatorial analysis, probability, and the theory of indivisibles (a forerunner of integral calculus). He developed the first commercial calculating machine. Isaac Newton (1642–1727) and Gottfried Leibniz (1646–1716) invented calculus. Thomas Reid discovered the first non-Euclidean geometry, long before the better-known discoveries by G. F. B. Riemann (1826–1866).

Immanuel Kant (1724–1804) reinterpreted the nature of mathematical truth. The received view was that mathematical truth was "out there," independent of human thought or discovery. But Kant grounded geometry's truths in the human faculty of sensibility. Also, geometrical truth and proof depended on procedures for constructing figures in a spatiotemporal process,

Figure 5.2. The first page of William of Ockham's text on logic. (This is Gonville and Caius College's manuscript 464/571, written in 1341, reproduced with kind permission from the Masters and Fellows of Gonville and Caius College, Cambridge.)

not just on seeing the completed figures. Although logic had been based on axioms since Aristotle, as had geometry since Euclid, Kant taught that arithmetic could not be axiomatized. But later, Giuseppe Peano (1858–1932) did axiomatize arithmetic.

For millennia the various branches of deduction – such as logic, arithmetic, and geometry – had been developed as separate and unrelated systems. Early great works aiming to unify logic and mathematics were the brilliant *Grundgesetze der Arithmetik* (*The Basic Laws of Arithmetic*) of Frege (1893; also see Lucas 2000 and Potter 2000) and the monumental *Principia Mathematica* of Whitehead and Russell (1910–13).

ELEMENTARY PROPOSITIONAL LOGIC

Propositional logic, also called statement calculus and truth-functional logic, is a rather elementary branch of deductive logic. Nevertheless, it is quite important because it pervades common-sense reasoning and scientific reasoning. Also, it is an integral component within many of the more advanced and more complex branches of logic.

A simple proposition or statement has a subject and a predicate. Hamilton (1978:1) gives examples, with underlined words being the subjects and the remaining words being the predicates: "Napoleon is dead," "John owes James two pounds," and "All eggs which are not square are round." A compound proposition is formed from simple propositions by means of connectives like "and," "or," and "if . . . then . . . ," such as "If the barometer falls, then either it will rain or it will snow."

Propositional logic considers only declarative statements. Accordingly, every simple proposition has the property of having one or the other of two possible truth-values: true (T) and false (F). Note that the truth-value applies to the proposition as a whole, such as "Napoleon is dead" is true. In propositional logic, as introduced in this section, there is no further analysis of the subject and predicate within a proposition. But in predicate logic, to be explained in a later section of this chapter, further analysis is undertaken. Hence, predicate logic is more complicated, subsuming propositional logic and adding additional concepts and analysis.

Proposition constants represent specific simple propositions and are denoted here by uppercase letters like A, B, and C (except that T and F are reserved to represent the truth-values true and false). For example, "The barometer falls" can be symbolized by B, "It will rain" by R, and "It will snow" by S. Then the compound sentence "If the barometer falls, then either it will rain or it will snow" can be symbolized by "If B, then R or S."

The most common connectives or operators are "not," "and," "or," "implies," and "equals" ("implies" is also expressed as "if . . . then . . . ," and

Table 5.1. Truth-table definitions for negation, conjunction, disjunction, implication, and equality

Assignments		Not	And	Or	Implies	Equals
A	B	$\sim B$	$A \wedge B$	$A \vee B$	$A \rightarrow B$	$A \equiv B$
T	T	F	T	T	T	T
T	F	T	F	T	F	F
F	T		F	T	T	F
F	F		F	F	T	T

"equals" as "if and only if"). These five connectives are denoted here by these symbols: "$\sim$," "$\wedge$," "$\vee$," "$\rightarrow$," and "$\equiv$." The meanings of these connectives are specified by a truth table (Table 5.1). "Not" is a unary operator applied to one proposition. If B is true, then $\sim B$ is false; and if B is false, then $\sim B$ is true. That is, B and $\sim B$ have opposite truth-values. The other connectives are binary operators applied to two propositions. For example, "A and B," also written as "$A \wedge B$," is true when both A is true and B is true, and is false otherwise. Simple propositions can be combined with connectives, such as $B \rightarrow (R \vee S)$ to symbolize the preceding compound proposition about a barometer. Parentheses are added as needed to avoid ambiguity. To simplify expressions, the conventions are adopted that negation has priority over other connectives and applies to the shortest possible subexpression, and parentheses may be omitted whenever the order makes no difference.

Proposition variables stand for simple propositions and are denoted here by lowercase letters like p, q, and r. Within a given argument, the proposition constant A stands for a particular simple proposition, such as "Apples are sweet." But the proposition variable p stands for any simple proposition, such as A or B or C.

Proposition expressions stand for proposition constants or variables and may also contain connectives. They are denoted by script letters like $\mathcal{A}$, $\mathcal{B}$, and $\mathcal{C}$. They are formed by one or more applications of two rules: (1) Any proposition variable is a proposition expression. (2) If $\mathcal{A}$ and $\mathcal{B}$ are proposition expressions, then $(\sim \mathcal{A})$, $(\mathcal{A} \wedge \mathcal{B})$, $(\mathcal{A} \vee \mathcal{B})$, $(\mathcal{A} \rightarrow \mathcal{B})$, and $(\mathcal{A} \equiv \mathcal{B})$ are proposition expressions. Hence, examples of proposition expressions are p, q, $(\sim p)$, and $(p \wedge q)$. By repeated application of these rules, more complex expressions can be formed, such as $(((\sim p) \wedge q) \rightarrow ((\sim q) \wedge r))$.

Any expression formed by these rules makes sense in that for any assignment of truth-values to its proposition variables, the expression has a truth-value of either true or false. But other sequences of symbols, not formable by these rules, such as ") $\sim (p(q,$" are nonsense and lack truth-values.

A proposition expression is a tautology, also called a logical truth, if it is true under every possible assignment of truth-values to its proposition variables, whereas it is a contradiction, also called a logical falsity, if it is false under every possible assignment. For example, $(p \lor (\sim p))$ is a tautology, always true, whereas $(p \land (\sim p))$ is a contradiction, always false. A proposition expression is a logical contingency if it is neither a tautology nor a contradiction, that is, if some assignment of truth-values to its variables makes the proposition true and another assignment makes it false, such as the proposition $p \land q$.

$\mathscr{A}$ logically implies $\mathscr{B}$ if $(\mathscr{A} \to \mathscr{B})$ is a tautology. $\mathscr{A}$ is logically equivalent to $\mathscr{B}$, denoted by $\mathscr{A} :: \mathscr{B}$, if $(\mathscr{A} \equiv \mathscr{B})$ is a tautology. For example, $(p \land q)$ logically implies p, and $\sim\sim p$ is logically equivalent to p. Equivalent expressions may replace each other wherever they occur. The following equivalence rules are useful (McKay 1989). These rules are also given abbreviated names to facilitate annotation in proofs, as will be useful momentarily.

Double Negation (DN)
$p :: \sim\sim p$

Conditional Exchange (CE)
$p \to q :: \sim p \lor q$

Commutativity (Comm)
$p \land q :: q \land p$
$p \lor q :: q \lor p$

Distribution (Dist)
$p \lor (q \land r) :: (p \lor q) \land (p \lor r)$
$p \land (q \lor r) :: (p \land q) \lor (p \land r)$

Associativity (Assoc)
$p \land (q \land r) :: (p \land q) \land r$
$p \lor (q \lor r) :: (p \lor q) \lor r$

Contraposition (Contra)
$p \to q :: \sim q \to \sim p$

DeMorgan's Rules (DeM)
$\sim(p \lor q) :: \sim p \land \sim q$
$\sim(p \land q) :: \sim p \lor \sim q$

Exportation (Exp)
$(p \land q) \to r :: p \to (q \to r)$

Redundancy (Red)
$p :: p \lor p$
$p :: p \land p$

Biconditional (Bic)
$p \equiv q :: (p \to q) \land (q \to p)$

An argument is a structured, finite sequence of proposition expressions, with the last being the conclusion (ordinarily prefaced by the word "therefore" or the symbol "$\therefore$"), and the others the premises. The premises are intended to support or prove the conclusion. For example, *modus ponens* is the argument $p; p \to q; \therefore q$. An argument is valid if under every assignment of truth-values to the proposition variables that makes all premises true, the conclusion is also true. That is, an argument is valid if the truth of all of its premises guarantees the truth of its conclusion. Otherwise, the argument is invalid. In other words, the argument $\mathcal{A}_1, \ldots, \mathcal{A}_N; \therefore \mathcal{A}$ is valid if and only if the proposition $((\mathcal{A}_1 \wedge \ldots \wedge \mathcal{A}_N) \rightarrow \mathcal{A})$ is a tautology.

There are several methods for proving that an argument is valid, or else invalid as the case may be. Different methods all give the same verdict, but one method may be easier to understand or use in a given instance than is another. The conceptually simplest method, directly reflecting the definitions of validity and invalidity, is to construct a truth table to determine whether or not each assignment of truth-values to the argument's proposition variables that makes all premises true also makes the conclusion true. An argument with N constants or variables has 2^N lines in its truth table. As an example, Table 5.2 analyzes the argument $A \rightarrow B; B \rightarrow (C \vee D); A \wedge \sim C; \therefore D$. An asterisk draws attention to the 1 assignment of the 16 assignments that makes all premises true. Because the conclusion is always true when all premises are true, this argument is proved valid. This example has just four variables, but when an argument has considerably more variables, obviously this truth-table method will become tedious.

Table 5.2. Truth-table proof for the valid argument $A \rightarrow B$; $B \rightarrow (C \vee D); A \wedge \sim C; \therefore D$

Assignments				Premises			Conclusion
A	B	C	D	$A \rightarrow B$	$B \rightarrow (C \vee D)$	$A \wedge \sim C$	D
T	T	T	T	T	T	F	T
T	T	T	F	T	T	F	F
T	T	F	T	T	T	T	T*
T	T	F	F	T	F	T	F
T	F	T	T	F	T	F	T
T	F	T	F	F	T	F	F
T	F	F	T	F	T	T	T
T	F	F	F	F	T	T	F
F	T	T	T	T	T	F	T
F	T	T	F	T	T	F	F
F	T	F	T	T	T	F	T
F	T	F	F	T	F	F	F
F	F	T	T	T	T	F	T
F	F	T	F	T	T	F	F
F	F	F	T	T	T	F	T
F	F	F	F	T	T	F	F

Note: An asterisk indicates the line for which all of the premises are true.

A second method for proving validity is to deduce the argument, expressed in the form of a tautological proposition, as a theorem from assumed axioms and rules. The axioms and rules are assumed without proof because the process of deducing propositions from other propositions cannot regress indefinitely, but rather must start with something. This method is especially valuable for arguments with numerous variables that would make the truth-table method cumbersome. Hence, one could begin with axiomatic statements, and then, by clever rearrangements, eventually deduce the tautological proposition $((A \rightarrow B) \wedge (B \rightarrow (C \vee D)) \wedge (A \wedge \sim C)) \rightarrow D$. Davis (1986:370–371) gives examples. For propositional logic, the following three axioms suffice (Davis 1986:369):

Identity (Iden)
$p \rightarrow p$

Excluded Middle (EM)
$p \vee \sim p$

Noncontradiction (Nonctr)
$\sim(p \wedge \sim p)$

Besides these three axioms assumed true, rules are needed to generate new propositions from old ones. The foregoing axioms and the following several inference rules suffice to prove any theorem of propositional logic (McKay 1989):

Modus ponens (MP)   *Modus tollens (MT)*
$p \rightarrow q$ $p \rightarrow q$
p $\sim q$
$\therefore q$ $\therefore \sim p$

Simplification (Simp)   *Conjunction (Conj)*
$p \wedge q$ $p \wedge q$ p
$\therefore p$ $\therefore q$ q
                                          $\therefore p \wedge q$

Disjunctive Syllogism (DS)   *Hypothetical Syllogism (HS)*
$p \vee q$ $p \vee q$ $p \rightarrow q$
$\sim p$ $\sim q$ $q \rightarrow r$
$\therefore q$ $\therefore p$ $\therefore p \rightarrow r$

Addition (Add)   *Constructive Dilemma (CD)*

p p $p \rightarrow q$

$\therefore p \vee q$ $\therefore q \vee p$ $r \rightarrow s$

                                                 $p \vee r$

                                                 $\therefore q \vee s$

A third method for proving an argument valid is to deduce the argument's conclusion from its premises, using the logic's rules. The following proof exemplifies this method for the same argument as in Table 5.2, namely, $A \rightarrow B$; $B \rightarrow (C \vee D)$; $A \wedge \sim C$; $\therefore D$. Each line of the proof is numbered on the left, and the justification for each step is noted on the right.

(1) $A \rightarrow B$ Premise 1
(2) $B \rightarrow (C \vee D)$ Premise 2
(3) $A \wedge \sim C$ Premise 3
(4) A 3 Simp
(5) B 1, 4 MP
(6) $C \vee D$ 2, 5 MP
(7) $\sim C$ 3 Simp
(8) $\therefore D$ 6, 7 DS

A fourth method for testing validity is the truth-tree method explained by Jeffrey (1991:21–34). It is beyond the scope of this chapter, however, to present that and other proof strategies. To acquire facility with propositional logic, one must study a standard text and work through numerous exercises.

Remarkably, however, the preceding few pages provide all of the resources needed to prove *any* propositional argument valid or invalid. Indeed, the simple truth-table method alone is sufficient, even if tedious for long arguments. As a representative example, consider the following argument:

If component A short-circuits, then circuit B will fail. If circuit B fails, then component C or D will burn out. Component A will short-circuit, but C will not burn out. Therefore, D will burn out.

Is this argument valid? Depending on one's logical prowess, the answer to this question may not be entirely or quickly obvious. But even before getting the answer, you can rest assured that the resources already provided are adequate for a definitive yes or no for this problem, as well as for any other propositional reasoning ever encountered in science. Actually, this particular word problem is an instance of the argument analyzed earlier, as the reader may have detected, so the answer is yes, this argument is valid.

FORMAL PROPOSITIONAL LOGIC

It is amazing that the preceding section provided, in a few pages, all the logical resources needed to prove any propositional argument either valid or invalid. And yet, one could make do with far fewer resources. For example, all of the preceding equivalences, such as $p \equiv \sim\sim p$, can be deduced from the axioms and rules, so they are unnecessary luxuries. They were given earlier merely for convenience, to make proofs easier. The following considerably smaller formal system L, drawing on Hamilton (1978:28), is sufficient for deducing any theorem of propositional logic.

(1) Language. The symbols used in L are as follows: $\sim$, $\rightarrow$, (,), p_1, p_2, p_3, ...

(2) Expressions. A well-formed formula (wff) is formed by one or more applications of two rules: (a) Each p_i is a wff. (b) If $\mathcal{A}$ and $\mathcal{B}$ are wffs, then $(\sim\mathcal{A})$ and $(\mathcal{A} \rightarrow \mathcal{B})$ are wffs.

(3) Axioms. For any wffs $\mathcal{A}$, $\mathcal{B}$, and $\mathcal{C}$, axioms are formed by the following axiom schemes:

Axiom Scheme 1.	$(\mathcal{A} \rightarrow (\mathcal{B} \rightarrow \mathcal{A}))$
Axiom Scheme 2.	$((\mathcal{A} \rightarrow (\mathcal{B} \rightarrow \mathcal{C})) \rightarrow ((\mathcal{A} \rightarrow \mathcal{B}) \rightarrow (\mathcal{A} \rightarrow \mathcal{C})))$
Axiom Scheme 3.	$(((\sim\mathcal{A}) \rightarrow (\sim\mathcal{B})) \rightarrow (\mathcal{B} \rightarrow \mathcal{A}))$

(4) Rule. One rule, *modus ponens*, says from $\mathcal{A}$ and $(\mathcal{A} \rightarrow \mathcal{B})$, infer $\mathcal{B}$.

(5) Interpretation. The symbols "$\sim$" and "$\rightarrow$" (and the associated parentheses needed to specify the order of operations) are the logical connectives negation and implication, and the symbols p_1, p_2, p_3, and so on, represent proposition variables having truth-values of either true or false. A proof is a sequence of wffs $\mathcal{A}_1, \ldots, \mathcal{A}_N, \mathcal{A}$ such that each wff either is an axiom or follows from two previous members of the sequence by application of *modus ponens*. The final wff, $\mathcal{A}$, is a theorem.

Even this quite frugal foundation for propositional logic can be reduced further. For starters, the infinite set of symbols ($p_1, p_2, p_3, \ldots$) can be replaced by sequences of just two symbols, say 0 and 1. In the current context, each symbol, 0 or 1, represents one binary bit of information, rather than one of the literal numbers zero and one. For example, computers often use a string of eight bits, called a byte, to represent an alphabet of 256 symbols. In practice, that usually will be a sufficient alphabet, but a longer binary sequence could be used for an arbitrarily larger alphabet, such as a string of 20 binary bits to specify over 1,000,000 symbols.

Greater frugality is also possible for the logical connectives. Table 5.1 specified four binary connectives, such as "and" with the truth-values TFFF (transposing columns in that table into rows in this paragraph in an obvious manner). Because a binary connective is specified by four choices of T or F, there are 2^4 or 16 possible connectives, but only some of these are useful enough to have recognized names. In ordinary English, "or" is used in two senses: the inclusive "or," meaning A or B or both, and the exclusive "or," meaning A or B but not both, which can be expressed more specifically by "A or else B." The "or" in Table 5.1 is the ordinary inclusive "or," with truth-values TTTF; so another binary connective is the exclusive "or," sometimes denoted by "xor," specified by the truth-values FTTF.

The foregoing formal system uses only two connectives, negation and implication ($\sim$ and $\rightarrow$), which are fewer than the five used in the earlier informal logic, because the others can be defined in terms of just these two connectives: $(A \land B) :: (\sim(A \rightarrow (\sim B)))$, $(A \lor B) :: ((\sim A) \rightarrow B)$, and $(A \equiv B) :: (\sim((A \rightarrow B) \rightarrow (\sim(B \rightarrow A))))$. Besides this set $\{\sim, \rightarrow\}$, other adequate sets of connectives are $\{\sim, \lor\}$ and $\{\sim, \land\}$.

Another binary connective is joint denial of A and B, expressed by "Neither A nor B," symbolized by "$A \downarrow B$," and specified by the truth-values FFFT. It is also named "Nor" because its FFFT is the negation of "or" with TTTF. Still another connective is alternative denial of A and B, expressed by "Either not A or not B" or equivalently "Not both A and B," symbolized by "$A \mid B$," and specified by the truth-values FTTT. It is also named "Nand" because its FTTT is the negation of "and" with TFFF. Remarkably, the singleton sets $\{\downarrow\}$ and $\{\mid\}$ are also adequate sets of connectives for defining all of the unary and binary connectives. Hence, the foregoing formal system could be written with only one logical connective, say $\downarrow$. The required replacements to define negation and implication are $(\sim A) :: (A \downarrow A)$ and $(A \rightarrow B) :: ((A \downarrow A) \downarrow B) \downarrow ((A \downarrow A) \downarrow B)$, adapted from Hamilton (1978:21). The two singleton sets, $\{\downarrow\}$ and $\{\mid\}$, are the only individually adequate connectives among the binary connectives (Hunter 1971:70–71). However, there are additional individually adequate connectives, not elaborated here, among trinary (and higher) connectives (Shankar 1994:29).

Yet another opportunity for frugality is to replace the three axiom schemes with only one. Leblanc and Wisdom (1976:339) have shown that they can be exchanged for the single axiom scheme $((((\mathscr{A} \rightarrow \mathscr{B}) \rightarrow (\sim \mathscr{C} \rightarrow \sim \mathscr{D})) \rightarrow \mathscr{C}) \rightarrow \mathscr{E}) \rightarrow ((\mathscr{E} \rightarrow \mathscr{A}) \rightarrow (\mathscr{D} \rightarrow \mathscr{A}))$. They have also explained that the three axioms in the foregoing formal system are not uniquely suitable, for numerous other choices are equivalent. However, these three axiom schemes are independent, meaning that were any one to be deleted, it could not be deduced from the remaining axiom schemes (Hunter 1971:122). So, neither fewer than these three axiom schemes, nor some shortened version of the one combined axiom scheme, could suffice to derive the theorems of propositional logic.

So, the foregoing frugal system could be exchanged for a yet more Spartan system that has only three symbols ($\downarrow$, 0, and 1, plus associated parentheses), one axiom, one rule, and the usual interpretation as propositional logic. From this, all of the theorems of propositional logic could be deduced. Any (propositional) argument could be proved valid or invalid. One would pay for such frugality, however, with long and complicated expressions. Accordingly, both informal and formal versions of logic are useful.

PREDICATE LOGIC

Predicate logic, also called predicate calculus and first-order logic, considers the internal structure and meaning of a sentence, as contrasted with the simpler propositional logic that merely considers the truth-value of a sentence construed as a whole. A quite simple example suffices to motivate this deeper analysis. Consider the world's most familiar argument: All men are mortal; Socrates is a man; therefore Socrates is mortal. From the simple perspective of propositional logic, this argument has three different statements arranged in the argument form p; q; $\therefore$ r (Hamilton 1978:45). Hence, in this analysis, the argument's conclusion does not follow from its premises – it is a *non sequitur*. Yet patently this familiar argument is valid. Its validity, however, depends not merely on the truth-functional relationships among this argument's statements, such as would be within the purview of propositional logic. Rather, its validity depends on relationships between parts of the statements, which requires predicate logic. The required logic for this simple example must analyze an argument that can be formalized as follows: All A's are B; C is an A; $\therefore$ C is B. To handle this deeper analysis, two elaborations are needed.

First, a statement's subject and predicate must be denoted separately. Recall that (simple) statements each have a subject and a predicate. The following rather standard notation is convenient (Hamilton 1978:46). Predicates are represented by uppercase letters, and subjects by lowercase letters. For example, $A(s)$ may represent "Socrates is a man" by having A stand for the predicate "is a man" and s stand for the subject "Socrates." Likewise, $A(n)$ could symbolize "Napoleon is a man" by having n stand for Napoleon. Similarly, $B(I)$ can symbolize "I write books." And letting C represent the predicate "is a cow," the negated term $\sim C(s)$ represents "Socrates is not a cow." Note that an entire statement, like $A(s)$ or $\sim C(s)$ or $A(s) \wedge A(n)$, makes an assertion and hence has a truth-value; but a constituent part of a statement, like A representing the predicate "is a man" or s representing the subject "Socrates," makes no assertion and hence has no truth-value.

Second, notation is needed for the universal quantifier "all" and the existential quantifier "some." Let the predicate symbols A and M represent "is a man" and "is mortal." How can "All men are mortal" be translated into symbols? A convenient device is to use a variable as an indeterminate subject, denoted by a lowercase italic letter such as x, to represent any object x in the universe (as contrasted with specific subjects like s representing Socrates). Then "All men are mortal" can be written as follows: For all x, $(A(x) \rightarrow M(x))$ (Hamilton 1978:46). (Of course, y or z would be an equally suitable variable.) The phrase "for all x" is symbolized by $(\forall x)$, so this statement can be written more compactly: $(\forall x)\,(A(x) \rightarrow M(x))$. The implication $A(x) \rightarrow M(x)$ is asserted for all objects x. If x happens to be a man, then by implication x is mortal; but if x is not a man, then the implication is still true because its antecedent is false (recalling the truth-table definition of implication).

The existential quantifier "some" has the meaning "there exists at least one object x such that" and is symbolized by $(\exists x)$ (Hamilton 1978:47). For example, $(\exists x)(A(x) \wedge M(x))$ means "There exists at least one object x such that x is a man and x is mortal." For comparison, $(\forall x)(A(x))$ means "All x are A" and $(\exists x)\,(A(x))$ means "Some x is A." For a more complicated example with two variables, $(\forall x\,)(\exists y)\,(A(x) \vee B(y))$ means "For all x and some y, x is A or y is B."

Although it is convenient to use universal and existential quantifiers, actually only one kind of quantifier is needed. Note that the two statements "There exists an x that has property P" and "It is not the case that all x's do not have property P" have exactly the same meaning (Hamilton 1978:48). Hence, $(\exists x)$ can be replaced by $\sim(\forall x\,)\sim$. For example, $(\exists x)(A(x))$ is equivalent to $\sim(\forall x) \sim (A(x))$. Likewise, $(\forall x)$ can be replaced by $\sim (\exists x)\sim$. So any statement using "$\forall$" and "$\exists$" symbols can be rewritten using only "$\forall$" symbols (or, alternatively, using only "$\exists$" symbols).

The following deductive system $K_{\mathscr{L}}$ is based on the work of Hamilton (1978:49–56, 71–72). For the following symbols, subscripts are indices that identify individual items, whereas superscripts indicate the number of places or arguments (such as "$\sim$," which has one argument, as in $\sim p$, and "$\wedge$," which has two arguments, as in $p \wedge q$). Hamilton (1978:129) provides a formal definition of a function that will not be reiterated here. Basically, a function gives a unique output value for any values of its inputs over some specified domain.

(1) Language. A first-order language $\mathscr{L}$ has as its alphabet of symbols the following: variables $x_1, x_2, \ldots$; the punctuation symbols left parenthesis, right parenthesis, and comma; the connectives "$\sim$" and "$\rightarrow$"; and the quantifier "$\forall$." Optionally, there may also be individual constants $a_1, a_2, \ldots$, predicate letters A_I^n, and function letters f_I^n.

(2) Expressions. A term in $\mathscr{L}$ is generated by one or more applications of two rules: (a) Each variable and each individual constant is a term.

(b) If f_l^n is a function letter in $\mathscr{L}$, and $t_1, \ldots, t_n$ are terms in $\mathscr{L}$, then $f_l^n(t_1, \ldots, t_n)$ is a term. Terms are to be interpreted as the objects that have properties – as the things regarding which assertions are made. If A_j^k is predicate letter in $\mathscr{L}$ and $t_1, \ldots, t_k$ are terms in $\mathscr{L}$, then $A_j^k(t_1, \ldots, t_k)$ is an atomic formula of $\mathscr{L}$. Atomic formulas are the simplest expressions in $\mathscr{L}$ to be interpreted as assertions. A well-formed formula (wff) is formed by one or more applications of two rules: (a) Every atomic formula of $\mathscr{L}$ is a wff of $\mathscr{L}$. (b) If $\mathscr{A}$ and $\mathscr{B}$ are wffs of $\mathscr{L}$, then $(\sim\mathscr{A})$, $(\mathscr{A} \to \mathscr{B})$, and $(\forall x_i)\mathscr{A}$ are also wffs, where x_i is any variable. In the wff $(\forall x_i)\ \mathscr{A}$, $\mathscr{A}$ is the scope of the quantifier. A variable x_i is bound if it occurs within the scope of a $(\forall x_i)$ in the wff or if it is the x_i in a $(\forall x_i)$; otherwise, the variable is free. A term t is free for x_i in $\mathscr{A}$ if x_i does not occur free in $\mathscr{A}$ within the scope of a $(\forall x_j)$, where x_j is any variable occurring in t.

(3) Axioms. For any wffs $\mathscr{A}$, $\mathscr{B}$, and $\mathscr{C}$, axioms are formed by the following axiom schemes:

Axiom Scheme 1. $(\mathscr{A} \to (\mathscr{B} \to \mathscr{A}))$

Axiom Scheme 2. $((\mathscr{A} \to (\mathscr{B} \to \mathscr{C})) \to ((\mathscr{A} \to \mathscr{B}) \to (\mathscr{A} \to \mathscr{C})))$

Axiom Scheme 3. $(((\sim\mathscr{A}) \to (\sim\mathscr{B})) \to (\mathscr{B} \to \mathscr{A}))$

Axiom Scheme 4. $((\forall x_i)\mathscr{A} \to \mathscr{A})$ if x_i does not occur free in $\mathscr{A}$

Axiom Scheme 5. $((\forall x_i)\ \mathscr{A}(x_i) \to \mathscr{A}(t))$ if $\mathscr{A}(x_i)$ is a wff of $\mathscr{L}$ and t is a term in $\mathscr{L}$ that is free for x_i in $\mathscr{A}(x_i)$

Axiom Scheme 6. $(\forall x_i)(\mathscr{A} \to \mathscr{B}) \to (\mathscr{A} \to (\forall x_i)\mathscr{B})$ if $\mathscr{A}$ contains no free occurrences of the variable x_i

(4) Rules. One rule, *modus ponens*, says from $\mathscr{A}$ and $(\mathscr{A} \to \mathscr{B})$, infer $\mathscr{B}$. A second rule, generalization, says from $\mathscr{A}$, infer $(\forall x_i)\ \mathscr{A}$, where $\mathscr{A}$ is any wff of $\mathscr{L}$ and x_i is any variable.

(5) Interpretation. The symbols of $\mathscr{L}$ are interpreted as predicate logic, as already intimated.

ARITHMETIC

In the contemporary vision of deductive systems, many branches of mathematics, such as arithmetic, are built on a foundation of predicate logic. To build a branch of mathematics on logic, two items must be added: an interpretation and some axioms. A formal language is abstract, and an interpretation attaches a particular meaning to some symbols of a formal language. Additional axioms are needed because often a mathematical statement is true (or false) because of the mathematical meanings of its terms, rather than merely the logical arrangement of its terms (Hamilton 1978:101). There are both logical truths and mathematical truths. Both require axioms.

An interpretation I of the formal language $\mathscr{L}$ is a non-empty domain D_I of objects over which the variables range, together with a collection of distinguished elements or individual constants $(\bar{a}_1, \bar{a}_2, \ldots)$, a collection of distinguished functions on the domain D_I $(\bar{f}_I^n, i > 0, n > 0)$, and finally a collection of distinguished relations on the domain D_I $(\bar{A}_I^n, i > 0, n > 0)$ (Hamilton 1978:57). For arithmetic, the following interpretation N is suitable (Hamilton 1978:57). Let $D_N = \{0, 1, 2, \ldots\}$, the set of natural numbers. The single distinguished element a_1 is interpreted as the number 0. The functions f_1^2 and f_2^2 are given the interpretations of addition and multiplication of natural numbers ("+" and "×"). And the successor function is the interpretation of f_1^1, where the successor of 0 is 1, the successor of 1 is 2, and so on. The successor of x, namely $x + 1$, will also be written here as x'. The relation "=" (equals) is the interpretation of the predicate letter A_1^2.

Because arithmetic is being built on predicate logic, which already has the 6 logical axioms given in the preceding section, the numbering for the additional required mathematical axioms continues here with 7. Three more axioms are needed to specify the mathematical concept of "equals." Though written here in a technical manner, following Hamilton (1978:102–103), the essential meaning of the following axioms is simply that any number equals itself and that any equals can be substituted for another equals.

Axiom Scheme 7. $A_1^2(x_1, x_1)$

Axiom Scheme 8. $A_1^2(t_k, u) \rightarrow A_1^2(f_i^n(t_1, \ldots, t_k, \ldots, t_n), f_i^n(t_1, \ldots, u, \ldots, t_n))$, where $t_1, \ldots, t_n, u$ are any terms, and f_i^n is any function letter of $\mathscr{L}$.

Axiom Scheme 9. $(A_1^2(t_k, u) \rightarrow (A_i^n(t_1, \ldots, t_k, \ldots, t_n) \rightarrow A_i^n(t_1, \ldots, u, \ldots, t_n)))$, where $t_1, \ldots, t_n, u$ are any terms, and A_i^n is any predicate symbol of $\mathscr{L}$.

The meaning of axiom 7 is that any number x_1 equals itself. To simplify notation, the relation "equals" will henceforth be written here using "=" rather than its more formal equivalent A_1^2. Finally, we also need seven more axiom schemes (Hamilton 1978:112–113). To make these arithmetic axioms somewhat more readable, liberty is being taken to substitute a set of symbols – "=," "0," "'"(prime), "+," and "×" – for their more formal equivalents (namely, $A_1^2, a_1, f_1^1, f_1^2,$ and f_2^2).

Axiom Scheme 10. $(\forall x_1) \sim (x_1' = 0)$

Axiom Scheme 11. $(\forall x_1)(\forall x_2)(x_1' = x_2' \rightarrow x_1 = x_2)$

Axiom Scheme 12. $(\forall x_1)(x_1 + 0 = x_1)$

Axiom Scheme 13.	$(\forall x_1)(\forall x_2)(x_1 + x_2' = (x_1 + x_2)')$
Axiom Scheme 14.	$(\forall x_1)(x_1 \times 0 = 0)$
Axiom Scheme 15.	$(\forall x_1)(\forall x_2)(x_1 \times x_2' = (x_1 \times x_2) + x_1)$
Axiom Scheme 16.	$\mathscr{A}(0) \to ((\forall x_1)(\mathscr{A}(x_1) \to \mathscr{A}(x_1')) \to (\forall x_1)$ $\mathscr{A}(x_1))$ for each wff $\mathscr{A}(x_1)$ in which x_1 occurs free.

For the most part, the meanings of these axioms should be fairly obvious. For example, axiom 12 says that any number plus 0 equals itself, and axiom 14 says that any number times 0 equals 0. The arithmetic resulting from these axioms is generally similar to Peano arithmetic (Hamilton 1978:113–116). Axiom 16 is essentially the principle of mathematical induction (Davenport 1992:14–16; Velleman 1994:245–283).

The severe simplicity of science's core logic has grand philosophical implications. It means that science's logic is relentlessly objective. Whether or not a given argument is valid or a given calculation is correct is determined by the rules of logic or mathematics in advance of any given application. For example, the premises "Magnesium or else iron is the enzyme's cofactor" and "Iron is not the enzyme's cofactor" point inexorably to one correct conclusion, for all times and places and persons. No special pleading or vested interests can change or thwart logic's inexorable verdict. Furthermore, the alternative situation would be unthinkable. Imagine a world in which each scientist has his or her own private version of logic and mathematics, so each person gets a different sum for $27 + 62$. In such a world, there would be no science! There would be no banks either! It is precisely because the rules of science's logic are fixed before the game starts that science is a meaningful, rational activity.

Lesser minds may miss the wonder, but Albert Einstein, shown in Figure 5.3, asked "How is it possible that mathematics, a product of human thought that is independent of experience, fits so excellently the objects of physical reality?" (Frank 1957:85). Likewise, he remarked that the most incomprehensible thing about the world is that it is comprehensible.

Potter (2000:17–18) expresses this wonder specifically as regards arithmetic, remarking that "it is not immediately clear why the properties of abstract objects [numbers] should be relevant to counting physical or mental ones.... One has only to reflect on it to realize that this link between experience, language, thought, and the world, which is at the very centre of what it is to be human, is truly remarkable.... The challenge of accounting for the applicability of arithmetic to the world evidently participates in this wider puzzle of explaining the link between experience, language, thought and the world."

Finally, it may be mentioned that by their choice of different axioms, every branch of deduction has nonstandard variants. The most familiar is

Figure 5.3. Albert Einstein wondered at the marvelous match between human thought and physical reality. (This photograph by Yousuf Karsh is reproduced with kind permission from Woodfin Camp and Associates.)

non-Euclidean geometry. But deductive logic also has variants, which is why the logic text by Cleave (1991:1) opens with the words "Logics, in the plural, are the subject of this book." Arithmetic provides a simple example. In standard arithmetic, $2 + 2 = 4$, but ring arithmetics give other answers (Niven, Zucherman, and Montgomery 1991:121–126). A ring arithmetic is based on a circular arrangement of its integers, such as an ordinary clock, with 1 to 12 written in a circle, so that at 3 hours after 11 o'clock the time is 2 o'clock, or $11 + 3 = 2$. In a smaller ring arithmetic with only the integers 0, 1, 2, and 3, the sum of interest becomes $2 + 2 = 0$. For the ring arithmetic with just 0, 1, and 2, the sum of interest becomes $2 + 2 = 1$. Other unusual arithmetics can offer yet other answers for the problem $2 + 2$. All that need be remarked here is that for counting physical objects and events, standard arithmetic is appropriate. Two apples plus two apples equals four apples.

COMMON FALLACIES

Ever since Aristotle's *Sophistical Refutations*, logicians have been providing helpful analyses and classifications of logical fallacies (Parry and Hacker 1991:409–480; Hansen and Pinto 1995:19–38). Likewise, science educators report that "all the standard logical fallacies, known since Aristotle's day, are routinely committed by science students" (Matthews 2000:331; also see Zeidler 1997).

Pirie (1985:vii–viii) introduces his book on logical fallacies by saying that "Fallacies are fun.... This book is intended as a practical guide for those who wish to perpetrate fallacies with mischief at heart and malice aforethought.... In the hands of the wrong person, this is more of a weapon than a book; and it was written with that wrong person in mind. It will teach such a person how to cheat, and even more than that.... Your... own dexterity with them [fallacies] will enable you to be... successful... as you set about the all-important task of getting your own way." That guise as a naughty sophist, combined with lavish use of cartoons, can make logic appealing even for those who would expect logic to be dull.

Three main factors explain the high frequency and inordinate appeal of bad logic. First, invalid arguments often closely resemble valid arguments in their logical forms, thereby taking on apparent but spurious legitimacy. Syllogisms are especially rich in this problem, because many of the 232 invalid forms resemble one of the 24 valid forms. Second, invalid arguments often feature slick rhetoric and appealing conclusions, which invite careless thought. Conversely, a perfectly valid argument may be widely disregarded because it reaches an unpopular conclusion. For better or for worse, persuasion has psychological as well as logical dimensions (Rips 1994). Third, logic can be subverted to the self-serving business of seeming to prove your own arguments while refuting your opponent's arguments, rather than the honest goal of assessing arguments regardless whose they are. Regrettably, this brief section can offer only a sketchy overview of merely seven kinds of invalid arguments.

(1) Invalid Variations on *modus ponens*. Consider the following four arguments. All four arguments have the same implication as the first premise, but they have different second premises and conclusions.

(A) If plants lack nitrogen, then they become yellowish.
     The plants are yellowish. So they lack nitrogen.
(B) If plants lack nitrogen, then they become yellowish.
     The plants do not lack nitrogen. So they do not become yellowish.
(C) If plants lack nitrogen, then they become yellowish.
     The plants lack nitrogen. So they become yellowish.
(D) If plants lack nitrogen, then they become yellowish.
     The plants are not yellowish. So they do not lack nitrogen.

Which of these arguments are valid, and which are invalid? Superficially, these arguments are similar enough to cause occasional confusion. Indeed, although some readers may find this exercise trivial, others may find that this little logic test requires rather careful thought.

In the implication $p \to q$, meaning "if p then q," the "if" part p is termed the antecedent, and the "then" part q is termed the consequent. Exactly four

arguments can be formed by making this implication the first premise and
adding a second premise that either affirms or denies either the antecedent
or the consequent (presuming that the right conclusion is asserted for valid
arguments or the most tempting conclusion is asserted for invalid arguments).
Of these four arguments, two are valid, and two are invalid. These arguments
are listed below in the same order as before, appending to each the verdict
on its validity.

(A) Affirming the consequent. $p \rightarrow q; q; \therefore p.$ Invalid
(B) Denying the antecedent. $p \rightarrow q; \sim p; \therefore \sim q.$ Invalid
(C) Affirming the antecedent. $p \rightarrow q; p; \therefore q.$ *modus ponens*
(D) Denying the consequent. $p \rightarrow q; \sim q \therefore \sim p.$ *modus tollens*

To understand and remember which of these arguments are valid, it may
be helpful to recall that the implication $p \rightarrow q$ is logically equivalent to the
disjunction $\sim p \vee q$. Given this disjunction's truth, it is rather obvious that if
p is true, then the first part of this disjunction $\sim p$ is false, so it must be the
second part q that is true; and likewise, if $\sim q$, then $\sim p$ follows. But asserting
$\sim p$ says nothing about q (for $\sim p \vee q$ is then true regardless whether q is true
or false), and asserting q says nothing about p.

This difference between valid and invalid arguments can also be explained
by truth tables. Tables 5.3 and 5.4 illustrate a valid argument, affirming the
antecedent (*modus ponens*), and an invalid argument, affirming the consequent. Asterisks draw attention to lines for which all premises are true.
Note that for the valid argument, when all premises are true, the conclusion
must be true. By contrast, for the invalid argument, when all premises are
true, the conclusion can be either true or false. The truth of its premises entails the truth of its conclusion for a valid argument, but not for an invalid
argument.

Table 5.3. Truth-table analysis for a valid
argument: affirming the antecedent

Assignments		Premises		Conclusion
A	B	A	$A \rightarrow B$	B
T	T	T	T	T *
T	F	T	F	F
F	T	F	T	T
F	F	F	T	F

Note: An asterisk indicates the line for which all of
the premises are true.

Table 5.4. Truth-table analysis for an invalid
argument: affirming the consequent

Assignments		Premises		Conclusion
A	*B*	*B*	*A* $\rightarrow$ *B*	*A*
T	T	T	T	T*
T	F	F	F	T
F	T	T	T	F*
F	F	F	T	F

Note: Asterisks indicate the lines for which all of
the premises are true.

    Clearly understand that the corrective for the foregoing invalid arguments is *not* just to declare the opposite conclusion. For example, changing the first of the foregoing arguments to the opposite conclusion ($\therefore \sim p$) also results in an invalid argument. The proper corrective is to declare no conclusion at all: $p \rightarrow q$; q; $\therefore$ [nothing]. In a given actual instance of this first argument, the proffered conclusion that the plants lack nitrogen may be true; but it still is not true that this argument provides legitimate support for that conclusion. Perhaps some other argument can support a solid conclusion. But in any case, whether its conclusion is actually true or false, an invalid argument offers no basis for knowing which conclusion obtains.

    When embedded in longer arguments, these fallacies can become less obvious. For example, the fallacy of affirming the consequent can be hidden in argument forms such as "If *p* then *q*, and if *p* then *r*; either *q* or *r*; therefore *p*." Likewise, the fallacy of denying the antecedent resides in the following: "If *p* then *q*, and if *r* then *q*; either not *p* or not *r*; therefore not *q*."

    Upon rejecting an invalid argument, however, certain overreactions must be avoided. The triumph of overthrowing a bad argument can entice one to think the argument's conclusion is false. But the formula that "Invalid argument *A* fails to support conclusion *X*, so *X* must be false," is wholly irrational. It is always possible to construct a bad argument that supposedly can support any proposition whatsoever, whether it be true or false. So the existence of some bad argument that does or does not conclude *X* is no evidence either for or against *X*. Likewise, equally unfair is the formula "Bad argument *A* fails to support *X*, so without even weighing them, it must be that arguments *B* and *C* also fail to support *X*." The real issue is whether some good argument concludes *X* (or else $\sim X$). An invalid argument counts neither for nor against its supposed conclusion. Again, the proper reaction to an invalid argument is to conclude nothing.

    While knowingly taking a little logic test, one's guard is up, helping to prevent blunders. But in hasty, informal discourse, blunders often pass undetected. For example, imagine encountering the following little example from Cederblom and Paulsen (1986:108) when your guard is down: "If the President does a poor job, the economy gets worse. The economy has gotten worse. So the President has done a poor job." Especially if you already believe this argument's conclusion, this logic may seem fine. But this argument affirms the consequent, which is invalid logic.

Bad logic has bad consequences. It prompts us to offer illegitimate conclusions lacking valid support, or to miss legitimate conclusions actually supported by the data. For example, a farmer making the first fallacy listed earlier, affirming the consequent, could be prompted to apply nitrogen fertilizer when nitrogen is not the actual problem, which both wastes money and diverts attention away from finding and fixing the real problem, thereby leaving the crop yellowish and unproductive. On the other hand, a farmer failing to grasp the import of the last valid argument listed earlier, denying the consequent or *modus tollens*, could miss the logical conclusion that the crop does not need nitrogen and thereby could be tempted to apply nitrogen fertilizer needlessly, which would reduce the profit margin. Bad logic causes unrealistic beliefs, needless ignorance, and ineffective actions. Bad logic means less food, worse medicines, costlier cars, and smaller profits. But some study of common fallacies can help one to reject bad logic.

(2) Fallacies of Composition and Division. The fallacy of composition inappropriately applies a property of individuals or constituent parts to a class or organized whole. Examples are "The players are excellent; the team is several players; therefore the team is excellent" and "Sodium and chlorine are poisonous; table salt is sodium chloride; therefore table salt is poisonous."

Conversely, the fallacy of division inappropriately applies a property of a class or organized whole to individuals or constituent parts. Parry and Hacker (1991:429) give an example that, despite its true conclusion, obviously has invalid logic: Cows are numerous; white cats are not numerous; therefore white cats are not cows. In the first premise, "cows" is used collectively, but in the conclusion it is used divisively (individually). Consider a more complex example: Dogs are common; albino spaniels are dogs; therefore albino spaniels are common (Parry and Hacker 1991:428). Here, "albino spaniels" is used divisively in the second premise but collectively in the conclusion, and "dogs" is used collectively in the first premise but divisively in the second premise.

For the sake of brevity, these examples were chosen to make these fallacies transparent. But other examples could be devised that would be more subtle and alluring.

(3) Fallacies of Ambiguity. Many logical blunders arise from unclear wordings, inconsistent meanings of different occurrences of the same term, and ambiguous logical relationships. The fallacy of equivocation trades in ambiguous words or phrases interpreted differently in its premises and conclusion: A plane is a carpenter's tool; a Boeing 747 is a plane; so a Boeing 747 is a carpenter's tool (Davis 1986:58). The fallacy of amphiboly involves a confusing statement with multiple meanings due to faulty grammar or punctuation: "On campus last night, Professor Merkin gave a factual report on students' sexual activity in the library; therefore we may assume that not all student activity in the library is intellectual in nature" (Parry and Hacker 1991:427).

(4) False Dilemmas. A false dilemma mentions fewer alternatives than actually exist, often only two (Hurley 1994:157–158). In the false dilemma "*A* or else *B*; not *A*; therefore *B*," the logical form is valid, but the first premise "*A* or else *B*" is false because, in a particular case, there exist additional alternatives, such as "*C*," "*D*," "*A* and *B*," or "*C* or *B*." For example, "Either apply nitrogen fertilizer or get yellowish plants" is a false dilemma for many reasons, including the possibilities that a particular soil already has adequate nitrogen without adding fertilizer, or that a virus causes yellowish plants despite adequate fertilizer.

A dishonest variant on the false dilemma is the straw-man argument (Cederblom and Paulsen 1986:104–105). The logical form is "*A* or else *B*; not *A*; therefore *B*," where *A* represents an opponent's position and *B* our own position. The premise "not *A*" is supported, however, by attacking the opponent's weakest evidence or a simplistic, straw-man representation of the opponent's reasons. If we are honestly trying to test our own position *B*, then the opponent's position *A* must be represented by its strongest case. Only if *A* fails at its best is the premise "not *A*" established with honesty and strength. Indeed, when you see persons attack a straw man, you do well to suspect that they feared to tackle the real man.

A subtle variant on the false dilemma is the argument from ignorance (*argumentum ad ignorantiam*) (Parry and Hacker 1991:477–479). This fallacy attempts to drive opponents to accept my argument unless they can find a better argument to the contrary. For example, an environmentalist might say "We cannot prove that this pesticide is safe, so we must assume that it is dangerous and outlaw its use." There may or may not be some other good arguments against this pesticide's use, but an argument from ignorance is not a good reason. Similarly, scientists sometimes confuse failure to reject a hypothesis with proof of a hypothesis. The implicit dilemma in an appeal to ignorance is "Give me a better argument or else accept my argument," but the unmentioned third option is to admit current inability to construct a better argument while still either rejecting the offered argument or suspending judgment.

On balance, however, ignorance can provide a legitimate basis for rational actions. For example, precisely because I am ignorant about whether or not

I will be in a car accident tomorrow, it is rational for me to buy car insurance today. Known possibilities as well as known certainties can motivate actions.

Of course, the opposite problem of false dilemmas, namely, believing that we have more choices at our disposal than reality actually offers, is equally common if not even more common. Some dilemmas are false, but others are real! Discerning which dilemmas are false and which are real is vital for logical, realistic thinking.

(5) Circular Reasoning. A circular argument, also called "begging the question," assumes what it intends to prove, amounting to arguing "p; $\therefore$ p." Of course, presented that obviously, anyone would reject such stupid reasoning. But by lengthening the argument before closing the circle, and by saying the same thing in different words, circular reasoning can look plausible. So "p; $\therefore$ p" will not convince, but "$q \vee r$; s; $t \rightarrow (s \vee p)$; p; $\sim r \vee s$; $u \equiv r$; $\therefore$ p" might. Likewise, "Whatever is less dense than water will float, because whatever is less dense than water will float" sounds stupid, but "Whatever is less dense than water will float, because such objects won't sink in water" might pass (Cederblom and Paulsen 1986:109). Anyone lacking imagination regarding how to dress up circular reasoning in 101 ways may sit at Aristotle's feet (Parry and Hacker 1991:443–457). Circular reasoning ranks among the most common logical blunders. As always, logical blunders are most enticing to listeners who already like and believe the conclusion.

(6) Genetic Fallacies. A genetic fallacy, also called an *ad hominem* argument (meaning literally "to the man"), attacks one's opponent rather than debating the issue (Hurley 1994:120–123). For example, a scientist might seek to discredit a research report by saying that the researcher is at a small college, rather than refuting the report's data or logic. An *ad hominem* attack is irrational because an argument's merit depends on its content, not on who says it. The *ad hominem* fallacy appeals to prejudices rather than reasons. The intent is to discredit an argument or conclusion by discrediting its proponent.

Imagine that we are third parties to a scientific dispute between Jones and Smith. Jones stoops to an attack saying that Smith's grandmother smokes cigars. How should we react to this? Well, if this is the worst that Jones can find against Smith, we may interpret this irrelevant attack as presumptive evidence that Smith's scientific theory has something going for it! Smith's theory may well merit an honest hearing and careful evaluation. In such a case, it may be unwise to rely on Jones's account of Smith's theory, but rather Smith's own report should be studied.

On balance, however, some situations justify inspecting a scientist's training and motivations, so it could be wrong to forego such an inspection automatically on the grounds that it would constitute an *ad hominem* attack. Particularly when a scientist's conclusion is uncertain or false, personal factors can be relevant. For example, an explanation for someone believing that "Winds cause tides" cannot possibly involve ordinary and objective scientific facts about the earth and oceans, but rather must involve some defect in this person's education or reasoning that instilled that false belief. Also, in cases where definitive scientific conclusions are not yet possible, such as some court case regarding whether or not an employment situation caused a medical condition or an insecticide poses unacceptable environmental risks, it is rational to assess scientists' expertise. The motivation should be to clarify the relative weight to place on various and possibly conflicting testimonies, not to discredit perfectly honest and competent testimony in order to avoid the truth.

As a historical note, the original and current meanings of "*ad hominem*" are completely different. Apart from this one paragraph, herein this term will always be used in its ordinary, current meaning of attacking a person instead of his or her argument. An argument is always directed to some intended audience and is applied to some particular domain or context of relevant instances. An argument's audience and domain are termed its metabasis (Parry and Hacker 1991:451–454). An argument binding on one audience may be pointless to another, and a principle proved in the domain of one science, like physics, may be quite suspect in another science, like sociology. The fallacy of illicit metabasis is to claim that a good argument for a given audience and domain also applies, without further justification, to a different or wider audience and domain. An argument can be directed to different audiences, including oneself (*ad seipsum*), persons specifically involved in a dialogue (*ad hominem*), persons with whom no meaningful dialogue is possible (*ad populum*), everyone (*ad omnes*), or the matter at hand rather than any particular persons (*ad rem*). The original meaning of an *ad hominem* argument was to press a man with the consequences or implications of his own principles or concessions, most often by pressing a contradiction between his current view and what appear to be his more fundamental or permanent principles. In this sense, an *ad hominem* argument is perfectly rational, not at all fallacious. Its only limitation is that an *ad hominem* argument addresses a smaller audience than a stronger *ad omnes* argument. For example, wanting to prove something to a Marxist, one may deduce a thesis from the Marxist's own principles, but the resulting *ad hominem* argument may have a limited metabasis and may not be a proof at all to others who are not Marxists. "Before an argument is a proof to any audience, that audience must accept both the truth of that argument's premises and its validity" (Parry and Hacker 1991:453). Rightly understood, this principle emphasizes that part of the responsibility for rational dialogue falls to the audience. A speaker's argument may have true premises and valid logic, and yet some audience may be unresponsive because their data are lacking and their logic is weak. Scientists have responsibilities when they listen, as well as when they speak. Incidentally, the ancient term corresponding to the modern sense of *ad hominem* was *adversus personam*, meaning a personal attack on the opponent. In review, an *argumentum ad hominem* or "argument to the man" originally meant an argument to the person with whom I disagree about a particular point and also agree about some deeper fundamentals that may provide a basis for persuasion and resolution, but currently it means a personal attack on an opponent rather than a rational rebuttal of his or her argument.

(7) Failure of Will. Most fallacious reasoning in science results from accidental blunders. Unexamined presuppositions, bad data, and invalid logic lead to wrong conclusions, despite the best of intentions. To be honest, however, it must be admitted that scientists are only human, so occasionally errors result not from failure of competence but from failure of will. Ordinarily, scientists' goal is to find the truth about physical reality. But sometimes reason is usurped by desire, so the goal becomes not to embrace reality, but to evade reality. Logic is enlisted in this dirty, insincere business of rationalizing.

Errors caused by lack of will to find the truth tend to be subtle. Often the action resides in what is not said, rather than what is said. When a wildly unrealistic verdict is greatly desired from science, science's stubborn hold on truth is often too strong to allow a downright falsehood, so instead the more subtle lie emerges that science cannot reach a solid verdict either way.

For example, I recall reading an article about smoking. The author was a scientist employed by the tobacco industry. The article said that scientists used to be dogmatic about their findings, but now enlightening developments in the modern philosophy of science teach us that science does not deliver absolute truth, but rather science pursues tentative, imperfect, shifting paradigms that must always be open to future revision in light of further progress. The obvious implication was that science had no real evidence of health hazards from tobacco, so smokers should feel free to continue their habit and governments should leave the tobacco industry alone. What I found most disturbing about that article was its love for, and comfort in, a weak version of science guaranteed in advance not to disturb anyone by delivering truth.

Failure of will is more reprehensible and less remediable than failure of competence. Persons who do not want truth should not complain when they cannot find it, just as those who shut their eyes should not complain when they cannot see. Bad faith subverts reason. To cherish ignorance rather than knowledge is to break faith with the scientific attitude. It is insincere. Failure of will is a moral failure, rather than an intellectual failure. That is all that need be said here.

Lastly, what is the worst fallacy? The worst fallacy of all is energetic pursuit to root out others' fallacies while ignoring or even cultivating one's own! The first and foremost purpose for training in fallacies is to hunt and kill one's own fallacies. And the first and foremost reward is a greater liberty in dealing with reality.

MATERIAL LOGIC

The logic explored thus far in this chapter has been formal logic, whose referents are abstract concepts, such as logical variables and operators. But manifestly, the logic needed for the natural sciences is material logic, whose referents are physical entities, such as cats and dogs. Accordingly, this section explains the justification of knowledge claims about physical reality, be they either certain or probable.

At the outset, two ideas must be recalled from the preceding chapter. The PEL model says that presuppositions, evidence, and logic combine to support scientific conclusions (and irrelevant knowledge is relegated to a dismissed archive). And the reality check says that the common-sense belief is certain that "Moving cars are hazardous to pedestrians." In Chapter 4, the purpose of the PEL model was to present scientific arguments with full disclosure, and the purposes of the reality check were to install science's presuppositions and to dismiss radically skeptical worldviews from science's worldview forum. In this chapter, these two resources plus some logic are combined for the purpose of developing a material logic.

Besides the reality check itself, additional beliefs have already been deemed equally certain here, including that "Table salt is composed of sodium and chlorine" and the experimental result that "There is a coin in the cup." Likewise, Wittgenstein judged "the existence of the apparatus before my eyes" to have the same certainty as "my never having been on the moon" (Ludwig Wittgenstein, in Anscombe and von Wright 1969:43e; Newell 1986: 68–69). Such thinking – that various beliefs are equally certain – is intuitively appealing and undoubtedly right. This section will formalize that intuition.

Precisely what philosophical reasons can be given for deeming various beliefs to be certain? How does the reality check's assumed certainty extend to other beliefs' demonstrated certainty? To make these questions more concrete, consider the belief that "There are elephants in Africa." What formal, philosophical reason could be given for judging this belief to be certain? Prior to reading the rest of this section, the reader might like to try his or her own hand at constructing a philosophically rigorous proof that "There are elephants in Africa" is true and certain. The method of justification offered here has three steps.

(1) Reality check. By faith and without possibility of proof, adopt as true and certain the reality check's belief, denoted by A, that "Moving cars are hazardous to pedestrians." A is certain.

(2) Certainty equivalence. Demonstrate that the belief, denoted by B, has certainty equivalence with the beliefs in the reality check, denoted by A. According to the PEL model, this demonstration requires (a) the same presuppositions, (b) equally admissible, relevant, and weighty evidence, (c) equally valid or correct logic, and (d) equally inert and dismissible archives. Then A and B have certainty equivalence.

(3) Rule for justification. From the preceding two beliefs, that "A is certain" and "A has certainty equivalence with B," infer that "B is certain."

The first step merely reiterates the reality check's claim of certainty. Although adopted ultimately by faith, the evidence by which (non-skeptical) adults have learned and understood the reality check is the familiar story of safety or disaster resulting from obeying or disobeying the reality check. The second step is based on the PEL model. The quality of an argument's output or conclusion is determined by the quality of its three active inputs and its inert archive. Equal inputs give an equally certain output. The third step is essentially an application of *modus ponens*, which is the argument form "A; A implies B; therefore B." The premise "A equals B" is a subcase of the premise "A implies B," so the rule *modus ponens* includes as a subcase the rule "A; A equals B; therefore B." The premise "A is certain" just means that "A is true and known to be true," or, more briefly, "A." And likewise, "A has certainty equivalence with B" has the logical content "A equals B." This three-step model of justification is summarized in Figure 5.4.

For example, this model can be applied to Africa's elephants as follows. Denote "Moving cars are hazardous to pedestrians" by A, and "There are elephants in Africa" by B. First, the reality check is asserted to be certain. A is certain. Second, the same presuppositions are necessary and sufficient to believe A or B; recent sightings or photographs of elephants in Africa are as admissible, relevant, and weighty as any evidence that could be adduced for the reality check; equally valid or correct logic works in both cases; and both cases generate equally inert and dismissible archives. Thus, A has certainty equivalence with B. Third and finally, applying the rule for justification to these two premises produces the conclusion "That 'There are elephants in Africa' is certain."

Naturally, ordinary common-sense reasoning reaches this conclusion in a quick, intuitive manner. For philosophical exactness and insight, however, this model of justification provides an explicit account of how beliefs other than the reality check can be deemed equally certain. The essential faith in the first step and the simple reasoning in the second and third steps combine to make rational the assertion that additional beliefs are also certain. Thereby, this model extends the reality check's certainty to other beliefs. Or, to say the same thing in negative terms, the only way to unsettle the conviction

---

A Model of Justification

Reality Check. Adopt as realistic, true, and certain the reality check's belief, denoted by A, that "Moving cars are hazardous to pedestrians."

Certainty Equivalence. Demonstrate that the belief, denoted by B, has certainty equivalence with the beliefs in the reality check, denoted by A. This demonstration requires:

    (1) the same presuppositions,

    (2) equally admissible, relevant, and weighty evidence,

    (3) equally valid or correct logic, and

    (4) an equally inert and dismissible archive.

Rule for Justification. From the above beliefs, that A is certain and that A has certainty equivalence with B, infer that B is certain.

Figure 5.4. A model for justifying scientific beliefs based on the reality check, certainty equivalence, and rule for justification.

that "There are elephants in Africa" would be to embrace such profound skepticism as would also unsettle the reality check that "Moving cars are hazardous to pedestrians."

This model of justification explains the simple case of deeming other beliefs to be certain. Much knowledge, however, is probabilistic, such as a forecast that rain is likely tomorrow. This model is easily extended to justify probabilistic as well as certain conclusions, but probabilistic reasoning is better left to Chapters 6 and 7 on probability and statistics. The crucial move for justifying a probabilistic conclusion is to accept common-sense presuppositions so that the conclusion faces only the ordinary challenge of imperfect evidence, but not the radical challenge of skeptical presuppositions.

SUMMARY

Logic distinguishes between good and bad reasoning. It concerns the relationships between evidence and hypotheses, or between premises and conclusions. A deductive argument is valid if the truth of its premises entails the truth of its conclusions, and is invalid otherwise. Formal deductive logic begins with a language, axioms, and rules, and then derives theorems. As applied in science, deductive logic argues with certainty from an assumed model to particular expected data. By contrast, inductive logic argues with probability from particular actual data to an inferred general model. Scientific thinking alternates deduction, reasoning from mind to world, and induction, reasoning from world to mind.

Propositional logic is presented here with a mere three axiom schemes (or just one rather long axiom scheme) and one rule, and predicate logic with six axiom schemes and two rules. Additional deductive systems can be built on this base of predicate logic. For example, a suitable interpretation and ten more axioms suffice to add arithmetic.

Common invalid arguments involve incorrect variations on *modus ponens*, fallacies of composition and division, fallacies of ambiguity, false dilemmas, circular reasoning, genetic fallacies, and failure of will. Fortunately, most blunders in scientific reasoning involve a fairly small number of common logical fallacies, so even a little study of logic can improve reasoning skills markedly.

The logic needed for the natural sciences is material logic, whose referents are physical entities. A model of justification for truth claims about material reality has three components. First, the reality check is asserted as being true and certain. Second, other beliefs have certainty equivalence with the reality check if they have the same presuppositions, equally weighty evidence, and equally correct logic (and an equally dismissible archive). Third and finally, the rule for justification is that from the reality check's certainty and the other belief's certainty equivalence, we infer that this other belief is also certain. This model is easily extended to justify probabilistic as well as certain conclusions, using concepts developed in the following two chapters on probability and statistics.

CHAPTER SIX

PROBABILITY

Probability and statistics are deductive and inductive tools, respectively, that are used to deal with uncertainty. Probability concepts and values are used in both deductive and inductive reasoning, and inductive problems often contain deductive subproblems, so probability theory is needed to handle diverse contexts. Probability is important because in many cases the best answers that science can deliver are more or less probable conclusions, rather than absolute certainties. Daily life and scientific research are full of unavoidable practical decisions that must be made on the basis of the best available information.

Correct probability reasoning is not easy. Indeed, probability blunders are among the most common kinds of blunders, even in professional, refereed scientific journals. Regulatory agencies and law courts can perpetuate such blunders, as they often rely on scientific findings, thereby inflicting wrongful losses and injustices. The challenge in probability reasoning is that it requires some precise distinctions that are critical for getting the correct results, but those distinctions are not intuitively obvious, and few scientists learn or teach them. Also, there are conflicting paradigms that can seriously affect research efficiency and technological progress, but precious few scientists understand the issues. So there is much to be gained from a proper understanding of probability and statistics, as pursued in this and the next chapters.

This chapter cannot possibly do what an entire book on probability would do: present a comprehensive treatment. But it can do what most such books fail to do: present the pivotal elements with philosophical clarity and scientific relevance. So this chapter can be regarded as a prolegomenon to probability. This prolegomenon is important, however, because in learning probability theory, the most frequent and disastrous mistakes are made right at the outset.

This chapter will give common-sense definitions and then precising definitions of probability. Two fundamental requirements will be stated prior to selecting probability axioms suitable for doing business with physical reality.

Then some probability theorems will be derived, most notably Bayes's theorem, which is the foundation for one of the major paradigms in statistics. Finally, two common blunders will be discussed.

PROBABILITY CONCEPTS

Probabilists and philosophers have written many thousands of pages about the nature and meaning of probability, and doubtless much still remains to be discovered or clarified. This section defines probability in two stages: first as a simple common-sense notion, and then as a precise scientific notion.

There are two primary concepts of probability, one pertaining to events, and one to beliefs. An *objective* or *physical* probability is the chance of an event occurring. For example, upon flipping a fair coin, the probability of heads is 0.5 (because there are two possible events or outcomes, namely, heads or tails, and they are equally likely). A *subjective* or *personal* or *epistemic* probability is the degree of belief in a proposition warranted by the evidence. For example, given today's weather forecast, one may believe that "It will rain today" has a 90% probability of being true. Of course, personal and physical probabilities are often interrelated, particularly because personal beliefs are often about physical events.

Concepts of probability are as old as our species, for humans have always had to deal with uncertainties and probabilities: "Aristotle classified events into three types: (1) certain events that happen necessarily; (2) probable events that happen in most cases; and (3) unpredictable or unknowable events that happen by pure chance," and similarly, "In discussing probability, Thomas Aquinas distinguishes among science or certain knowledge, opinion or probable knowledge, and the accidental or chance" (Hald 1990:30).

The concepts of probability are philosophically rich and scientifically complex. Accordingly, we do well to begin our exploration of probability simply, because "the ordinary common-sense notion of probability is capable of precise and consistent treatment when once an adequate language is provided for it" (Jeffreys 1983:ix). Similarly, Kyburg (1970:3–4) began with informal interpretations of probability, saying that " 'Probability' has many partial synonyms in ordinary usage: the probable is that which we distinguish from the certain; probability is likelihood, degree of confirmation, degree of factual support; the probable is the credible, that which is supported by the evidence, that which the chances favor. . . . 'Probability' is a word that finds many uses in the sciences and in everyday life as well as in philosophy. . . . It would, of course, be too much to expect that in all of these occurrences the word 'probability' had exactly the same meaning."

This section will first give an extensional definition of probability, and then an intensional definition. (For those not familiar with this distinction, an extensional definition gives a list, partial or complete, of instances of the term to be defined, whereas an intensional definition describes salient, diagnostic properties sufficient to recognize appropriate instances of the term's use.) Consider the following eight common-sense usages of the concept of probability.

(1) A fair coin has a probability of 0.5 of heads, and likewise 0.5 of tails; so the probability of tossing two heads in a row is 0.25.

(2) There is a 10% probability of rain tomorrow.

(3) There is a 10% probability of rain tomorrow according to the weather forecast.

(4) Fortunately there is only a 5% probability that her tumor is malignant, but this will not be known for certain until the surgery is done next week.

(5) Smith has a greater probability of winning the election than does Jones.

(6) I believe that there is a 75% probability that she will want to go out for dinner tonight.

(7) I left my umbrella at home today because the forecast called for only a 1% probability of rain.

(8) Among 100 patients in a clinical trial given drug A, 83 recovered, whereas among 100 other patients given drug B, only 11 recovered; so new patients will have a higher probability of recovery if treated with drug A.

All eight examples use the same word, "probability." To a first approximation, the meaning of "probability" is the same in all of these examples. On the level of common sense, other words could be used with essentially the same meaning, such as "chance" or "likelihood." The common-sense meanings of these examples should be entirely clear to everyone. The following informal intensional definition captures many of the general features present in the eight examples.

Informal Intensional Definition of Probability

"Probability" is the propensity for an event to occur or be actualized or for a proposition to be true. Most events and propositions with low probabilities are not or will not be actualized or true, whereas most with high probabilities are or will be actualized or true.

That a common-sense concept of probability exists and that it is roughly as stated here can be proved by the simple observation that all dictionaries provide definitions of "probability" along these lines. Furthermore, even a young child easily comprehends simple probability statements, such as "We probably will visit grandmother tomorrow."

On balance, however, it must be admitted that even on the level of a common-sense discussion, our eight examples express a variety of distinguishable concepts. Likewise, this brief, intensional definition does not capture all of the subtleties and variations contained in those examples.

Example 1 is essentially a definition or theoretical description of what is meant by a "fair" coin, that it has equal chances of landing heads or tails. It then states the deductive implication regarding tossing two heads in succession. This example may appear to be talking about actual physical coins, but it does not necessitate or imply that any actual coin has this property of fairness, at least not exactly. Indeed, the simple fact that heads and tails have different depictions (shapes) suggests that actual coins are likely to be unfair to at least some minute degree (as is often confirmed by tedious experiments with enormous numbers of tosses). So this first example is of an essentially theoretical character, in contrast to being basically empirical.

Example 2 is solidly empirical, obviously purporting to convey information about the physical world, namely, about the probability of rain. Example 2 differs from Example 1 in that the first example's event (a coin toss) is a repeatable event, both in theory and in practice, whereas the second example's event (rain in a particular place and day) is a singular, nonrepeatable event. Hence, concepts of probability based on the long-run frequencies of events apply naturally to Example 1, but not to Example 2. Incidentally, Example 1 expresses its probabilities as numbers within the range of 0 to 1, whereas Example 2 multiplies such values by 100 to yield a percentage, but this cosmetic difference is not particularly significant.

Example 3 is based on Example 2, but adds an explicit statement about the evidence in support of its assertion. Hence, the probability in Example 2 appears to be a function of only one thing or argument, the event of rain, whereas the probability in Example 3 appears to be a function of two arguments, the event of rain and the evidence of the weather forecast. Clarification will be needed regarding whether the concept of probability can take various numbers of arguments or must always take the same number (although informal statements may sometimes fail to express all of the arguments explicitly). More generally, clarification is needed regarding the relationship, or lack thereof, between the concept of probability and the concept of evidence.

Example 4 expresses a 5% probability of malignancy. In fact, however, the tumor is either benign or malignant – it is *not* 5% malignant and 95% benign. It is one or the other. Hence, this probability value of 5% must refer to the present state of knowledge, as contrasted with the actual status of the tumor. This interpretation is reinforced by the suggestion that further knowledge will modify this probability, hopefully to a zero probability of malignancy, rather than to the dreaded 100%. Hence, the concept of probability must be capable of handling not only the presence of evidence (as in Example 3) but also the addition of new evidence to an existing body of old evidence (as in Example 4). This example also hints at some relationship between the concepts of probability and certainty, which will require further clarification.

Example 5 informs us that Smith has a better chance of winning the election than does Jones, but it does not tell us whether Smith's chance is great (say, over 0.5) or small. The problem is that no information is offered about the presence or absence of other candidates. If Smith and Jones are the only candidates, then Smith does have a good chance of winning. For all we know, however, both Smith and Jones may be obscure candidates with virtually no chance of defeating the overwhelming favorite, Adams. Hence, the probability in this example is a somewhat less informative quantity than are the probabilities in the preceding examples. Later this difference will be expressed by distinguishing the technical meanings of "probability" and "likelihood," wherein probability considers all possible outcomes, whereas likelihood considers a possibly incomplete roster of outcomes (as in Example 5). Also note that this example uses no number to express its probability (or likelihood), but rather merely expresses a relationship of one probability being "greater" than another. This raises a question regarding whether the concept of probability is inherently or necessarily expressed numerically or whether a more general sense of probability also exists, with relations like "equal," "greater," and "less."

Example 6 is the first that explicitly recognizes the existence and role of the person expressing a probability judgment, "*I* believe that . . . " Furthermore, this example is deliberately chosen to be highly subjective – evidently other persons might come up with a figure considerably different from 75%, particularly if they happen to have some special knowledge about this woman's plans or commitments for that evening. On the other hand, someone else with no knowledge whatsoever of this woman's plans might pick a probability of 50% to represent maximal uncertainty or ignorance. This suggests that the concept of probability needs to be able to model ignorance as well as knowledge. More generally, this example raises the issue whether or not the concept of probability should have or does have a subjective aspect relating to personal beliefs, as contrasted with an objective aspect relating to physical events.

Example 7 shows probability taking a role not only in personal inferences and beliefs but also in personal decisions and actions. It weighs the personal costs or bother of carrying an umbrella against the potential benefit of not getting soaked. This example's concern with decisions finds probability in a more complex context than a simpler concern with just inference, as in Example 3. This complex setting of human decisions and actions is tremendously important, however, because most of our thinking about probabilities occurs in exactly this setting. We rarely spend time assessing or manipulating probabilities unless motivated to do so by the need to make practical decisions.

Example 8 uses data on two recovery rates to derive a conclusion about a probability judgment. Its logical progression is therefore the reverse of that in the first example. Example 1 is representative of deductive thinking that begins with a model or theory (about a fair coin) and then derives conclusions regarding expected observations (of two heads in succession). Such thinking could be elaborated, for example, to calculate the probability of throwing three heads and five tails in a set of eight tosses. By contrast, Example 8 is representative of inductive thinking that progresses in the reverse direction. It begins with specific actual data (regarding 83 and 11 recoveries) and then derives general conclusions (about two drugs' relative merits). Hence, the concept of probability is used in both deductive and inductive settings.

Clearly, an informal definition of probability leaves much unclear that a precise, scientific definition must specify. Does the concept of probability concern events or beliefs or both? Does it pertain just to repeatable events, or also to singular events? Is probability to be expressed only by exact numbers, or also by inexact comparative relations? Is probability a function of one argument (such as an event) or of two arguments (such as an event and some evidence), or either? How can old and new evidence be combined in reaching probability judgments? Can ignorance as well as knowledge affect probability judgments? How do theoretical and empirical aspects of probability combine, and likewise how do deductive and inductive applications interconnect? Can a single, unified probability theory cover all common-sense and scientific applications?

But equally clearly, it must also be emphasized that any useful scientific definition of probability must respect and build on the common-sense notion of probability. The intent of a precising definition is to do better than common sense by adding substantial science, not to do worse than common sense by adding weird philosophy.

A philosopher, Hacking (1965:227), has concluded that the informal common-sense concept of probability is useful and well defined: "[Nearly] every adult speaker of English can use the word 'probable' correctly. No account of probability could be sound if it did not examine what the English word 'probable' means. Some say the word and its cognates are entirely vague, but our ability to use them regularly suggests they must be governed by pretty stern regularities, which mark out the concept or concepts of probability." A statistician, Barnett (1982:12), expressed similar sentiments.

Jeffreys (1983:15) cited the following simple case as exemplifying in principle most of the subject of probability reasoning: "Suppose that I know that

Smith is an Englishman, but otherwise know nothing particular about him. He is very likely, on that evidence, to have a blue right eye. But suppose that I am informed that his left eye is brown – the probability is changed completely. This is a trivial case, but the principle in it constitutes most of our subject-matter [of probability]. It is a fact that our degrees of confidence in a proposition habitually change when we make new observations or new evidence is communicated to us by somebody else, and this change constitutes the essential feature of all learning from experience."

TWO FUNDAMENTAL REQUIREMENTS

So many superb books on probability theory have been written that one faces an embarrassment of riches. For the purpose of providing a general audience of scientists with an elementary account of probability, however, an informative summary of one good book seems better than a superficial treatment and comparison of many books. Hence, the following formal account of probability largely follows Jeffreys (1983), whose book is remarkably wise, clear, and practical. It progresses from two fundamental requirements to eight general rules to several axioms and finally to various theorems.

The motivation for constructing a careful theory of probability arises mainly from the centrality of probability within the important larger subjects of inductive logic and learning. Consequently, a theory of probability will be most helpful if developed in a manner that explicitly anticipates and recognizes, from beginning to end, this larger setting of induction. Accordingly, an adequate theory of induction must satisfy two fundamental requirements:

(1) Generality. An adequate theory of probability and induction must provide "a general method," that is, it must be suitable for all of its intended applications: "If the rules are not general, we shall have different standards of validity in different subjects, or different standards for one's own hypotheses and somebody else's" (Jeffreys 1983:7).

(2) Impartiality. An adequate theory of induction must be impartial: "the principles of the method must not of themselves say anything about the world" (Jeffreys 1983:7). The purpose of induction is to discover things about the world in light of observations and data, and that purpose would be thwarted if the theory of induction itself contained pronouncements or presuppositions in advance of looking at the world. For example, induction could not be used to *discover* electrons if the theory of induction already *presupposed* that electrons either did or did not exist. Jeffreys (1983:7) explains: "If the rules [of induction] of themselves say anything about the world, they will make empirical statements independently of observational evidence, and thereby

limit the scope of what we can find out by observation. If there are such limits, they must be inferred from observation; we must not assert them in advance."

In summary, an adequate theory of induction must be general and impartial. Generality will ensure applicability to all cases and persons. Impartiality will ensure fairness in evaluating various hypotheses, and there must be no presuppositions to favor one hypothesis over another before one looks at the world to gather relevant data. Remarkably, this is all that need be said to ground a good theory of probability and to exclude bad theories.

EIGHT GENERAL RULES

Following in the spirit of his two fundamental requirements, and prior to selecting his probability axioms, Jeffreys presents eight general rules. These rules serve three roles. They detail the implications of the initial two fundamental requirements, they structure the subsequent choice of axioms, and in the end they serve as judges of the adequacy or inadequacy of a theory of probability and induction.

(1) Explicit. All presuppositions or axioms must be stated explicitly, and subsequent theory and conclusions must follow from them (Jeffreys 1983:8).

(2) Coherent. The theory of probability and induction must be coherent or self-consistent. That is, "it must not be possible to derive contradictory conclusions from the postulates [axioms] and any given set of observational data" (Jeffreys 1983:8).

(3) Practical. The rules for induction must be practical, applicable to real experiences and experiments. For example, "The existence of a thing or the estimate of a quantity must not involve an impossible experiment" (Jeffreys 1983:8).

(4) Revisable. The theory must provide for occasional revisions of incorrect scientific inferences: "It is a fact that revision of scientific laws has often been found necessary in order to take account of new information – the relativity and quantum theories providing conspicuous instances – and there is no conclusive reason to suppose that any of our present laws are final. But we do accept inductive inference in some sense; we have a certain amount of confidence that it will be right in any particular case, though this confidence does not amount to logical certainty" (Jeffreys 1983:9). One must be able to recover from a mistake.

(5) Empirical. Scientific conclusions must be dominated by empirical evidence, not philosophical presuppositions: "The theory must not

deny any empirical proposition *a priori*; any precisely stated empirical proposition must be formally capable of being accepted ... given a moderate amount of relevant evidence" (Jeffreys 1983:9).

(6) Parsimonious. The number of axioms should be small, "because if we choose the postulates [axioms] so as to cover the subject with the minimum number of postulates we thereby minimize the number of acts of apparently arbitrary choice" (Jeffreys 1983:9).

(7) Human. Probability theory must be compatible with our humanness and imperfection: "While we do not regard the human mind as a perfect reasoner, we must accept it as a useful one and the only one available" (Jeffreys 1983:9). The theory need not capture every possible detail, but should be correct in outline. It may have both descriptive and prescriptive aspects. Among statisticians, there may be some unresolved disagreements that are deemed less than crucial: "I disagree utterly with many arguments produced by the chief current schools of statistics, but I rarely differ seriously from the conclusions; their practice is far better than their precept. I should say that this is the result of common sense emerging in spite of the deficiencies of mathematical teaching" (Jeffreys 1983:10).

(8) Not perfectionistic. Induction is more general and more complex than deduction, and yet even erudite theories of deduction have apparent limitations and imperfections (Jeffreys 1983:10–15). Therefore, we cannot expect a theory of probability and induction to be perfect in every way, so we cannot be expected to respond with complete surprise and abject dismay if our theory can do only 99.9% of what scientists (or philosophers) would like. Likewise, we must accept the fact that the experimental data that enter into inductive analyses are also imperfect, containing some measure of error and uncertainty for even the most precise experiments.

PROBABILITY AXIOMS AND RULES

The probability theory sought here is to be practical and suitable for science, which is why Jeffreys (1983) precedes axiom choice with two fundamental requirements and eight general rules (whereas more formal or abstract treatments usually begin directly with the axioms). The desired probability axioms are to be not only mathematically coherent but also physically sensible. As Miller (1987:272–273) remarks, "axioms ... are meant to fit sound vernacular reasoning about probabilities." Momentarily, you will be presented several probability axioms and will then need to make your own personal judgment regarding whether or not these axioms are sensible – whether or not they reflect daily and scientific experiences of the physical world.

As a warm-up exercise, here is a little probability test. Judge for yourself whether or not the following three probability statements, which concern two tosses of a fair coin, seem sensible: (1) The probability of getting two heads is $1/4$. (2) The probability of getting any outcome, namely, two heads or two tails or else one heads and one tails, is 1. (3) The probability of getting two heads, or else one heads and one tails, is $1/4 + 1/2 = 3/4$.

Does this make sense to you? Does it match your experience of the world? If so, apparently you already hold, at least implicitly, the probability axioms introduced by Kolmogorov (1933; 1956:2; Berry 1996:107). Furthermore, in line with the fundamental requirements of generality and impartiality, additional reasoning about probabilities is to fit with these simple examples.

Accordingly, those three statements can be restated in generality to yield three probability axioms: (1) The probability of getting a particular outcome can be expressed as a value greater than or equal to 0. (2) The probability of getting any among all possible outcomes is 1. (3) The probability of getting either of two mutually exclusive outcomes equals the sum of their individual probabilities.

One final question completes this little warm-up exercise. Do the following three statements all make sense to you? (1) The probability of an ideal fair coin landing heads is $1/2$. (2) The probability of an actual fair coin landing heads is nearly $1/2$. (3) The probability that my belief, "The fair coin will land heads," will be true is $1/2$.

If all three of these statements make sense to you, then you should feel comfortable applying probability concepts to abstract entities, actual events, and personal beliefs (although the Kolmogorov axioms happen to have been expressed in a language of events and outcomes). Thus, a single unified probability theory serves all ordinary and scientific applications, just as does a single unified arithmetic theory. A broad and unified concept of probability fits various usages of the term "probability," as in this chapter's opening eight examples. It also respects Jeffreys's two fundamental requirements, that probability theory be general and impartial.

What one needs and obtains from an axiomatized theory of probability is "an impersonal criterion [or theory] that will enable an individual to see whether, in any particular instance, he is following the rules that other people follow and that he himself follows in other instances" (Jeffreys 1983:15). By contrast, some philosophers, such as O'Hear (1989:155–176), are inclined to sharply contrast and strongly oppose different concepts or applications of probability. But that seems needless and counterproductive.

Now if these three little axioms and their broad applications to events and beliefs make sense to you, then remarkably the whole subject of probability theory is already settled in principle. Everything else is just a collection of theorems derivable from these axioms. No more fundamental choices are required.

Some preliminaries are needed before stating probability axioms in more formal notation. Recall the notations that "~" means "not," "∧" "and," "∨" "or," "→" "implies," and "≡" "equals," as defined in Table 5.1. Also, probability theory is built on predicate logic and arithmetic, which required 16 axioms (and two rules). So the following probability axioms could resume with the numbering 17. But it suffices instead merely to mention the inheritance of logic and arithmetic and to number this chapter's axioms starting with 1.

The following axioms are expressed in a language of abstract, mathematical sets, although the foregoing discussion has emphasized the intended applications to events and propositions as well. The members of a set are listed in brackets: $A = \{1, 2\}$ is the set composed of the integers 1 and 2. $A \cup B$ designates the union of A and B, the set containing everything that belongs to A or B or both. $A \cap B$ designates the intersection of A and B, the set containing everything belonging to both A and B. For example, if $A = \{1, 2\}$ and $B = \{2, 3, 4\}$, then the union is $A \cup B = \{1, 2, 3, 4\}$, and the intersection is $A \cap B = \{2\}$. A is a subset of B if every member of A is also a member of B. For example, the set $\{2\}$ is a subset of the set $\{1, 2, 3\}$ but is not a subset of $\{3, 4\}$. Two sets are mutually exclusive if they have no members in common, such as the sets $\{2\}$ and $\{3, 4\}$.

The conditional probability $P(H \mid D)$, meaning the probability of H given D, is defined as the simple probability $P(H \cap D)$ divided by the simple probability $P(D)$, provided that $P(D) \neq 0$. This definition is depicted in Figure 6.1. For the present, these symbols "H" and "D" are merely arbitrary labels, but they have been chosen deliberately to facilitate subsequent interpretations of hypothesis and data. For example, assume there are 100 objects, of which 40 have property D, 27 have property H, and 6 have both properties $H \cap D$. Then the unconditional probability that an object has property H, $P(H)$, is 27 out of 100, or 0.27. But the conditional probability that an object has property H given that it has property D, $P(H \mid D)$, is 6 out of 40, or 0.15. Incidentally, in Figure 6.1, $P(H \cap D)$ has 15% of the area of $P(D)$, so this figure fits this particular example.

For another example, assume that a class has 6 girls with blue eyes and 4 with brown, and 5 boys with blue eyes and 8 with brown. The conditional probability that a student has blue eyes, given that the student is a girl, is $6/(6 + 4)$, or 0.6. For comparison, the conditional probability that a student has blue eyes, given that the student is a boy, is $5/(5 + 8)$, or about 0.38. Likewise, the unconditional probability that a student has blue eyes is $(6 + 5)/(6 + 4 + 5 + 8)$, or about 0.48.

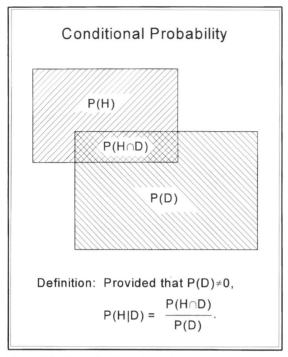

Figure 6.1. The definition of conditional probability.

For generality, the axioms given here use conditional probabilities. However, to cover simple probabilities also, the conditioning event A may be taken to be a tautology (such as $X \lor \sim X$) and then removed. Similar axioms are given by Jeffreys (1983:15–26), but the following convenient axioms follow Salmon (1967:59–60). For comparison, Gustason (1994:102, 157–158), following Burks (1977:40–41), has an interesting approach with somewhat different axioms, but it is not suitable here because of the additional burden of presuming knowledge of modal logic. Incidentally, as an intellectual exercise rather than a scientific tool, Burks (1977:99–164) also develops several nonstandard probability theories from nonstandard axioms. The preceding three Kolmogorov axioms and the following four axioms are equivalent and exchangeable, but the following axioms make it easier to derive theorems:

Axiom 1. $P(B|A)$ is a single number $0 \le P(B|A) \le 1$.
Axiom 2.   If A is a subset of B, then $P(B|A) = 1$.
Axiom 3.   If B and C are mutually exclusive, then
                    $P(B \cup C \mid A) = P(B|A) + P(C|A)$.
Axiom 4. $P(B \cap C|A) = P(B|A) \times P(C|A \cap B)$.

Salmon offers a brief and lucid explanation of the meaning of each of these four axioms:

> "Axiom 1 tells us that a probability is a *unique* real number in the interval zero to one (including these endpoints)" (Salmon 1967:59). A common use of probabilities is to express the propensity of a statement to be true. In this setting, the probability endpoint 0 represents false, the endpoint 1 represents true, and a probability of 0.5 means that a statement is equally likely to be true or false, such as "The next coin toss will give heads." Likewise, a probability of 0.9 means that a statement has a 90% chance of being true (and a 10% chance of being false). The endpoints 0 and 1 are exceptional probabilities in that they represent certainty, namely, certain falsehood and certain truth, respectively. Intermediate probability values represent some degree of uncertainty.
>
> "Axiom 2 states that the probability of an A being a B is one if every A is a B" (p. 59).
>
> "Axiom 3 tells us, for example, that the probability of drawing a black card from a standard deck equals the probability of drawing a club plus the probability of drawing a spade. It does not apply, however, to the probability of drawing a spade or a face card, for these two classes are not mutually exclusive" (p. 59).
>
> "Axiom 4 applies when, for example, we want to compute the probability of drawing two white balls in succession from an urn containing three white and three black balls. We assume for illustration that the balls are all equally likely to be drawn and that the second draw is made without replacing the first ball drawn. The probability of getting a white ball on the first draw is one half; the probability of getting a white ball on the second draw *if the first draw resulted in white* is two fifths, for with one white ball removed, there remain two whites and three blacks. The probability of two whites in succession is the product of the two probabilities – i.e., one fifth. If the first ball drawn is replaced before the second draw, the probability of drawing two whites is one fourth" (pp. 59–60).

Finally, probability axioms serve in essence to enforce coherence among a set of probability assignments and to derive certain probabilities given others, but they do little to provide probability assignments. Instead, that job is done by some probability rules (Gustason 1994). For instance, the most basic rule is the so-called "straight rule," which says that "If n As have been examined and m have been found to be Bs, then the probability that the next A examined will be a B is m/n" (Earman 2000:22). Of course, the sample must be representative (unbiased) and sizable in order for this fraction m/n to be reliable and repeatable. Statisticians have developed fancier rules that provide probabilities that the true value of this fraction lies within a specified interval.

PROBABILITY THEOREMS

Theorems are statements derivable deductively from axioms. Consequently, probability theorems express probability beliefs or relationships that are coherent with the axioms. The function of theorems is to address a rich variety of probability cases beyond those covered explicitly and immediately by the axioms themselves. For example, given a probability of 0.17 that a light bulb will fail in its first thousand hours of use, what is the probability that it will not fail? None of the preceding four probability axioms answers this question directly (and neither do any of the 16 inherited axioms and two rules of logic and arithmetic). But after deriving Theorem 1 momentarily, we will be well equipped to answer this question.

Just five theorems are given in this section to convey the flavor of probability theorems, leaving thorough treatment to probability texts.

Theorem 1. $P(\sim B|A) = 1 - P(B|A)$.

Informal Proof: By Axiom 2, $P(B \cup \sim B|A) = 1$, because every member of A is a member of either B or $\sim B$, and hence A is a subset of $B \cup \sim B$. Then by Axiom 3, because B and $\sim B$ are mutually exclusive, $P(B \cup \sim B|A) = P(B|A) + P(\sim B|A)$, which by the first result also equals 1. Therefore, $P(B|A) + P(\sim B|A) = 1$. Finally, inherited arithmetic procedures allow subtraction of $P(B|A)$ from both sides to obtain Theorem 1. Incidentally, applying Theorem 1 to the light-bulb problem, we obtain the solution that the probability of not failing is $1 - 0.17 = 0.83$, which is obviously sensible.

Theorem 2. If $P(A|E) \geq P(B|E)$ and $P(B|E) \geq P(C|E)$, then $P(A|E) \geq P(C|E)$.

Informal Proof: Theorem 2 follows from the transitivity property of real numbers, made applicable to probabilities by Axiom 1.

Theorem 3. $P(A|E) = (P(B|E) \times P(A|E \cap B)) + (P(\sim B|E) \times P(A|E \cap \sim B))$.

Informal Proof: Following Salmon (1967:60), the set A equals the things that are both A and B or else both A and $\sim B$, so $P(A|E) = P((A \cap B) \cup (A \cap \sim B)|E)$. Because B and $\sim B$ are mutually exclusive, so are $A \cap B$ and $A \cap \sim B$, and therefore Axiom 3 yields $P((A \cap B) \cup (A \cap \sim B)|E) = P(A \cap B|E) + P(A \cap \sim B|E)$. Axiom 4 gives $P(A \cap B|E) = P(B|E) \times P(A|E \cap B)$, and likewise

$P(A \cap \sim B \,|\, E) = P(\sim B \,|\, E) \times P(A \,|\, E \cap \sim B)$. Combining these results gives Theorem 3.

Two additional probability theorems are given next without proof. Incidentally, note that Theorem 5 generalizes Axiom 3, giving the probability of $A \cup B$ regardless of whether or not A and B are mutually exclusive, because Theorem 5 reduces to Axiom 3 if A and B are mutually exclusive, so that $P(A \cap B \,|\, E) = 0$, but Theorem 5 makes the necessary adjustment of subtracting $P(A \cap B \,|\, E)$ if A and B are not mutually exclusive.

Theorem 4.   If A is a subset of B, then $P(A \,|\, E) \leq P(B \,|\, E)$.
Theorem 5. $P(A \cup B \,|\, E) = P(A \,|\, E) + P(B \,|\, E) - P(A \cap B \,|\, E)$.

It is important for a scientist to know the content of probability theorems, as in the foregoing examples, but an equally important objective of this section is to explain the nature and function of probability axioms and theorems. Accordingly, let us review Theorem 1, that $P(\sim B | A) = 1 - P(B | A)$, with the goal of really understanding its meaning. Its meaning can be better appreciated by also considering and contrasting various other (wrong) renditions of Theorem 1.

Assume that the probability of rain today, given some weather forecast or other evidence, is 0.3. Then what is the probability of no rain today? By Theorem 1, the probability of no rain is simply $1 - 0.3 = 0.7$.

But now let us examine the thinking of Mr. Quandary, who, sad to say, has suffered the grave misfortune of never having studied probability theory. Mr. Quandary knows that there is a 0.3 probability of rain today. He is in a quandary, however, concerning what to believe about the probability of no rain today. Sometimes he is inclined to think it is 0.8, and at other times he thinks it is 0.6. At yet other times, he just gives up, thinking that the probability of no rain is unknowable.

The salient feature of Mr. Quandary's probability beliefs is that they generate incoherence when compared with the probability axioms. Consider, for example, his beliefs that the probability of rain is 0.3 and the probability of no rain is 0.8. Those two events, rain and no rain, are mutually exclusive, so by Axiom 3 we have P(rain or no rain) $= 0.3 + 0.8 = 1.1$. But this result is incoherent with Axiom 1, which restricts probability values to the range 0 to 1 inclusive. Likewise, these beliefs are incoherent with Theorem 1, which was derived from the axioms, for it assigns the total of the probabilities of all outcomes a value of 1. Hence, it is not possible to assert coherently these probabilities of 0.3 for rain and 0.8 for no rain *and* the probability axioms. Something must give. Likewise, Mr. Quandary's other belief, that there is a 0.6 probability of no rain, fares no better with the probability axioms. Furthermore, his opinion that the probability of no rain is unknowable is irrational with respect to a probability theory based on the foregoing axioms, because the answer to this simple problem is readily forthcoming.

A hallmark of coherent, rational beliefs is that they give thought some structure and predictability. Thereby, rational thinking can be public. If both you and I use probability theory, then I can expect you also to deduce that the probability of no rain is 0.7. But if you do not know or respect probability theory, then anything goes, and I cannot presume to predict what you are likely to think about the probability of no rain. That the scientific community exists is clear evidence that scientists have a shared and sensible probability theory, for otherwise this community could neither have been formed nor endure.

Of course, even with probability theory in place, mistakes are still possible. However, probability axioms and reasoning make mistakes remediable. For example, presume that you and I both adhere to the probability axioms and that we both believe that the probability of rain today is 0.3. Furthermore, assume that upon being asked about the probability of no rain today, you respond 0.7, but I respond 0.5. Naturally, we are both disconcerted by our disagreement. However, presumably a brief discussion will show me the error of my ways, I will cheerfully retract my error, and we will happily agree on 0.7. The situation for probability mistakes is thus really no different from that for arithmetic mistakes. In either case, because of the underlying axioms, be they understood formally or used informally, errors are correctable. However, without a coherent probability theory, errors could not be detected nor corrected.

To understand that the probability axioms are the basis for deriving probability theorems and for correcting probability errors, as discussed thus far, is only to begin to understand the enormous significance and role of the axioms. More fundamentally, the axioms are necessary for any probability statement to have any meaning whatsoever. For example, consider the statement that the probability of an event A, given evidence E, is known to be $P(A \mid E) = 0.3$. What does this mean?

Given the simultaneous, coherent assertion of the probability axioms, this probability statement is meaningful. By the scaling convention of Axiom 1, we understand that this probability of 0.3 is somewhat closer to 0 (representing total impossibility) than to 1 (representing total certainty). More specifically, an obvious understanding is that among 10 cases with $P(A \mid E) = 0.3$, we would expect, on average, that A would occur in 3 cases, and $\sim A$ in 7 cases. Furthermore, by Theorem 1, which is deducible from the axioms, we realize that $P(\sim A \mid E) = 0.7$.

Far otherwise would be our situation, however, if the naked assertion that $P(A \mid E) = 0.3$ were made in the absence of probability axioms. Then what would this assertion mean? Nothing! Absolutely nothing! Apart from some scaling convention (as has been provided by Axiom 1), the probability value 0.3 could mean anything from total impossibility to total certainty. Because it could mean anything, it does mean nothing. Likewise, apart from some relationship between an event A and its opposite $\sim A$ (as has been provided by Theorem 1), there is no implied consequence about the probability of $\sim A$. For example, knowing the probability of rain then would imply nothing whatsoever about the probability of no rain. Such thinking is meaningless. It would erode science hopelessly. Incidentally, the same situation would obtain for a naked arithmetic assertion, such as "There are 7 apples," were it separated from arithmetic axioms. Without its ordinary context and implications, such as 7 apples resulting from 4 apples plus 3 apples, this simple arithmetic assertion would be meaningless too.

In order for any statement about probability to have a meaning, that statement must reside within the larger context of a coherent interpretation based on the probability axioms. This larger context may be informal and implicit, as in common sense, or may be formal and explicit, as in probability theory. If a statement such as "There is a 30% probability of rain today" has any meaning to you, it is because and only because you interpret this statement by invoking the coherent and meaningful context provided by the probability axioms. Formal probability theory recognizes these axioms explicitly.

BAYES'S THEOREM

The preceding section derived several theorems from the probability axioms, and this section derives another theorem that merits special attention: Bayes's theorem. Several alternative forms are given that are convenient for solving various kinds of problems.

This extraordinarily important theorem is nothing less that "the chief rule involved in the process of learning from experience," and it provides us "a formal rule in general accordance with common sense, that will guide us in our use of experience [or data] to decide between hypotheses" (Jeffreys 1983:28–29). Bayes's theorem can even be applied to the process of learning itself, resulting in learning about learning (Morris H. DeGroot, in Bernardo et al. 1980:391).

Bayes (1763) stated several probability axioms and then deductively derived his famous theorem. That paper has been reproduced by Barnard (1958) and Press (1989:185–217). Jeffreys (1983:30–34) has provided a helpful summary, as well as an essentially equivalent restatement in terms of modern probability theory. Bayes's theorem has been derived from the foregoing probability axioms by Salmon (1967:61; also see Burks 1977:65–66 and Berry

1996:147–157). The simpler derivation given here follows trivially from the definition of conditional probability (Howson and Urbach 1993:28).

The following is a simple form of Bayes's theorem. The symbol "$\propto$" means "proportional."

Bayes's Theorem, Simple Form
$P(H \mid D) \propto P(D \mid H) \times P(H)$, or
Posterior $\propto$ Likelihood $\times$ Prior

> *Informal Proof:* Rearranging the definition of conditional probability gives $P(H \cap D) = P(H \mid D) \times P(D)$. A second application of this same definition, with H and D reversed, gives $P(D \cap H) = P(D \mid H) \times P(H)$. But $H \cap D$ and $D \cap H$ are the same set, so $P(H \cap D) = P(D \cap H)$, and these two formulas can be equated to obtain $P(H \mid D) \times P(D) = P(D \mid H) \times P(H)$. Finally, ignoring the normalizing constant $P(D)$, and hence substituting proportionality for equality, we get Bayes's theorem. Hence, Bayes's theorem is an extremely simple theorem that is basically just a rearrangement of the definition of conditional probability.

Three conventional terms are useful in discussing Bayes's theorem. The first term, $P(H \mid D)$ in the foregoing equation for Bayes's theorem, is called the "posterior." The middle term, $P(D \mid H)$, is the "likelihood." And the last term, $P(H)$, is the "prior." To gain an initial understanding of this theorem to build on later, first grasp the basic meanings of its three components individually, and then assemble them together into a functioning whole. In this simplified introduction, these components are described as probabilities, but in general they are represented by probability distributions that necessitate calculus to solve induction problems. A single probability value is the simplest kind of probability distribution, which simplifies the calculations.

(1) Prior. The term "prior" is short for the phrase "prior probability" applied to a proposition, belief, hypothesis, model, or event. The term "prior" is synonymous with what might otherwise be called "previous," "initial," "old," or "original." Hence the prior is the initial probability of a proposition's truth or an event's occurrence, evaluated prior to collecting some particular data or evidence.

(2) Likelihood. The likelihood summarizes the data's impact on the probabilities of the hypotheses. A (poor) hypothesis that gives low probability to the newly observed data is weakened in its credibility, whereas a (good) hypothesis that gives high probability to the newly observed data is strengthened. Hence, under the influence of the data, the hypotheses will shift in terms of their credibilities, in accordance with how poorly or how well they fit the data.

(3) Posterior. The term "posterior" is short for "posterior probability," again applied to the same proposition, belief, hypothesis, model, or event as in the prior. The term "posterior" is synonymous with "final," "new," or "conclusion." If there is ongoing research, with plans for additional data collection, "posterior" can also be synonymous with "current." In any event, the term "posterior" means posterior to having utilized the new information considered in the likelihood.

Robert (1994:8) has offered an insightful explanation of the meaning of Bayes's theorem in statistical applications, saying that it "appears as a major step in the history of Statistics, being the first *inversion* of probabilities." The foregoing equation contains three items, but the equation's central feature is the presence of $P(H|D)$ on the left and $P(D|H)$ on the right, with the roles of H and D reversed or inverted. Bayes's theorem solves the inverse or inductive problem of calculating the probability of a hypothesis given some data $P(H|D)$ from the probability of some data given a hypothesis $P(D|H)$.

This simple form of Bayes's theorem mentions only one hypothesis, but hypotheses always have competitors. Accordingly, additional forms of this theorem are useful to clarify the expression and simplify the solution of various inference problems. The following form expresses Bayes's rule for an inquiry with two hypotheses, H_1 and H_2, that are mutually exclusive and jointly exhaustive:

Bayes's Theorem, Two-Hypotheses Form

$$P(H_1|D) = \frac{P(D|H_1) \times P(H_1)}{(P(D|H_1) \times P(H_1)) + (P(D|H_2) \times P(H_2))}$$

Of course, a similar equation holds for the other posterior $P(H_2|D)$. However, because we already have $P(H_1|D)$, and these two hypotheses are mutually exclusive and jointly exhaustive, the easier way to calculate the other result is to use the equation $P(H_2|D) = 1 - P(H_1|D)$. This formula for two hypotheses is easily extended to three mutually exclusive and jointly exhaustive hypotheses by adding the term $P(D|H_3) \times P(H_3)$ to the denominator, and so on for more hypotheses.

Sometimes probabilities are expressed as ratios or odds, such as saying that the odds that H_1 and H_2 are true are 3:1, meaning that H_1 is three times as likely as H_2 to be true. If H_1 and H_2 are mutually exclusive and jointly exhaustive, their odds can be converted to probabilities if desired, such as 0.75 and 0.25 in this instance. But if the situation is otherwise and more complex, particularly because additional unspecified hypotheses may need to be considered, further information would be required to move from odds to probabilities. The ratio form of Bayes's rule is as follows:

Bayes's Theorem, Ratio Form

$$\frac{P(H_1|D)}{P(H_2|D)} = \frac{P(D|H_1)}{P(D|H_2)} \times \frac{P(H_1)}{P(H_2)}$$

This equation may be read as the posterior odds equal the likelihood odds times the prior odds. For example, if initial considerations give hypotheses H_1 and H_2 prior odds of 3:1 and a new experiment gives them likelihood odds of 1:27, then the posterior odds of 1:9 reverse the initial situation to favor H_2.

Incidentally, although the usual focus is on obtaining the posterior probabilities from the data and the prior probabilities, other options are possible. For instance, the focus could be on determining the range of prior probabilities, given the data, that would give H_1 greater posterior probability than H_2 (Gustason 1994:125).

There are many additional forms of Bayes's equation, some of which use calculus and represent probabilities by distributions rather than single values. Such elaborations are readily available in many statistics texts, but such technicalities are beyond this book's objectives and needs.

PERMUTATIONS AND COMBINATIONS

It is convenient to accompany probability theory with some elementary combinatorial theory for counting permutations and combinations of objects. Many probability problems can be solved by counting the number of ways that a given event can occur.

For R events or experiments such that the first event has N_1 possible outcomes, and for each of those the second event has N_2 possible outcomes, and so on up to the Rth with N_R outcomes, there is a total of $N_1 \times N_2 \times \cdots \times N_R$ possible outcomes. For example, how many different license plates could be made with three digits followed by three letters? The solution is $10 \times 10 \times 10 \times 26 \times 26 \times 26 = 17{,}576{,}000$.

A "permutation" is a distinct ordered arrangement of items. For example, for the set of letters A and B and C, all possible permutations are ABC, ACB, BAC, BCA, CAB, and CBA – which number 6, because the first choice has 3 options, the second 2, and the third 1, for a total of $3 \times 2 \times 1 = 6$ permutations. The general rule is that for N entities, there are $N \times (N-1) \times (N-2) \times \cdots \times 3 \times 2 \times 1$ permutations. This number is called "N factorial," and is denoted by $N!$ For example, $1! = 1$, $3! = 6$, and $5! = 120$. Also, by definition $0! = 1$.

Sometimes the entities are not all unique, as in the set A, A, B, and C, which has two letter A's that are alike. For N objects, of which N_1 are alike, N_2 are alike, and so on up to N_R alike, there are $N! / (N_1! \times N_2! \times \cdots \times N_R!)$ permutations. For example, this set of letters has $4! / (2! \times 1! \times 1!) = 24/2 = 12$ permutations.

A "combination" is a particular number for each of several different entities or outcomes for which the order does not matter. Ross (1994:6) gives the example of selecting 3 items from the 5 items A, B, C, D, and E. There are 5 ways to select the first item, 4 for the second, and 3 for the third, so there are $5 \times 4 \times 3 = 60$ permutations that distinguish different orderings. But every group of 3 items, such as A and B and C, gets counted $3! = 6$ times, as explained earlier. So the number of combinations, which do not distinguish different orderings, of 3 items selected from 5 is $60/6$ or 10. The general rule is that N objects taken R at a time have $N!/((N-R)! \times R!)$ combinations. For $R \leq N$, this value is symbolized by $\binom{N}{R}$, meaning the number of possible combinations of N objects taken R at a time. For example, 4 horses are to be chosen from 8. How many combinations are possible? The solution is $\binom{8}{4}$, which equals $(8 \times 7 \times 6 \times 5)/(4 \times 3 \times 2 \times 1) = 1{,}680/24 = 70$.

For a somewhat more ambitious example, given a fair coin, what is the probability that 100 tosses will produce 45 heads and 55 tails? The total number of possible outcomes is 2^{100} and all outcomes are equally likely. The relevant number of combinations of 45 heads out of 100 tosses is $\binom{100}{45}$. Doing the required arithmetic and dividing the relevant outcomes by all outcomes will yield the answer, 0.048474. So there is about a 5% chance of getting this particular outcome.

COMMON BLUNDERS

The following examples can illustrate two of the most common blunders in probability reasoning. The first involves an ignored prior, and the second a reversed conditional probability.

Ignored Prior

Stirzaker (1994:25) has posed the following simple medical problem: "You have a blood test for some rare disease which occurs by chance in 1 in every 100 000 people. The test is fairly reliable; if you have the disease it will correctly say so with probability 0.95; if you do not have the disease, the test will wrongly say you do with probability 0.005. If the test says you do have the disease, what is the probability that this is a correct diagnosis?" Bourne et al. (1986:294) remarked regarding a similar example that most people would answer that the probability of disease was about 95% (Faust 1984:59–61, 90–94; Tarantola 1987:105–108; Mendenhall and Beaver 1994:161; Isaac 1995:31–34; Royall 1997:1–3). Indeed, a study conducted in a leading American hospital found that 80% of those questioned gave wrong answers (Howson 2000:52–54; *Economist*, 1999). But when we plug the numbers into the two-hypotheses form of Bayes's theorem, surprisingly the correct answer is $(0.95 \times 0.00001)/((0.95 \times 0.00001) + (0.99999 \times 0.005))$, or only about 0.2%. This is *drastically* different from 95%, and it strongly supports the opposite conclusion! As Stirzaker (1994:26) remarked, "Despite appearing to be a pretty good test, for a disease as rare as this the test is almost useless." For every real instance of the disease detected by that test, there would be more than 500 false positives, so the results could hardly be taken seriously. At best, such a test might offer economical screening before administering another more expensive, definitive test.

What went wrong to give that incorrect answer? Knowledge of the general population gives odds for diseased:healthy of 1:100,000. And knowledge of a positive test gives likelihood odds of 95:5. The common mistake is to base one's conclusion on the likelihood, and to ignore the prior, contrary to what Bayes's theorem says to do. It is alarming and potentially dangerous that most people in general and most doctors in particular make this common blunder in probability reasoning.

Reversed Conditional Probability

As an example to sharpen probability reasoning, consider the genders of children in a family. Make the simplifying assumption that the probabilities of boys (B) and of girls (G) are equal, $P(B) = P(G) = \frac{1}{2}$. Also assume that the gender of each child is an independent factor. Actually, large samples show that boys are slightly more numerous than girls, and gender is not an independent factor in the rare case of identical twins. The topic of this chapter, however, is probability, not reproductive biology, so these simplifying assumptions are expedient.

Consider the question posed by Stewart (1996): "Suppose Mr. and Mrs. Smith tell you they have two children, one of whom is a girl. What is the probability that the other is a girl?" You might want to reach your own conclusion before reading further. Figure 6.2 depicts this problem, with one child obscured by the family dog, so that this child's gender is not apparent. As Stewart remarks, the usual response is that the other child is a boy or a girl with the same probability of $\frac{1}{2}$ for either. But that is incorrect.

There are four possible gender sequences – BB, BG, GB, and GG – where B and G denote boy and girl, and the letters are arranged in order of birth. All sequences are equally likely, and hence each has a probability of $\frac{1}{4}$. Because we know that the Smiths have at least one girl, the sequence BB can be eliminated. That leaves three equally likely cases, and in just one of those cases (GG) is the other child also a girl. So the probability that the other child is a girl is actually $\frac{1}{3}$. Hence, far from it being equally likely that the other child is a boy or a girl, actually it is twice as likely that the other child is a boy.

Figure 6.2. Children in a family. The Smith family has two children. One is a girl, but the other is obscured by the family dog. Reasoning with conditional probability can calculate the probability that the hidden child is also a girl. (This drawing by Susan Bonners is reproduced with her kind permission.)

As Stewart also shows, this argument can be repeated more formally using the definition of conditional probability, which underlies Bayes's theorem. Let X denote that at least one of two children is a girl, and let Y denote that the other child is also a girl. Recall that "$\cap$" is the symbol for "and" and that "$|$" is the symbol for "conditional on" or "given." Then $P(X) = \frac{3}{4}$ and $P(X \cap Y) = \frac{1}{4}$. Finally, by the definition of conditional probability, $P(Y|X) = P(X \cap Y)/P(X) = (\frac{1}{4})/(\frac{3}{4}) = \frac{1}{3}$, just as was shown by the informal proof in the preceding paragraph. Incidentally, this example involving children is analogous to other stories involving coins or cards, and such stories have a long history (Galton 1894; Ball and Coxeter 1987:44).

Why is confusion about reversed conditionals so common? Two conjectures seem plausible. First, for some other operators, order does not matter, and that may tempt people to think that order does not matter for conditional probabilities. For the operators "and" and "or," order is irrelevant, because $P(X \cap Y) = P(Y \cap X)$ and $P(X \cup Y) = P(Y \cup X)$. But, instead, $P(X|Y) \neq P(Y|X)$.

Second, informal remarks about probabilities often use undefined and vague terms, the biggest offender being "when" (Howson 1997b). Recall this subsection's question, which can be expressed as "For a family with two children, what is the probability that the other child is a girl, given that one is a girl?" This is a proper probability question. But now change just one key term, obtaining "For a family with two children, what is the probability that the other child is a girl, when one is a girl?" This is an improper question precisely because P(X when Y) is *not* defined. Most people interpret P(X when Y) as P($X|Y$), the probability of X given Y. But it can also be construed as P($Y|X$), the probability of Y given X, or as P($X \cap Y$), the probability of X and Y. So, in careless usage, P(X when Y) equivocates between three meanings: P($X|Y$), P($Y|X$), and P($X \cap Y$). Always, these three probabilities have different meanings, and ordinarily they also have different numerical values, sometimes wildly different. Consequently, an indispensable prerequisite for rational thinking about probability and statistics that will correctly match with experiences of physical reality is a sharp distinction among these three probabilities that is implemented and reinforced by precise and consistent terminology.

Finally, this discussion of the genders of children in a family photograph might seem to be only a cute device for explaining conditional probabilities, without larger significance. But Stewart (1996) draws out implications for the courtroom. One of his remarks concerns DNA profiles, which produce a "match probability" between a DNA sample related to a crime and another taken from a suspect: "The prosecutor's fallacy refers to a confusion of two different probabilities. The 'match probability' answers the question 'What is the probability that an individual's DNA will match the crime sample, given that he or she is innocent?' But the question that should concern the court is 'What is the probability that the suspect is innocent, given a DNA match?' The two queries can have wildly different answers" (Stewart 1996).

For instance, suppose that a blood stain collected at a crime scene and a blood sample taken from the suspect are submitted to DNA profiling, and they match at all genetic markers tested. The prosecutor establishes that the probability of this DNA match, given that a person is innocent, is only one in a million. Should the jury accept that as strong evidence of guilt?

The proffered evidence concerns P(match|innocent). But the question before the jury concerns the probability that the suspect is innocent, given the DNA match, P(innocent|match). To estimate this relevant quantity, additional information is required to specify prior probabilities, and that information is about the genetic history and structure of the population of possible perpetrators. When that additional genetic information is incorporated into the analysis, typically the evidence is about as strong as the evidence one would get from a match that used half as many genetic markers and that ignored population structure. Because the evidence from the markers combines multiplicatively, the resulting adjustment equates to taking a square root. So if P(match | innocent) is one in a million, typically P(innocent | match) is about one in a thousand. And if the crime occurred in a locality with a strong ethnic identity, and many persons there had common ancestors a century or two ago, then the adjustment could be larger than average, leading to a probability of, say, one in a hundred. Obviously, that increase in the probability of innocence, from one in a million to one in a hundred, might convince some jurors that the evidence was not clear beyond a reasonable doubt.

Anyway, the salient lesson in this example, apart from the details about DNA tests and population structures, is that P(match | innocent) and P(innocent | match) are different quantities that should not be confused. Again, this is a particular instance of the difference between $P(D | H)$ and $P(H | D)$, as well as another illustration that the usual quantity of interest in science (and law) is $P(H | D)$.

Scientists want to know which hypothesis is true (or most likely to be true), given the data. The business of inductive logic, as pursued in the next chapter, is to specify the evidential bearing of data on hypotheses. But that business must begin by not confusing a conditional probability with its reverse.

Sadly, an innocent person can go to jail or a guilty person can go free because a jury misunderstands conditional probabilities and thereby misinterprets the evidence. Accordingly, Stewart (1996) suggests that "Of course, it would be silly to suggest that every potential juror should take (and pass) a course in Bayesian inference, but it seems entirely feasible that a judge could direct them on some simple principles." A quick review to avoid confusion between a conditional probability and its reverse could discourage lawyers from attempting fallacious arguments and claims.

The problems illustrated here also occur in medicine and other areas. Patients can receive detrimental treatments just because a care-giver has faulty reasoning about conditional probabilities. The arenas in which probability reasoning applies can be coin tossing or weather forecasting, but they can also be law courts and medical establishments.

SUMMARY

Probability is the propensity for an event to occur or for a proposition to be true. A convenient scaling for probabilities is values from 0 to 1, with 0 representing certain falsity, 1 representing certain truth, and intermediate values representing uncertainty.

To do business with physical reality, probability theory must meet two fundamental requirements: generality and impartiality. These requirements can be specified in greater detail by eight general rules. These requirements and rules then inform the selection of appropriate probability axioms.

Probability theory is built on predicate logic and arithmetic, requiring the 16 axioms and 2 rules supplied in Chapter 5, and it needs just the four new axioms given in this chapter. Several probability theorems have been derived from those axioms. A particularly important one is Bayes's theorem, which is the basis for an important statistical paradigm that will be explored in Chapter 7. Simple rules for counting permutations and combinations have also been given, as they are useful in solving many probability problems.

Finally, two common probability blunders have been discussed. One involves ignoring prior information when reaching a probability conclusion. The other consists in confusing a conditional probability with its reverse conditional probability.

CHAPTER SEVEN

INDUCTIVE LOGIC AND STATISTICS

The logic that is so essential for scientific reasoning is of two basic kinds: deductive and inductive. Chapter 5 reviewed deductive logic, and Chapter 6 probability, which is a branch of deductive logic. This chapter reviews inductive logic, with "statistics" being essentially the term meaning applied inductive logic.

For better or for worse, statisticians are not unified. Rather, there are two major paradigms for induction: Bayesian and frequentist statistics. At stake are scientific concerns, seeking efficient extraction of information from data to answer important questions, and philosophical concerns, involving rational foundations and coherent reasoning.

In order for scientists to judge for themselves which paradigm can best address their research needs, they must understand exactly which questions these two paradigms are asking and exactly what data or information these paradigms require in order to provide answers. By understanding the relative merits of different statistical paradigms, scientists have much to gain in the way of efficiency and productivity.

This chapter cannot possibly do what entire books on statistics do – present a comprehensive treatment. But it can provide a prolegomenon that will clarify the most basic and pivotal issues, which are precisely the aspects of statistics that scientists generally comprehend the least. The main objectives are to depict and contrast the Bayesian and frequentist paradigms and to explain why inductive logic or statistics often can function just fine despite imperfect data, models, and scientists.

AWESOME RESPONSIBILITIES

Modern research in agriculture, medicine, engineering, and other fields imposes on statistical practice a great burden of decision and responsibility that is nothing less than harrowing. Accordingly, this chapter's issues cannot be approached properly apart from healthy alarm at the significant losses that result from inappropriate or second-rate statistics, as well as an appreciation of the potential benefits from first-rate statistics.

For example, writing poignantly of his experience as a statistical consultant in wartime research, Pearson (1947) said that "decisions had to be reached on statistical data far less ample than could be wished." For instance, he discussed an experiment in which 12 naval shells of one kind were tested, of which 2 failed to penetrate an armor plate, and 8 shells of another kind were tested, of which 5 failed. Was that significant evidence that the first kind of shell was superior, or not? There a statistician's advice may have influenced a navy's decision and affected sailors' ability to defend themselves.

Such grave responsibility moved Pearson to ask himself "What weight do we actually give to the precise value of a probability measure when reaching decisions of first importance?" In turn, that practical question motivated a deeper theoretical or philosophical question about the adequacy of "the logical basis of the reasoning process" used in statistics. Furthermore, such questions invoke a moral responsibility in every individual who uses statistics: "each individual who hopes to use his own judgement to the full in drawing conclusions from the statistical analysis of sampling data, must decide for himself what he requires of probability theory." Accordingly, this chapter will endeavor to present the Bayesian and frequentist paradigms with sufficient clarity that readers can discharge their personal responsibilities to make informed decisions about which paradigm best suits their research needs.

Certainly, we all hope not to be caught up in a world war, as Pearson was. But in any case, we cannot avoid the war against hunger, the war against cancer, and other such wars. As scientists see statistics informing their conclusions, which in turn influence food supplies and medical treatments and other critical matters, an awareness of such weighty responsibilities should prompt keen interest in the logical basis of statistics.

INDUCTION AND DEDUCTION

Chapter 5 distinguished deduction and induction in terms of three contrasts. First, the conclusion of a deductive argument is already contained implicitly in its premises, whereas the conclusion of an inductive argument goes beyond the information in its premises. Second, given the truth of all of its premises, the truth of a valid deductive argument's conclusion follows with certainty, whereas given the truth of all of its premises, the truth of an inductive argument's conclusion follows with at most high probability. Third, deduction argues from a given model's general principles to specific cases of expected data, whereas induction argues in the opposite direction, from actual data to an inferred model.

The distinction between deduction and induction can be recapitulated, in the current context of (deductive) probability and (inductive) statistics, by comparing the following two problems. For the sake of simplicity, define a "fair coin" as a coin that gives heads with probability 0.5, and tails 0.5, and an "unfair coin" as one that gives heads with probability 0.6, and tails 0.4.

> *Problem 1.* Given that a coin is a fair coin. What is the probability that 100 tosses of the coin will produce 45 heads and 55 tails?

> *Problem 2.* Given that 100 tosses of a coin produce 45 heads and 55 tails. What is the probability that the coin is a fair coin?

Is Problem 1 deductive or inductive? Likewise, which is Problem 2? At a superficial reading, these two problems may seem quite similar. But in actuality, they are fundamentally different, having opposite reasoning directions. By the foregoing three criteria, Problem 1 is deductive, whereas Problem 2 is inductive. Problem 1 is a probability problem, whereas Problem 2 is a statistics problem.

Why is induction so pervasive and critical in science? Because despite the appearance of being strictly about data and experiments, science is actually almost entirely about unobservables, or, more specifically, about things and times outside the database of actual observations. For example, chemists report that iron melts at 1,535°C. Obviously, that report is based on actual observations of iron made with particular samples at particular times. Yet that is taken to be a general fact, applicable to any iron at any place at any time, though it is strictly unobservable, because scientists cannot possibly observe all of the iron in the universe over all time. Without induction, there would be no general conclusions whatsoever, and science would perish.

HISTORICAL PERSPECTIVE ON INDUCTION

This section gives a brief history of induction from Aristotle to John Stuart Mill. Later sections will continue the story with the positive proposal of Thomas Bayes, the relentless attack of David Hume, and the relatively recent methodology of Sir Ronald Fisher.

Aristotle (384–322 B.C.) offered three methods of induction (Pérez-Ramos 1988:201–211; Losee 1993:7–8). The unifying concept is that induction (ἐπαγωγή) is a method for establishing the premises in a deductive argument that reaches a law-like or general conclusion in the context of ordinary human discourse about the world. Cicero (106–43 B.C.) translated Aristotle's ἐπαγωγή as the Latin *inductio* in his treatises on rhetoric, and that became the root for the English word "induction."

The first type of ἐπαγωγή is dialectical induction, in the *Topics*. It is not particularly germane in science, but it is especially revealing of the fundamental notions in Aristotle's induction. It is a method of discourse between partners, such as a teacher and a pupil, that leads from questions to answers in order to disclose and support the argument's premises (πρότασεις). The dialectician advancing an inductive argument needs to adduce as much particular evidence (καθ' ἕκαστα) as possible, which may be certain or only probable. Aristotle's example is as follows: "If a skilled pilot is the best pilot and the skilled charioteer the best charioteer, then, in general, the skilled man is the best man in any particular sphere" (Pérez-Ramos 1988:204).

The second type of ἐπαγωγή is enumerative induction, in the *Prior Analytics*. In simple enumeration, statements about individual objects provide the basis or premises for a general conclusion. For example, from observing numerous adult humans with 32 teeth, an inductive argument could conclude that all humans have 32 teeth (or a more cautious version could reduce that claim to cover most adult humans, at least prior to any tooth loss).

The third and final type of ἐπαγωγή is intuitive induction, in the *Posterior Analytics*. It involves direct intuition of the general principles exemplified in the data. Aristotle gave the example of a scientist who notices repeatedly that the bright side of the moon is always turned toward the sun and who concludes that the moon shines because of reflected sunlight.

Aristotle had a broad conception of induction (Chakrabarti 1995:65–82). Primarily, induction is reasoning from particular instances to general conclusions. That is ampliative reasoning from observed to unobserved, from part to whole, from sample to population. Induction also concerns co-presence. If two things or properties are observed to be together in some or many instances, one may generalize that such is the case frequently or perhaps always, such as objects being gold and being dense. Likewise, induction can concern co-absences, as well as concomitant variation. Aristotle's example of concomitant variation was that if pleasure is good is a true generalization, then a greater pleasure should be a greater good.

A special type of co-presence (or co-absence) is what Aristotle called inference from signs. A sign is an evident, empirical indicator of something else that may not be evident. Aristotle taught that "it is necessary to use evident things [φανεροῖς] in witness [μαρτυρίοις] of nonevident things [ἀφανῶν]" (Asmis 1984:216). Hence, smoke is a sign of fire, and a smile is a sign of inward happiness. He warned that legitimate signs should not be confused with mere chance concurrences or coincidences. He limited signs to universal generalizations, distinguishing signs from mere probabilities.

Aristotle cautioned against hasty generalizations and noted that a single counterexample suffices to nullify a universal generalization. He carefully

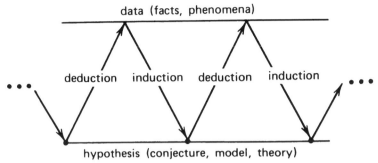

Figure 7.1. Science's iterative learning process that alternates deduction and induction. (Reprinted from Box et al., 1978:2, with kind permission from John Wiley & Sons.)

distinguished induction from deduction, analogy, and isolated examples (παράδειγμα). Curiously, Aristotle claimed that deductive proofs were more convincing to intellectual persons, whereas inductive evidence was more compelling to the unintellectual masses.

One of Aristotle's most influential contributions to the philosophy of science was his model of scientific logic or reasoning, the inductive-deductive method. Scientific inquiry alternates inductive and deductive steps (Losee 1993:6–9). Indeed, as shown in Figure 7.1, even current explanations of the scientific method feature Aristotle's iterative process as the central core (Box and Tiao 1973:5; Box et al. 1978:2). From observations, induction provides general principles, and with those principles serving as premises, deduction predicts or explains observed phenomena. Overall, there is an advance from knowledge of facts to knowledge of an explanation for the facts.

To understand Aristotle's scientific method well is to grasp its relationship to his view of truth. Aristotle's correspondence theory of truth features correspondence between a mental model and external reality. His inductive-deductive method for scientific inquiry seeks truth in successive arguments from actual data to model and from model to expected data, eventually bringing the mental model into correspondence with the data and thereby with physical reality.

Epicurus (341–271 B.C.) discussed the very fundamental role of induction in forming concepts and learning language for kinds of things in his doctrine of πρόληπσις, subsequently translated into Latin as *anticipatio* (Urbach 1987:37–38; Chakrabarti 1995:83–92): From repeated sense perceptions, a general idea or image is formed that combines the salient, common features of the objects, such as the concept of a horse derived from numerous observations of horses. Once stored in memory, this concept, or "anticipation" to use Epicurus's term, acts as an organizing principle or convention for discriminating which perceptions or objects are horses and for stating truths about horses.

Robert Grosseteste (c. 1168–1253) affirmed and refined Aristotle's inductive-deductive method, which he termed the Method of Resolution and Composition for its inductive and deductive components, respectively. But he added to Aristotle's methods of induction (Crombie 1962:61–90; Losee 1993:31–38). His purposes were to verify true theories and to falsify false theories. Causal laws were suspected when certain phenomena were frequently correlated, but natural science (*naturalis*) sought robust knowledge of real causes, not accidental correlations. "Grosseteste's contribution was to emphasize the importance of *falsification* in the search for true causes and to develop the method of verification and falsification into a systematic method of experimental procedure" (Crombie 1962:84). His approach used deduction to falsify proposed but defective inductions. As mentioned in the earlier chapter on deduction, Grosseteste's Method of Verification deduced consequences of a theory beyond its original application and then checked those predictions experimentally. His Method of Falsification eliminated bad theories by deducing implications known to be false.

Grosseteste clearly understood that his optimistic view of induction required two metaphysical presuppositions about the nature of physical reality: the uniformity of nature and the principle of parsimony (Crombie 1962: 85–90). Without those presuppositions, there is no defensible method of induction in particular or method of science in general.

In essence, at Oxford in 1230, Grosseteste's new scientific method – with its experiments, Method of Resolution and Composition, Method of Verification, Method of Falsification, emphasis on logic and parsimony, and common-sense presuppositions – was the paradigm for the design and analysis of scientific experiments. Science's goal was to provide humans with truth about the physical world, and induction was a critical component of scientific method.

Roger Bacon (c. 1214–1294) promulgated three prerogatives of experimental science, as mentioned in Chapter 2. Of those, two concerned induction. His first prerogative was that inductive conclusions should be submitted to further testing. That was much like his predecessor Grosseteste's Method of Verification. His second prerogative was that experiments could increase the amount and variety of data used by inductive inferences, thus helping scientists to discriminate between competing hypotheses.

John Duns Scotus (c. 1265–1308), at Paris, reflected Oxford's confidence about inductive logic. He admired Grosseteste's commentaries on Aristotle's *Posterior Analytics* and *Physics*, but disagreed on some points. Duns Scotus admitted that, ordinarily, induction could not reach evident and certain knowledge through complete enumeration, and yet he was quite optimistic that "probable knowledge could be reached by induction from a sample and, moreover, that the number of instances observed of particular events being correlated increased the probability of the connexion between them being a truly universal and causal one.... He realized that it was often impossible to get beyond mere empirical generalizations, but he held that a well-established empirical generalization could be held with certainty because of the principle of the uniformity of nature, which he regarded as a self-evident assumption of inductive science" (Crombie 1962:168–169).

Building on an earlier proposal by Grosseteste, Duns Scotus offered an inductive procedure called the Method of Agreement. "The procedure is to list the various circumstances that are present each time the effect occurs, and to look for some one circumstance that is present in every instance" (Losee 1993:33–34). For example, if circumstances $ABCD$, ACE, $ABEF$, and ADF all gave rise to the same effect x, then one could conclude that A could be the cause of x, although Duns Scotus cautiously refrained from the stronger claim that A must be the cause of x. The Method of Agreement could promote scientific advances by generating plausible hypotheses that merited further research to reach a more nearly definitive conclusion.

Henry of Ghent (c. 1217–1293), in contrast to Duns Scotus, believed that real scientific knowledge had to be about necessary things, not contingent things (Vier 1951:117). Had his view prevailed, science in general and induction in particular would now be held in low philosophical esteem.

William of Ockham (c. 1285–1347) further developed inductive logic along lines begun earlier by Grosseteste and Duns Scotus. He added another inductive procedure, the Method of Difference. "Ockham's method is to compare two instances – one instance in which the effect is present, and a second instance in which the effect is not present" (Losee 1993:34–35). For example, if circumstances ABC gave effect x, but circumstances AB did not, then one could conclude that C could be the cause of x. But Ockham was cautious in such claims, especially because he realized the difficulty in proving that two cases differed in only one respect. As a helpful, although partial, solution, he recommended comparing a large number of cases to reduce the possibility that an unrecognized factor could be responsible for the observed effect x.

Nicholas of Autrecourt (c. 1300–1350) had the most skeptical view of induction among medieval thinkers, prefiguring the severe challenge that would come several centuries later from David Hume. "He insisted that it cannot be established that a correlation which has been observed to hold [in the past] must continue to hold in the future" (Losee 1993:42). Indeed, if the uniformity of nature is questioned in earnest, then induction is in big trouble. Recall that Grosseteste had recognized that induction depended on the uniformity of nature.

Francis Bacon (1561–1626) so emphasized induction that his conception of scientific method is often known as Baconian induction (Urbach 1987:25–58;

Pérez-Ramos 1988:239–285; O'Hear 1989:12–34; Losee 1993:66). He criticized Aristotelian induction on three counts: haphazard data collection without systematic experimentation; hasty generalizations, often later proved
false; and simplistic enumerations, with inadequate attention to negative instances.

Bacon discussed two inductive methods. The old and defective procedure
was the "anticipation of nature," with "anticipation" reflecting its Epicurean
usage, which led to hasty and frivolous inductions. The new and correct procedure was the "interpretation of nature." Inductions or theories that were acceptable interpretations "must encompass more particulars than those which
they were originally designed to explain and, secondly, some of these new
particulars should be verified," that is, "theories must be larger and wider
than the facts from which they are drawn" (Urbach 1987:28). Good inductive theories would have predictive success.

Bacon's basic view of induction, as systematic tabulation of presuppositionless observations, was untenable, as was explained in Chapter 4, on science's presuppositions. So René Descartes (1596–1650) attempted to invert
Bacon's scientific method: "But whereas Bacon sought to discover general
laws by progressive inductive ascent from less general relations, Descartes
sought to begin at the apex and work as far downwards as possible by a deductive procedure" (Losee 1993:74). Of course, that inverted strategy shifted the
burden to establishing science's first principles, which had its own challenges.
At any rate, a wholesome and enduring aspect of Baconian induction was its
balanced search for both positive and negative evidence so as to minimize
errors.

Isaac Newton (1642–1727) developed an influential view of scientific
method that was directed against Descartes's attempt to derive physical laws
from metaphysical principles. Rather, Newton insisted on careful observation and induction, saying that "although the arguing from Experiments and
Observations by Induction be no Demonstration of general Conclusions, yet
it is the best way of arguing which the Nature of Things admits of" (Losee
1993:85). Newton affirmed Aristotle's inductive-deductive method, which
Newton termed the "Method of Analysis and Synthesis" for its deductive
and inductive components, respectively. "By insisting that scientific procedure should include both an inductive stage and a deductive stage, Newton
affirmed a position that had been defended by Grosseteste and Roger Bacon
in the thirteenth century, as well as by Galileo and Francis Bacon at the
beginning of the seventeenth century" (Losee 1993:85).

In Newton's scientific method, induction was extremely prominent, being
no less than one of his four rules of scientific reasoning: "In experimental philosophy we are to look upon propositions collected by general induction from
phænomena as accurately or very nearly true, notwithstanding any contrary

hypotheses that may be imagined, till such time as other phænomena occur, by which they may either be made more accurate, or liable to exceptions" (Williams and Steffens 1978:286). Nevertheless, he did not exaggerate induction's power, but rather held that scientific theories were tentative and potentially revisable in light of future research and experimentation.

John Stuart Mill (1806–1873) wrote a monumental *System of Logic* (Nagel 1950; Broad 1968:139–143; Losee 1993:154–165) that covered deductive and inductive logic, with a subtitle proclaiming a connected view of the principles of evidence and the methods of scientific investigation. Like Francis Bacon, Mill recommended a stepwise inductive ascent from detailed observations to general theories. He defined induction as "the operation of discovering and proving general propositions" and held that "every deductive inference is at bottom an inductive one" (Nagel 1950:xxxviii). That latter sentiment is reminiscent of Aristotle's use of induction to supply premises for deductive arguments.

Mill had four (or five) inductive methods for discovering scientific theories or laws that were essentially the same as those of Grosseteste, Duns Scotus, and Ockham. Despite his enthusiasm for induction, Mill recognized that his methods could not work well in cases of multiple causes working together to produce a given effect. Mill wanted not merely to discover scientific laws, but also to justify and prove them, while carefully distinguishing real causal connections from merely accidental sequences. But his justification of induction has not satisfied subsequent philosophers of science.

More recent developments in inductive logic will be discussed later in this chapter. During the past century, induction has picked up a common synonym: statistics. Statistics *is* inductive logic. The historically recent advent of statistical methods, digital computers, and enormous databases has stimulated and facilitated astonishing advances in induction.

PRESUPPOSITIONS OF INDUCTION

An earlier chapter discussed science's presuppositions, and the following chapter will discuss parsimony, so the presuppositions of induction can be quickly dispatched here. Medieval scholars mastered induction's presuppositions, and induction's death struggle with skepticism clearly reveals those presuppositions. Nevertheless, some people still think about induction or statistics essentially as did Francis Bacon, as the collection and analysis of presuppositionless facts. Consequently, at least a brief account is needed.

In a sense, to work in science, induction needs all of the presuppositions listed in Chapter 4. For example, our inductions presuppose our human nature. Scientists individually and the scientific community collectively are human, so perceptions and inductions are possible and are shared – unlike the imaginary situation were one scientist a human and another a rock or a river. But to be more germane, induction has two principal presuppositions (Caldin 1949:51–70; Nagel 1950:181–184; Keynes 1962:251–264; Broad 1968; O'Hear 1989:25–34; Kornblith 1993), one obvious, and one subtle.

The obvious presupposition of induction is the uniformity or regularity of nature, also expressed by saying that the universe is governed by general laws. "The method of induction, the inference from yesterday to tomorrow, from here to there, is of course only valid if regularity exists" (Hans Hahn et al., in Neurath and Cohen 1973:313).

The subtle presupposition of induction is parsimony, construed in a distinctively ontological sense. Keynes (1962:251) expresses it as the law of the limited variety of nature (Caldin 1949:57–58), and Kornblith (1993:13–57) as natural kinds. For example, consider the fact that iron melts at 1,535°C. How can such a statement be true? Well, were nature's variety so unlimited that each atom was unique, not merely in its identity but also in its properties, then there would be no such thing as iron or oxygen (or humans!). But in fact, although the immense number of atoms in the universe exhausts human imagination, the number of chemical elements is limited to about 100. Among atoms, there is limited variety. Hence, there exists such a thing as iron, and properties determined for iron samples also apply to other specimens of iron. Induction applies within classes of things that resemble one another in relevant respects, and such classes exist because of the limited variety or parsimony of nature. In statistics, replicates are classes of similar things, and legitimate statistical inferences are limited to those things for which the replicates are representative.

BAYESIAN EXAMPLE

Bayes's theorem was derived from probability axioms in Chapter 6, which also mentioned that this theorem is the basis for the Bayesian paradigm in statistics. Statistical procedures, Bayesian and otherwise, have been developed for a variety of purposes, such as designing experiments, estimating the values of quantities of interest, and testing hypotheses (Berger 1985; James O. Berger, in Raftery, Tanner, and Wells 2002:275–290; Bradley P. Carlin and Thomas A. Louis, in Raftery et al. 2002:312–318). Here the Bayesian paradigm is introduced by a simple example involving tests of hypotheses.

Envision joining an elementary statistics class during a lesson in Bayesian inference. The professor shows the class an ordinary fair coin, an opaque urn, and some marbles identical except for color, being either blue or white. Two volunteers, students Juan and Beth, are appointed as experimentalists. Juan receives his instructions and executes the following: He flips the coin without showing it to anyone else. If the coin toss gives heads, he is to place in the urn 1 white marble and 3 blue marbles. But if the coin toss gives tails, he is

---

Marble Experiment: Problem

Setup

    Flip a fair coin.

    If heads, place in an urn 1 white and 3 blue marbles.

    If tails, place in an urn 3 white and 1 blue marble.

Hypotheses

    H_B: 1 white and 3 blue marbles (WBBB).

    H_W: 3 white and 1 blue marble   (WWWB).

Purpose

    To determine which hypothesis, H_B or H_W, is probably true.

Experiment

    Mix the marbles, draw a marble, observe its color, and replace it, repeating this procedure as necessary.

Stopping Rule

    Stop when a hypothesis reaches a posterior probability of 0.999.

Figure 7.2. A marble experiment's setup, hypotheses, and purpose.

to place in the urn 3 white marbles and 1 blue marble. Juan knows the urn's contents, but the remainder of the class, including Beth and the professor, know only that exactly one of two hypotheses is true: either H_B, that the urn contains 1 white marble and 3 blue marbles, or else H_W, that it contains 3 white marbles and 1 blue marble.

The class is to determine which hypothesis, H_B or H_W, is probably true, by means of the following experiment: Beth is to mix the marbles, draw 1 marble, show its color to the class, and then replace it in the urn. That procedure is to be repeated as necessary. The stopping rule is to stop when either hypothesis reaches or exceeds a posterior probability of 0.999. In other words, there is to be at most only 1 chance in 1,000 that the conclusion will be false. This marble problem is summarized in Figure 7.2.

How shall we analyze the data emerging from this experiment? Before developing a careful scientific approach, it is instructive first to formulate and compare a common-sense perspective. Three facts are transparent to common sense. First, the basic story is clear: If after a large number of draws the yield has been mostly blue marbles, the urn probably contains the 1 white and 3 blue marbles; and the reverse outcome supports the opposite conclusion. Second, the exact story is unclear. Common sense knows that it does not know exactly what suffices as a "large" number of draws, nor just how "probably" true the conclusion is. Are 10 draws enough, or 25, or 100, or what? And does the conclusion have just 1 chance in 100 of error, or 1 in 1,000, or what? Third, common sense can judge that the exact story could be useful. Lacking a quantitative basis, experiments based only on common-sense reasoning often play it safe by collecting lots of data, say from 100 draws. But is that much data really necessary? In this context of drawing marbles, the observations are quick and inexpensive, so the penalty for excessive experimentation is small. But in a medical trial that costs $18,000 for each observation, not to mention considerable discomfort for each patient, unnecessary data are wasteful and undesirable. So it would be nice to be able to calculate the exact strength of a conclusion and to avoid collecting either too much or too little data. Also, it could be helpful to be able to forecast in advance the approximate amount of data that would be needed to produce results with some preselected confidence of truth. Then scientists and administrators could weigh the likely costs, benefits, and risks of a proposed experiment in advance, before actually paying for and conducting it.

Another reward from scientific analysis is greater objectivity. For example, assume that 20 draws gave an outcome of 6 white and 14 blue marbles, favoring the conclusion that the urn probably contains 1 white and 3 blue marbles. Ten common-sense thinkers could offer rather widely divergent assessments of that conclusion's strength. Also, those individuals could easily disagree about whether or not more data would be needed to make sure that the conclusion had no more than 1 chance in 1,000 of being false. But ten scientific thinkers would offer one exactly correct assessment, namely, 99.98476% confidence in that conclusion's truth, and they would all know and agree that no more data would be required to achieve 99.9% confidence.

For this marble problem, the ratio form of Bayes's rule is easiest. Here it is recalled, with the earlier generic hypothesis labels "1" and "2" replaced by more informative labels, namely, "B" meaning mostly blue marbles (1 white and 3 blue) and "W" meaning mostly white marbles (3 white and 1 blue).

$$\frac{P(H_B \mid D)}{P(H_W \mid D)} = \frac{P(D \mid H_B)}{P(D \mid H_W)} \times \frac{P(H_B)}{P(H_W)}$$

Table 7.1 gives the data from an actual experiment with blue and white marbles and analyzes the data using this equation. To begin, we can deduce the prior odds of H_B:H_W from all of our initial knowledge about the marble

Table 7.1. Bayesian analysis for a marble
experiment, assuming prior odds for H_B:H_W of 1:1

Draw	Result	Posterior H_B:H_W	Posterior $P(H_B|D)$
	(Prior)	1:1	0.500000
1	White	1:3	0.250000
2	Blue	1:1	0.500000
3	White	1:3	0.250000
4	Blue	1:1	0.500000
5	Blue	3:1	0.750000
6	Blue	9:1	0.900000
7	Blue	27:1	0.964286
8	Blue	81:1	0.987805
9	Blue	243:1	0.995902
10	White	81:1	0.987805
11	Blue	243:1	0.995902
12	White	81:1	0.987805
13	Blue	243:1	0.995902
14	Blue	729:1	0.998630
15	Blue	2187:1	0.999543

Note: The final conclusion, at 15 draws, is that the probability that H_B is true is 0.999543, and that H_W is true is
0.000457. These values mean that H_B can be accepted with
99.9543% confidence of truth, or that the conclusion that
H_B is true has a chance of error of only 1 in about 2,200.

experiment prior to executing this experiment. From the setup with a flip
of a fair coin, these odds are 1:1, as shown in the first line of this table, and
accordingly the prior probabilities for H_B and H_W are $P(H_B) = P(H_W) = 0.5$.

The likelihood odds $P(D \mid H_B)/P(D \mid H_W)$ arising from each possible empirical outcome of drawing a blue or a white marble are as follows. Recalling
that H_B has 3 of 4 marbles blue, but H_W has only 1 of 4 marbles blue, probability theory shows that a blue draw is three times as probable given H_B as it
is given H_W. Because $P(\text{blue} \mid H_B) = \frac{3}{4} = 0.75$ and $P(\text{blue} \mid H_W) = \frac{1}{4} = 0.25$,
a blue draw contributes likelihood odds of 0.75:0.25 or 3:1 for H_B:H_W, favor-
ing H_B. By similar reasoning, a white draw contributes likelihood odds of 1:3
against H_B. Furthermore, because each draw is an independent event after
remixing the marbles, probability theory shows that individual trials combine
multiplicatively in an overall experiment. (This rule is just a simplification
of Axiom 4 in Chapter 6 for this special case of independent events.) For
example, two blue draws will generate likelihood odds in favor of H_B of 3:1
times 3:1, which equals 9:1. Thus, in a sequential experiment, each blue draw
will increase the posterior odds for H_B:H_W by 3:1, whereas each white draw
will decrease it by 1:3.

Applying this analysis to the data in Table 7.1, note that the first draw is a white marble, contributing likelihood odds of 1:3 against H_B. Multiplying those likelihood odds of 1:3 by the previous odds (the prior) of 1:1 gives posterior odds of 1:3, decreasing the posterior probability for H_B to $P(H_B|D) = 0.25$, where the data at this point reflect one draw. In this sequential experiment, the posterior results after the first draw become the prior results at the start of the second draw. That is, after having drawn a marble once, our total knowledge prior to a second draw becomes the original prior knowledge about the setup with a coin flip plus the previous experimental knowledge about a single white draw. The second draw happens to be blue, contributing likelihood odds of 3:1 favoring H_B, bringing the posterior probability back to the initial value of 0.5.

Moving on to the sixth draw, for example, the previous odds are 3:1, and the current blue draw contributes likelihood odds of 3:1, resulting in posterior odds of 9:1 favoring H_B and hence a posterior probability of $P(H_B) = 0.9$. Finally, after 15 draws, the posterior probability happens to exceed the stopping rule's preselected value of 0.999, so the experiment stops, and hypothesis H_B is accepted with more than 99.9% confidence. Incidentally, in this particular instance of a real marble experiment, the conclusion was indeed correct, because the urn actually did contain 3 blue marbles and 1 white marble, as could have been demonstrated easily by some different experiment, such as drawing out all 4 marbles at once.

Table 7.1 illustrates a very general feature of data analysis: that results become more nearly conclusive as an experiment becomes larger. During the first 6 draws, H_B has two wins, H_W has two wins, and there are two ties, so the results are quite inconclusive, and the better-supported hypothesis never reaches a probability beyond 0.75. Indeed, at only 1 draw and again at 3 draws, this experiment gives mild support to the false hypothesis! But draws 7 to 15 all give the win to H_B, which is true, finally with a probability greater than 0.999.

A few comments may now be offered regarding the expected length of the marble experiment, were it to be repeated a number of times. This one particular experiment required 15 draws to reach a conclusive result (at the predetermined level of 99.9% confidence). But how long would this experiment run on average?

A relatively simple analysis, using two insights, provides a good approximation. First, from the general principle that a blue draw multiplies the prior odds H_B:H_W by the likelihood odds 3:1, whereas a white draw multiplies the prior odds by 1:3, as exemplified in Table 7.1, clearly a blue-white pair leaves the posterior odds unchanged. Consequently, the posterior probabilities of H_B and H_W depend only on the margin or difference between the numbers of blue and white draws. Let M be the margin of blue draws over white, so

$M = 3$ means 3 more blue draws than white draws. Then the posterior odds $H_B:H_W$ equal $3^M:1$. These odds exceed 999:1 or 99.9% confidence favoring H_B when $M = 7$, or exceed 1:999 favoring H_W when $M = -7$. Consequently, an experiment can be scored very simply by starting at $M = 0$ and adding 1 for a blue draw or subtracting 1 for white, stopping when $M = \pm 7$. For example, the experiment in Table 7.1 progresses through the scores $-1, 0, -1, 0, 1, 2, 3, 4, 5, 4, 5, 4, 5, 6$, and finally 7 on draw 15, which declares H_B the winner and stops the experiment.

Second, 4 draws will give, on average, 3 draws that support the true hypothesis and 1 draw that supports the false hypothesis (such as 1 white and 3 blue when H_B is true). So, on average, for 4 draws, 2 draws will cancel and 2 will support the true hypothesis. Because half the data cancel and half count, the length L required to achieve a score of $\pm M$ averages about $2M$. The current preselected confidence of 99.9% requires $M = \pm 7$, so the implied average length of an experiment is about $2 \times 7 = 14$ draws. Hence, the particular experiment in Table 7.1, having 15 draws, is about average.

A formal analysis, which is too complicated to merit explanation here, can provide exact answers to this question about the length of experiments, but information on sequential probability ratio tests for Bernoulli data is available (Siegmund 1985). For M equal to 2, 3, 4, or 5, the average length L is 3.2, 5.6, 7.8, or 9.9 draws, but thereafter the approximation that $L \approx 2M$ is quite accurate.

As this example draws to a close, let us turn from technical details to the big picture. The remarkable features of this typical example are that it follows from the probability axioms and that it matches with physical reality. For example, Bayesian theory says that a margin of 7 draws will reach a conclusion with posterior odds of 2,187:1, meaning a $2,187/(2,187 + 1)$ or 99.9543% probability of truth and a 0.0457% probability of error.

Is this theory just an irrelevant deduction from unrealistic probability axioms, or is it genuinely relevant to real marble experiments? This question might sound philosophical or unanswerable, but clearly you could put it to the test for yourself with real marbles to get a concrete answer. This particular case of a margin of 7 would require, however, much tedious work to get accurate results.

Accordingly, anyone who wanted to put this Bayesian theory to the test with easier physical experiments could save work by instead checking results for a margin of 2 draws, because such experiments' average length is only 3.2 draws, and the larger error rate of 10% is much easier to detect and quantify. The outcome of all such checks will be to confirm the realism and accuracy of this application of probability axioms and theory. It works. It works great. It works invariably. Just as an arithmetic theory based on a few axioms fits countless physical experiences, so does probability theory.

BAYESIAN INFERENCE

The marble example was specifically chosen to represent a definite advance beyond mere common-sense reckoning, and yet to be simple enough for easy comprehension and for actual implementation with readily available materials. But it was also carefully chosen to sidestep various statistical problems and philosophical controversies.

This section considers six variations on the theme of the marble experiment that require adjustments or raise difficulties. Its purpose is to stimulate insight into why science often works just fine even when the data, model, and scientists are all imperfect. Bayesian inference is robust.

(1) Different Preselected Confidence. The marble experiment's preselected confidence of a true conclusion was an error rate of no more than 1 chance of error in 1,000 experiments, which is first attained for a margin of 7 more blue than white draws (or the reverse), giving posterior odds of 1 chance of error in 3^7 or 2,187 experiments, and hence a confidence of truth of 99.9543%, which is reached after about 14 draws on average. What would happen to the average length of experiments if we were willing to tolerate only 1 chance of error in 1,000,000? How much more work would be needed to achieve that greater confidence?

The required margin M that makes 3^M large enough is 13, resulting in posterior odds of 1,594,323:1 in favor of the true hypothesis, or a 99.999937% confidence of truth. Using the earlier approximation that $L \approx 2M$, that would require about 26 draws on average, or about twice as much work as before. For another example, if we tolerated 1 chance of error in 100, a smaller margin of 5 would suffice for 99.5902% confidence and would require only about 9.9 draws on average (using the more exact calculation mentioned earlier).

The general pattern makes sense: More data will reduce the error rate; more data will give greater confidence. But this scientific analysis makes exact an important relationship that common sense apprehends only vaguely. It shows that the exact relationship between experimental work and evidential weight is exponential in this particular case. This relationship contrasts strongly with other possibilities, such as a linear relationship between work and weight in which twice as much work gives twice as much weight to the conclusion. For example, as the average length of experiments is increased from 14 to 26 draws, the work increases by a factor of about 2, but the weight of the evidence for diagnosing the true model increases (on average) not by a factor of 2, but rather by about 1,000.

Because of the exponential relationship between work and weight that obtains in this case, a reasonable effort can produce virtually certain results. For example, a mere one chance of error in a trillion (10^{12}) experiments would constitute virtual certainty, and yet its average experimental length would be only a very manageable 52 draws, which would require only a few minutes of work with inexpensive materials. Given enough data and a favorable relationship between work and weight, inductive logic can deliver virtual certainty. To obtain absolute certainty, however, one would need to switch to some different experiment, such as drawing out all 4 marbles at once.

(2) Different Priors. Knowledge of the setup in the original marble experiment provides clear, exact, and correct prior odds. A fair coin is flipped, and heads gives WBBB, whereas tails gives WWWB, so the prior odds of $H_B:H_W$ before collecting any data are precisely 1:1. However, four situations could generate a different or problematic prior: (1) The prior could simply be different from 1:1, such as 9:1, because of a different setup. (2) We could have ignorance rather than knowledge of the setup, and hence not know the correct prior. (3) The prior could be controversial, with different investigators believing or preferring different priors. (4) The prior could be extremely strong, nearly or completely eliminating some hypothesis from consideration.

First, the prior could be different because of a different experimental setup. For example, instead of determining the composition of the urn's contents (WBBB or WWWB) by flipping a fair coin, consider the following procedure. The professor takes 10 identical small slips of paper, writes WBBB on 9 slips and WWWB on 1, places them in a hat, and mixes them up. Then the student, Juan, is asked to draw a slip, with WBBB directing Juan to put 1 white and 3 blue marbles in the urn, or else WWWB meaning 3 white marbles and 1 blue marble. Now the prior odds of $H_B:H_W$ are 9:1, rather than the original 1:1, because of this change in the setup procedure. Assuming that the stopping criterion remains the same, that the probability of a true conclusion be at least 99.9%, how does this change in the prior affect the experiment?

Originally a margin of 7 sufficed to declare a conclusion and stop. But now H_B starts off with a 9:1 advantage, which is analogous to the original experiment after encountering a margin of 2 blue draws (as is evident in Table 7.1). Consequently, now a margin of only $7 - 2 = 5$ more blue than white draws suffices for H_B to win, because prior odds of 9:1 multiplied by likelihood odds of $3^5:1$ or 243:1 gives the required posterior odds of 2,187:1. On the other hand, because H_W starts off with a 9:1 disadvantage, similar logic requires $7 + 2 = 9$ more white than blue draws for H_W to win. So this prior information about slips of paper in the setup can be incorporated into the experimental procedure by adjusting the stopping rule: accept H_B at 5 more blue than white draws, or else accept H_W at 9 more white than blue draws.

This new stopping rule shortens the experiment's length. The probability that H_B is true is 0.9, and then the average experiment's length is $L \approx 2 \times 5 = 10$ draws, whereas the probability of H_W is 0.1, and its length is $L \approx 2 \times 9 = 18$, so the overall average length is about $L \approx (0.9 \times 10) + (0.1 \times 18) = 10.8$ draws. The original average length was about 14.

Second, the prior could be problematic because we have ignorance rather than knowledge of the prior. The original setup enjoyed a great simplification and objectivity from selecting WBBB or WWWB on the basis of a fair coin toss. That caused the prior probabilities to have definite values, namely, $P(H_B) = P(H_W) = 0.5$, agreed upon by all sensible persons.

What would happen if instead the professor said only that either 1 white and 3 blue marbles or else 3 white marbles and 1 blue marble were placed in the urn, but revealed nothing about the process or reasons determining that choice? Depending on the process in the unknown setup, $P(H_B)$ could have any value from 0 to 1. It is unfortunate not to know the real prior precisely because the prior does make a difference, as we just saw, in regard to the stopping rule, experiment length, and conclusions.

One plausible reaction to our ignorance about the professor's choice is to consider H_B and H_W equally likely. Indeed, if we have no reason to favor either hypothesis, then this choice of priors seems quite sensible. This approach leads to $P(H_B) = P(H_W) = 0.5$, which happens to be the same as before in the original experiment with a coin toss in its setup. Obviously, this same choice of priors will give the same numerical analysis as already discussed for the original setup. Nevertheless, definite knowledge of a probability of 0.5 is not the same as an ignorant guess of 0.5. Consequently, despite its numerical similarities, the new analysis requires a different interpretation and justification.

What makes $P(H_B) = P(H_W) = 0.5$ a sensible and rational choice when we are modeling ignorance rather than knowledge of the setup? This choice is unique in optimizing some desirable properties, of which three may be mentioned here. First, this choice of 0.5 minimizes the maximum possible error. The actual prior is somewhere between 0 and 1, so a value of 0.5 is incorrect by at most 0.5, whereas any different prior could have a larger error, such as 0.3, which could be off by as much as 0.7. Second, priors of 0.5 make the analysis most responsive to the data. Recall that the posterior conclusion is affected by the likelihood and the prior. The more the prior is indefinite or based on ignorance, the more the analysis should weight the likelihood in preference to the prior. For the current experiment, the choice $P(H_B) = P(H_W) = 0.5$ constitutes the choice that satisfies this criterion. Third, the priors $P(H_B) = P(H_W) = 0.5$ lead to the most conservative estimates of the confidence of a true conclusion. Recall that these priors and a required confidence of 99.9% lead to the original stopping rule, with a margin of 7 and an average experiment length of about 14 draws, regardless whether H_B or H_W is true, and likewise the original error rate of 0.046% still obtains. If some unknown process should cause $P(H_B)$ to be, say, 0.9 or 0.103, rather than the 0.5 selected out of ignorance, then knowledge of the true prior could be exploited to refine the stopping rule and thereby get more confidence with less work. So the penalty for not knowing the real prior is experimental inefficiency. However, we can still achieve an equally confident conclusion without the real prior, but at a cost of somewhat more data.

Third, the prior may be controversial. When we are not given the prior probabilities of the various hypotheses, we are left to our own devices to come up with some values, and it is possible that different individuals may adopt different priors, making the priors controversial. For example, various students might give H_B a prior probability $P(H_B)$ ranging from, say, 0.2 to 0.9. What happens to the analysis now that the priors are controversial?

Consider the unfavorable case of a small prior being given to what is actually the true hypothesis, such as $P(H_W) = 0.1$ when H_W is true. Prior odds of 9:1 will require an additional likelihood odds factor of 1:9 to move the posterior odds back to the 1:1 starting point of the original setup, which entails on average about 4 draws. Hence, this prior will increase the experiment's average length from the original 14 draws to a new total of $14 + 4 = 18$ draws. So a prior that is unfavorable to the truth results in more work to get at the truth.

The general pattern is that controversial, different priors require an ex-periment to be somewhat longer in order for its conclusion to still satisfy everyone. The problem is that given different priors, often someone will as-sign an unusually small prior to what happens to be the true hypothesis, and overcoming that person's doubts will necessitate some extra work. However, because of the exponentially increasing weight of the current evidence, the data can rapidly swamp the negative effect of even an exceedingly small prior placed on the true hypothesis. So the problem of different individuals' priors can be resolved by considering the worst case (which arises from the smallest prior assigned by any person to any hypothesis, for at the outset any hypoth-esis might turn out to be true) and adjusting the planned experiment's length to satisfy even that worst possible case. The result will then satisfy everyone.

Fourth and finally, the prior may disfavor some hypothesis extremely strongly. The most extreme case is represented by two hypotheses with pri-ors of 0 and 1. Such priors make experimentation and evidence irrelevant, because all the data in the world cannot move or overcome a prior of 0, for 0 multiplied by any value is still 0. Having your mind already made up is fine if your conclusion is true, but it is problematic should your conclusion be false. Strongly held errors are hard to fix.

Sometimes a prior is very strong, yet not at an extreme limit of 0 or 1. For example, assume that some different experimental setup gives $H_B:H_W$ strong prior odds of 10,000:1 and that our objective remains to accept a conclusion with at least 99.9% confidence of truth. Then no experiment is needed! Just declare H_B true with 10 times the required confidence. Indeed, the original experiment always stops at a margin of 7 more blue than white draws, or the reverse, so it always generates likelihood odds of either 2,187:1 or 1:2,187, which multiplied by the prior odds of 10,000:1 favors H_B in either case, so this small amount of data cannot possibly overcome this prior's strong influence and change the conclusion to favor H_W. Only if greater confidence were required and longer experiments were allowed would there be any call for collecting data.

It makes sense for strong priors to discourage small experiments. If you already know the answer with sufficient confidence, why pay for a weak experiment that cannot change your mind anyway? Surely you have other questions that better merit your attention.

On balance, however, although this small original experiment cannot possibly make H_W win, it might at least elevate H_W to the level of serious reconsideration. Assume that a small experiment gives 7 more white than blue draws. Then prior odds of 10,000:1 against H_W are multiplied by likelihood odds of 1:2,187 for H_W, resulting in posterior odds of only about 5:1 against H_W. H_B is still favored, but by small rather than large odds. Consequently, the alternative hypothesis H_W has gained enough possibility of truth to merit consideration.

The lesson here is that current odds for or against various hypotheses influence not only scientists' conclusions about what is true or false but also their decisions about which potential research questions are worthwhile or else pointless. Less evidence is needed to identify a plausible research topic than to pronounce a definite conclusion. For example, weak odds of 5:1 for H_B:H_W do not make either "H_B is true" or "H_W is true" a good answer, but rather make "Is H_B or H_W true?" a good question.

(3) Messy Data. In the original experimental procedure, the student, Beth, faithfully showed the class the marble resulting from each draw, ensuring the collection of accurate, quality data. However, now suppose instead that we have a fickle experimentalist. Suppose that Beth works alone in the laboratory, rather than having the class as witnesses, and that she observes the drawn marble's color accurately half of the time, but the other half is too lazy to bother looking at the marble and just says "blue" or "white" at random with even odds. Now we have messy data, with half good and half junk. What problems result? Is it still possible to use these messy data to decide between H_B and H_W with confidence?

Note that H_B with 1 white and 3 blue marbles gives blue draws a probability of 0.75. But these draws are now intermixed with an equal number of draws with a probability of reporting blue of 0.5. So the modified expectation of blue draws with messy data becomes the average of these two probabilities, namely, 0.625. Similarly, the modified probability of a white report becomes 0.375. Also, by the same logic, the alternative hypothesis H_W reverses these two probabilities, having 0.375 for blue reports and 0.625 for white reports.

Consequently, messy data move the original hypotheses closer together. H_B and H_W originally gave blue draws probabilities of 0.75 and 0.25, but now 0.625 and 0.375, which is only half the original separation. As explained in the next point, to maintain the original confidence of a true conclusion at 99.9%, this change increases the average experiment's length from about 14 to 56 draws. Incidentally, in addition to increasing the average length of such experiments, this change also increases the variability around the average, so occasional experiments may be annoyingly longer than the average of 56 draws, such as more than 100 draws for 7.5% of the experiments.

So the bad news with these messy data is that we now require about four times as much data in order to still reach a conclusion with 99.9% confidence. However, the good news is that the problem is still soluble. Fortunately, sometimes quantity can compensate for quality (Tarantola 1987:108–109; Giere 1984:212–214).

(4) Different Hypotheses. Recall that the original setup had hypothesis H_B with 1 white and 3 blue marbles, and H_W with 3 white marbles and 1 blue marble, so the probabilities of a blue draw were 0.750 for H_B and 0.250 for H_W. What happens if we replace those hypotheses with different ones? For example, consider the new hypotheses H_1, with 3 white and 5 blue marbles, and H_2, with 5 white and 3 blue marbles. Accordingly, the probabilities of a blue draw are 0.625 for H_1 and 0.375 for H_2. Now the separation between the hypotheses is only half the original separation. Accordingly, intuition or common sense may (and should) suggest that the closer the hypotheses become, the harder it will be to distinguish between them, so experiments must generate more data.

Briefly, a blue draw will contribute likelihood odds of only 5:3 or 1.66667:1 in favor of H_1, so a margin of 14 blue draws will be required to achieve 99.9% confidence (and likewise a margin of 14 white draws for H_2). Furthermore, on average, every 8 draws will contain 5 supporting the true hypothesis and 3 supporting the false hypothesis, of which 6 draws (3 for and 3 against the true hypothesis) will cancel out, and 2 good draws will remain to improve the posterior odds in favor of the true hypothesis, so on average every 4 draws will add 1 to the margin. Hence, the average length of the experiments is about $14 \times 4 = 56$ draws.

As an opposite example with more widely separated hypotheses, consider the new hypotheses that there are either 1 white and 2,000 blue marbles, or else the reverse. The probabilities of a blue draw are now separated by about 0.999, which is almost twice the original separation. Logically, this wider separation poses an easier problem requiring fewer data. Indeed, a large enough posterior odds to satisfy the 99.9% confidence stopping rule arises from merely one draw.

The main lesson here concerning different hypotheses is eminently sensible: Rather similar hypotheses require relatively more work, whereas rather different hypotheses require relatively less work. Fortunately for us, in general, the more similar two hypotheses become, the less we care about which is true. For example, if two brands of light bulbs that cost the same last for 500 and 1,000 hours, we care about that, but fortunately relatively few data will suffice to reveal that difference. But if they last for 999 and 1,000 hours, we do not care much at all, and virtually impossible quantities of experimental data would be needed to detect that tiny difference.

(5) Different Statistical Frameworks. Different scientists adopt different statistical frameworks, particularly frequentist or Bayesian statistics, and persistent debates between those schools seem to resist resolution. Such debates have rather little significance for this marble experiment, but can be important in other cases. How do statistical debates affect science? Can scientists get the same answers even if they follow different statistical schools?

The short answer is that small experiments generating few data can leave scientists from different statistical schools with different conclusions about which hypothesis is most likely to be true. Rather frequently, scientists will have only quite limited data, so the choice of a first-rate, efficient statistical procedure is important. However, as more data become available, the influence of statistical differences will diminish, and eventually everyone will come to the same conclusion, even though they differ in terms of the particular calculations used and the exact confidence attributed to the unanimous conclusion.

(6) Paradigm Shifts. A scientific paradigm is a set of generally accepted beliefs about some physical object or process, including opinions about open questions inviting further research and examples of typical experiments and methods that have promoted further progress. Ordinary science works within a successful paradigm and makes progress by asking sensible questions, collecting relevant data, and drawing reliable conclusions. But sometimes ordinary science breaks down, when the reigning paradigm proves fundamentally flawed and unrealistic. Old questions and old methods get into trouble. Then substantial further progress requires some fundamental correction and repair of the paradigm – a paradigm shift (Kuhn 1970).

The most penetrating way to understand or depict a paradigm shift is to say that at a troubled moment in science's development, scientists are asking the wrong questions. The simplest scenario that signals the need for a paradigm shift is that, unbeknownst to the scientists, they are asking a question with a hypothesis set that does not include the truth.

Such a situation, requiring a paradigm shift, could emerge in the context of the marble experiment. Recall that the original marble experiment had two competing hypotheses having equal prior probabilities, H_B with 1 white and 3 blue marbles, and H_W with 3 white marbles and 1 blue marble, and 99.9% confidence of a true conclusion required a margin of 7 more blue than white draws for H_B or 7 more white than blue draws for H_W. What if the experimental outcome were 50 white and 57 blue draws?

Well, on the face of it, these data simply declare H_B true with 99.9% confidence, because there is a margin of 7 blue draws. Superficially, 50 white and 57 blue draws have the same evidential import as, say, 5 white and 12 blue draws. However, because the average experiment's length is only 14 draws, even common sense may suggest that such a long experiment, though possible, is peculiar. Indeed, calculations show that a length of 107 or more draws has a chance of happening less than one in ten million.

So, we should worry. Perhaps something went wrong in the experiment's setup. We might doubt that either H_B or H_W is true, suspecting rather that the truth is something else, such as H_E, with 2 white and 2 blue marbles. To sing a familiar refrain, the needed remedy is more data.

In particular, we might decide to run this little marble experiment 20 or 30 times, instead of just once. If the suspicion is correct that H_E is really true, rather than the originally expected H_B or H_W, then more data will supply clear evidence on at least three counts: (1) Most obviously, the frequency of blue (or white) draws for the pooled data will be nearly 0.50, rather than the expected 0.75 or 0.25. Given the exponential increase in the weight of the evidence, such a large quantity of data will render both H_B and H_W unimaginably improbable. (2) The experiments will not repeatedly confirm the same hypothesis, but rather will declare H_B true about half of the time, and H_W the other half. That is quite disturbing, because we expect to get a clear verdict from repeated experiments, with true conclusions dominating false ones in the proportions 2,187:1. (3) If H_E is true, then the average experiment's length will be about 49 draws, rather than 14. But an average near 49 draws for 20 or 30 experiments is unimaginably improbable given the original setup.

Incidentally, were "Is H_B or H_W true?" the question when H_E is true, a single run of the experiment might happen to be fairly short and thus raise no alarm. For example, although the average length is 49 draws, a sizable 27% of experiments would stop at 21 or fewer draws, which is not alarmingly larger than the originally expected average of 14 draws. Hence, there is insurance value in running this marble experiment several times even if the first run seems ordinary and fine. An unsuspected problem may escape detection after just one run, but probably not after three or four runs, and almost certainly not after 20 or 30 runs. This illustrates how perfectly honest and competent science can reach wrong conclusions at an early stage when few data of limited kinds are available.

The good news is that even if scientists are collecting data to answer a misdirected question, the data are likely, at least eventually, to embarrass the faulty paradigm and thus to precipitate a needed paradigm shift. That is why science works so often: Even rather severe mistakes can be remediable. Scientific discovery is like a hike in the woods: You can go the wrong way for a while and yet still arrive at your destination at the end of the day.

To summarize the foregoing six points, numerous problems can be overcome by the simple expedient of collecting more data, assuming that that option is not too expensive or difficult. This happy outcome is especially likely when the weight of evidence increases exponentially with the amount of evidence.

BAYESIAN DECISION

The distinction between inference, as discussed in the preceding section, and decision, as explained in this section, is that inference problems pursue true beliefs, whereas decision problems pursue good actions. Clearly, inference and decision problems are interconnected, because beliefs inform decisions and influence actions. Accordingly, decision problems incorporate inference subproblems.

Many decisions are too simple or unimportant to warrant formal analysis, but some decisions are difficult and important. Formal decision analysis provides a logical framework that makes an individual's reasoning explicit, divides a complex problem into manageable components, eliminates inconsistencies in a person's reasoning, clarifies the options, facilitates clear communication with others also involved in a decision, and promotes orderly and creative problem-solving. Sometimes life requires easy and rather quick decisions, but at other times it demands difficult and carefully deliberated decisions. Accordingly, formal decision methods are supplements to, not replacements for, informal methods. On the one hand, even modest study of formal decision theory can illuminate and refine ordinary informal decisions. On the other hand, simple common-sense decision procedures provide the only possible ultimate source and rational defense for a formal theory's foundations and axioms.

Decision theory has normative and descriptive dimensions. Philosophers and statisticians study the normative behavior of ideally rational agents, whereas psychologists and social scientists describe the actual decisions of ordinary mortals. The same distinction applies to arithmetic, where mathematicians study the sum $18 + 17$ to know what 18 potatoes plus 17 potatoes really total, whereas educators study children's actual sums to discover common errors and hence needed refinements in arithmetic instruction. This section emphasizes normative theory, but concludes with some descriptive results and prescriptions for better decision-making.

The basic structure of a decision problem is as follows. Decision theory partitions the components or causes of a situation into two fundamentally different groups on the basis of whether or not we have the power to control a given component or cause. What we can control is termed the "action" or choice. Obviously, to have a choice, there must exist at least two possible actions at our disposal. What we cannot control is termed the "state," or, to use a longer phrase, the state of nature. Each state-and-action combination is termed an "outcome," and each outcome is assigned a "utility" or "consequence" that assesses the value or benefit or goodness of that outcome, allowing negative values for loss or badness, and assigning zero for indifference. These possible consequences can be written in a consequences matrix, a two-way table with columns labeled with states and rows labeled with actions. There is also information on the probabilities of the states occurring, resulting from an inference subproblem with its prior probabilities and likelihood information. If the state of nature were known or could be predicted with certainty, determining the best decision would be considerably easier; having only probabilistic information about the present or future state causes some complexity, uncertainty, and risk. Finally, the information on consequences and probabilities of states is combined in a decision criterion that assigns values to each choice and indicates the best action.

Drawing on ideas from Barnett (1982:249), Jeffrey (1983:1–6), Giere (1984: 319–333), Berger (1985:6–7), French (1986:168–175), and Rubinstein (1986: 4–6), Figure 7.3 presents a simple example of a farmer's cropping decision. There are three possible states of nature, which are outside the farmer's control: good, fair, or bad weather. There are three possible actions among which the farmer can choose: plant crop A, plant crop B, or lease the land.

Beginning at the lower left portion of Figure 7.3, we know something about the probabilities of the weather states. We possess old and new data on the weather, summarized in the priors and likelihoods. For example, the old data could be long-run frequencies based on extensive historical climatological records, indicating prior probabilities of 0.30, 0.50, and 0.20 for good, fair, and bad weather. The new data could be a recent long-range weather forecast that happens to favor good weather, giving likelihoods of 0.60, 0.30, and 0.10 for good, fair, and bad weather. Bayesian inference then combines the priors and likelihoods to derive the posterior probabilities of the weather states, as shown near the middle of the figure. Multiplying each prior by its corresponding likelihood gives values of 0.18, 0.15, and 0.02, for a total of 0.35, and division of those three values by their total yields the posterior

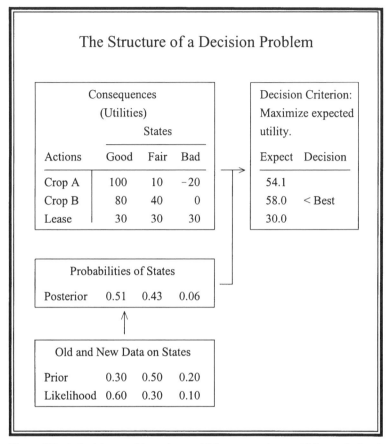

Figure 7.3. A decision problem about which crop to plant, which concludes that crop B is the best choice.

probabilities, namely, approximately 0.51, 0.43, and 0.06 for good, fair, and bad weather. So far, this is a standard inference problem. But a decision problem is more complicated, with two additional components, as explained next.

The upper left portion of Figure 7.3 shows the matrix of consequences or utilities. The outcome for any given growing season is specified by its particular state-and-action combination. The three possible states are good, fair, and bad weather, and the three possible actions are to plant crop A, plant crop B, or lease the land, for a total of $3 \times 3 = 9$ possible outcomes. The consequences matrix shows the utility or value of each possible outcome, using a positive number for a utility or gain, a negative number for a loss, or a zero for indifference. For example, in a given year the outcome might be fair weather for crop B, which has a utility of 40, where this number represents profit in dollars per acre or whatever.

Finally, the upper right portion of Figure 7.3 specifies a decision criterion, which will maximize the expected utility. The expected utility is the average or predicted utility, calculated for each possible action by multiplying the utility for each state by its corresponding probability and summing over the states. For example, the expected utility for crop A is $(100 \times 0.51) + (10 \times 0.43) + (-20 \times 0.06) \approx 54.1$. Likewise, the expected utility for crop B is 58.0, and that for leasing is 30.0. The largest of these three values is 58.0, indicating that planting crop B is the best decision to maximize the expected utility.

This example illustrates a frequent feature of decision problems: that different penalties for different errors can cause the best decision to differ from the best inference. Bayesian inference gives the greatest posterior probability of 0.51 to good weather, and good weather favors the choice of crop A, with its utility of 100. But Bayesian decision instead chooses crop B, with its largest expected utility of 58.0, primarily because fair weather is rather likely and will involve a tremendous reduction in crop A's utility.

For the sake of brevity and clarity, elementary Bayesian decision has been introduced here by a simple, common-sense example. However, this subject can be developed instead in a formal, mathematical manner. Incidentally, the earlier section on inductive inference inherited the axioms from earlier chapters on deduction and probability, but required no additional axioms. Probability and statistics have the same axioms. They differ only in reasoning directions and questions.

Remarkably, Bayesian decision theory follows from the addition of a mere one axiom to the three Kolmogorov probability axioms given earlier. So decision theory requires just one more axiom than inference theory. Recalling that "$\wedge$" symbolizes conjunction ("and"), that "$\vee$" symbolizes disjunction ("or"), and using "prob" to denote probability and "des" to denote desirability or utility, Jeffrey (1983:80–81) has stated the required axiom, and similar systems are given by Burks (1977:247–335), Berger (1985:47–50), and French (1986:39–56):

Axiom of Desirability
If $\mathrm{prob}(X \wedge Y) = 0$ and $\mathrm{prob}(X \vee Y) \neq 0$, then $\mathrm{des}\,(X \vee Y) =$
$(\mathrm{des}\,X\,\mathrm{prob}\,X + \mathrm{des}\,Y\,\mathrm{prob}\,Y) \,/\, (\mathrm{prob}\,X + \mathrm{prob}\,Y)$.

This axiom connects probabilities and desirabilities, specifically probabilities of states and utilities of outcomes. It says that the utility of an action equals the average of the utilities for its various outcomes weighted by their probabilities. This axiom was used in the foregoing cropping decision to calculate the expected utilities, such as 54.1 for crop A.

Using the elementary example of the cropping decision as a point of departure, several elaborations may be mentioned briefly. First, because of inexact knowledge or uncertain predictions, the quantities in a decision problem might be better modeled by distributions than by numbers. For example, Figure 7.3 specifies the utility of crop B in fair weather as 40, but fluctuations in fertilizer costs and crop values might change that utility somewhat, so a better model might be a normal distribution with mean 40 and standard deviation 5. However, solving that fancier decision problem would require calculus instead of arithmetic.

Second, because of different attitudes toward risk, decision criteria other than maximized expected utility may be appropriate and preferable (Rescher 1983). For example, one might prefer to maximize the worst possible utility, which in this case would favor leasing the land (as the worst possible utility from leasing would be 30, whereas crop A could be as bad as –20, and crop B as bad as 0). Sometimes the response to the expected utility is nonlinear, such as strong response to utilities below some minimum needed for survival, but mild response to differences among utilities above that amount that merely distinguish various levels of luxury. Furthermore, decisions can be evaluated in terms of not only their average but also their variability around that average, with large variability implying much uncertainty and risk. Sometimes a relatively minor compromise in the average can gain a substantial reduction in the variability, which, incidentally, is the basis for the insurance industry.

Third, decisions may have several criteria to be optimized simultaneously, probably with some complicated trade-offs and compromises. For example, a farmer might want to optimize income, as in Figure 7.3, but also want to rotate crops to avoid an epidemic buildup of pest populations and want to diversify crops to stagger the work load during busy seasons. Those other constraints might result in a decision, say, to plant 60% crop B and 40% crop A, which would reduce the expected utility slightly to $(0.6 \times 58.0) + (0.4 \times 54.1) \approx 56.4$.

Fourth and finally, a decision may be made by a group, such as a committee or jury, rather than by a single individual, such as one farmer. Those and other elaborations are beyond the scope of this study, but are treated in texts on decision theory (Berger 1985; Joyce 1999).

Although decision problems are more complex than inference problems, in practice they often are easier than inference problems because the necessity to take some action can allow even small probability differences to force sensible decisions. For example, other things being equal, even a slightly higher probability that a particular medicine is effective or a particular airplane is safe will suffice to generate strong preferences. So odds of merely 60:40 can force practical decisions. On the other hand, for an inference problem that aims at truth but has no particular practical implications, it is difficult to evaluate just how strong a probability will justify a conclusion. Comparatively much stronger odds of 30:1 in favor of a given hypothesis might satisfy one scientist, but not another, especially if the latter scientist's prior opinion has been challenged by the current experiment. Because most probability reasoning is motivated by the practical need to make good decisions, not merely by theoretical interest, even rather weak data and small probability differences can still significantly inform and influence decisions.

Finally, some common traps result in bad decisions. Extensive research by Russo and Schoemaker (1989) has shown that most bad decisions result from ten dangerous mistakes: (1) hastily gathering information and reaching decisions before stepping back to organize the problem carefully, (2) solving the wrong problem and overlooking the best options and important objectives, (3) defining the problem in only one way, thereby restricting possible solutions, (4) failing to gather the facts because of overconfident assumptions and opinions, (5) relying excessively on readily available and simplistic information, (6) shooting from the hip, skipping systematic decision procedures, (7) assuming that many smart people automatically produce a good group decision, (8) ignoring feedback from past decisions and outcomes, (9) failing to keep systematic records to track past decisions and lessons, and (10) failing to audit decision procedures, so that the earlier mistakes continue at an undiminished rate. Another problem is that one may fail to appreciate the logical implications of revising a belief or changing a decision, such that one isolated problem may be fixed, but not everything else connected to it, and consequently the overall objectives will remain compromised (Levi 1986, 1991; Gärdenfors 1992).

THE FREQUENTIST PARADIGM

Because the frequentist paradigm is better known than the Bayesian paradigm among contemporary scientists, its description here can be somewhat briefer. Particularly such concepts as Type I and II errors and p-values are quite well known.

Historically, the Bayesian paradigm preceded the frequentist paradigm by about a century and a half, so the latter was formulated in reaction to perceived problems with its predecessor. Specifically, the frequentist paradigm sought to eliminate the Bayesian prior because it burdened scientists with the search for additional information that often was unavailable, diffuse, inaccurate, or controversial. Frequentists wanted to give scientists a paradigm with greater objectivity.

In 1956, one of the principal innovators of the frequentist paradigm, Sir Ronald A. Fisher, acknowledged that the Reverend Thomas Bayes had developed "the first serious attempt ... to give a rational account of the process of scientific inference" (Fisher 1973:8). Given an exactly known prior, there was no problem, and "the method of Bayes could properly be applied" (p. 20). Fisher gave an example of a mouse experiment with known priors from genetic knowledge of the mice (pp. 18–20; Gustason 1994:123–125) that was analogous to my example of the marble experiment with known priors from the initial toss of a fair coin. In such cases, everybody uses the same unproblematic prior knowledge, and everybody gets the same, objective answer. Rather, the problem that troubled frequentists concerned Bayesian priors based on ignorance: "if knowledge of the origin of the mouse tested were lacking, no experimenter would feel he had warrant for arguing as if he knew that of which in fact he was ignorant, and for lack of adequate data Bayes' method of reasoning would be inapplicable" (Fisher 1973:20).

Fisher's tale of brown and black mice was a moral tale that waxed sermonic in its conclusion that "It is evidently easier for the practitioner of natural science to recognize the difference between knowing and not knowing than this seems to be for the more abstract mathematician" (Fisher 1973:20). That is, the statistician should follow the scientist in discerning clearly whether the prior probabilities of the hypotheses are known or unknown. Perhaps "the data are such as to allow us to apply Bayes' theorem, leading to statements of probability," but in other cases "we may be able validly to apply a test of significance to discredit a hypothesis the expectations from which are widely at variance with ascertained fact" (p. 37).

Jerzy Neyman and Egon S. Pearson modified and developed Fisher's ideas. Their hypothesis tests emphasized falsification (rejection), rather than accepting or proving hypotheses. At the same time, independently of those statisticians, Sir Karl Popper was formulating his falsificationist view of theory testing, which fit with the emerging frequentist paradigm (Howson 2000:94–108). So the time was right for a paradigm shift from Bayesian to frequentist statistics among the majority of practicing scientists.

In frequentist statistics, among the hypotheses under consideration, one is designated as the null hypothesis. Ordinarily, the null hypothesis is that there is no effect of the various treatments, whereas one or more alternative hypotheses express various possible treatment effects. For example, given an old treatment and a new treatment for some disease, the null hypothesis says that there is no difference between them, whereas the alternative hypotheses say that the old treatment is better or that the new treatment is better. A null hypothesis is either true or false, and a statistical test either accepts or rejects the null hypothesis, so there are four possibilities, as shown in Figure 7.4.

A Type I error is to reject a true null hypothesis, whereas a Type II error is to accept a false null hypothesis. By contrast, an inference is a success if it accepts a true null or rejects a false null. Incidentally, some statisticians prefer the precise wording "fail to reject" to the briefer wording "accept" because of the falsificationist mentality of frequentist hypothesis tests. However, to reject a null is tantamount to accepting an alternative, and the latter term has an advantage of conciseness.

	True	False
Accept	Success	Type II Error
Reject	Type I Error	Success

Figure 7.4. Type I and Type II error events.

The basic idea of frequentist hypothesis testing is that a statistical procedure with low Type I and II error rates provides reliable learning from experiments. Type I errors can be avoided altogether merely by accepting every null hypothesis regardless what the data show, and Type II errors can be avoided by rejecting every null. Hence, there is an inherent trade-off between Type I and II errors, so some compromise must be struck.

The ideal way to establish this compromise is to evaluate the cost or penalty for Type I errors and the cost for Type II errors and then balance those errors so as to minimize the overall expected cost of errors of both kinds. In routine practice, however, scientists tend to set the Type I error rate at some convenient level and not to be aware of the accompanying Type II error rate, let alone the implied overall or average cost of errors.

Another important quantity is the p-value, defined as the probability of getting an outcome at least as far from what is expected as is the actual observed outcome under the assumption that the null hypothesis is true. To calculate the p-value, one envisions repeating the experiment an infinite number of times under the assumption that the null hypothesis is true and finds the probability of getting an outcome as extreme as or more extreme than the actual experimental outcome. Hence, the p-value is the probability of a Type I error for a given experiment. The smaller the p-value, the more strongly a frequentist test rejects the null hypothesis. It has become the convention in the scientific community to call rejection at a p-value of 0.05 a "significant" result, and rejection at the 0.01 level a "highly significant" result.

To illustrate the calculation of a p-value, Table 7.2 reanalyzes, from a frequentist paradigm, the marble experiment that was analyzed in Table 7.1 from a Bayesian paradigm. Barnett (1982:28–63) and Berger and Berry (1988) have also given examples of parallel analyses by both paradigms. Let the null hypothesis be H_W, that the urn contains 3 white marbles and 1 blue marble, and let the alternative hypothesis be H_B, that it contains 1 white and 3 blue marbles. (In this case, neither hypothesis corresponds to the idea of no treatment effect, so H_W has been chosen arbitrarily to be the null hypothesis, but the story would be the same had H_B been designated the

Table 7.2. Frequentist analysis for a marble experiment, assuming that the null hypothesis H_W is true and the experiment stops at 15 draws

Blue draws	Probability	p-value
0	0.01336346101016	1.00000000000000
1	0.06681730505079	0.98663653898984
2	0.15590704511851	0.91981923393905
3	0.22519906517118	0.76391218882054
4	0.22519906517118	0.53871312364936
5	0.16514598112553	0.31351405847818
6	0.09174776729196	0.14836807735264
7	0.03932047169656	0.05662031006068
8	0.01310682389885	0.01729983836412
9	0.00339806545526	0.00419301446527
10	0.00067961309105	0.00079494901001
11	0.00010297168046	0.00011533591896*
12	0.00001144129783	0.00001236423850
13	0.00000088009983	0.00000092294067
14	0.00000004190952	0.00000004284084
15	0.00000000093132	0.00000000093132

Note: The conclusion, marked by an asterisk for the experimental outcome of 11 blue draws, is to reject H_W at a highly significant p-value of 0.000115. This p-value means that if experiments with H_W true were to be conducted numerous times, a result as extreme as 11 or more blue draws would occur with a frequency of only 0.000115, or only 1 in about 8,700 experiments.

null hypothesis instead.) Recall that the experiment had 15 draws, of which 4 were white marbles and 11 were blue marbles. Also recall, for comparison momentarily with the new frequentist analysis, that the Bayesian conclusion from the experimental evidence was that the probability that H_B was true was 0.999543, and that H_W was true was 0.000457. But unlike the Bayesian analysis, with its initial coin toss that supplied the prior probabilities of H_W and H_B, assume now that we know nothing about the prior.

Table 7.2 has three columns of numbers. The first column lists, for an experiment with 15 draws, the 16 possible outcomes, namely, 0 to 15 blue draws (and, of course, correspondingly 15 to 0 white draws). This analysis takes H_W as the null hypothesis, and under the assumption that the urn contains 3 white marbles and 1 blue marble, the probability of a blue draw is 0.25. So experiments with 15 draws will average $15 \times 0.25 = 3.75$ blue draws. Accordingly, were this experiment repeated many times, outcomes of about 3 or 4 blue draws would be expected to be rather frequent, whereas 14 or 15 blue draws would be quite rare. To upgrade this obvious intuition with an exact calculation using the probability theory explained in the preceding chapter, an outcome of b blue draws and w white draws from a total of $n = b + w$ draws can occur with $n!/(b! \times w!)$ permutations, and the probability of each such outcome is $0.25^b \times 0.75^w$.

For example, the probability of 5 blue and 10 white draws is $[15!/(5! \times 10!)] \times 0.25^5 \times 0.75^{10} \approx 0.165146$. These probabilities, for all possible outcomes from b values of 0 to 15, are listed in the second column of Table 7.2. Of course, these 16 values sum to 1 because all possible experimental outcomes are listed here. Finally, the third column is the p-value, obtained for b blue draws by summing the probabilities for all outcomes with b or more blue draws. For example, the p-value for 0 blue draws is 1, because it is the sum of all 16 of these probabilities, whereas the p-value for 14 blue draws is the sum of the last two probabilities. For the particular marble experiment considered here, the actual outcome was 11 blue draws, and an asterisk draws attention to the corresponding p-value of 0.000115. The conclusion, based on this very small p-value, is to reject H_W as a highly significant result.

The values in Table 7.2 can serve to define test procedures that will control Type I errors at a given rate. For instance, the procedure of rejecting the null hypothesis H_W if there are 7 or more blue draws results in a Type I error rate of about 0.0566. To obtain a significant result at the 0.05 level, the appropriate procedure is to reject the null hypothesis H_W if there are 8 or more blue draws, as that is the first outcome that reaches a p-value of 0.05 or less. Or, for a highly significant result at the 0.01 level, reject the null if there are 9 or more blue draws.

Unlike Bayesian analysis, which requires specification of prior probabilities in order to do the calculations, the frequentist analysis requires no such input, and thereby it seems admirably objective. So even if we know nothing about the process whereby the urn receives either the 1 white and 3 blue marbles or the reverse, we can still carry on unhindered with this wonderfully objective analysis! Or, so it seems.

Most scientists who have read this section thus far probably have not sensed anything ambiguous or misleading in this frequentist analysis. It all seems so sensible. Besides, this statistical paradigm pervades contemporary scientific research, so it hardly seems suspect. Nevertheless, on four counts, scientists need to understand exactly which question a frequentist hypothesis test is asking and answering. Exactly what are the meanings of Type I and II error rates and of p-values?

(1) The Meaning of Frequentist Hypothesis Tests. The results of frequentist hypothesis tests are expressed by statistics such as Type I and II error rates and p-values. Accordingly, the first necessity for a scientist using these values is to understand their precise meanings with absolutely no ambiguity. It should help to consider a concrete example (Figure 7.5) that parallels the definitions of Type I and II errors shown in Figure 7.4. Such numbers could

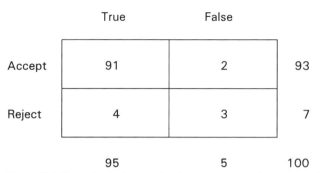

	True	False	
Accept	91	2	93
Reject	4	3	7
	95	5	100

Figure 7.5. Hypothetical example of error events and rates. A null hypothesis is either true or false, and a test, involving experimental data and a statistical inference, is used to accept or reject the null hypothesis, so there are four possible outcomes, with counts as shown. Also shown are row totals, column totals, and the grand total of 100 tests. To reject a true null hypothesis is a false-positive error event, whereas to accept a false null hypothesis is a false-negative error event. The Type I error rate α is P(reject | true) $= 4/95 \approx 0.0421$. The Type II error rate β is P(accept | false) $= 2/5 = 0.4$, and the power is $1 - \beta = 0.6$. The Type A error rate A is P(true | reject) $= 4/7 = 0.5714$, and the Type B error rate B is P(false | accept) $= 2/93 = 0.0215$. Type I and II error rates are probabilities of data given hypotheses, $P(D \mid H)$, whereas Type A and B error rates are probabilities of hypotheses given data, $P(H \mid D)$.

result when a diagnostic test accepts or rejects a null hypothesis of no disease and subsequently a definitive test determines for sure whether the null is true or false.

What is the Type I error rate for this example? As Schield (1996) has remarked for a similar example, that seemingly clear question can elicit three different answers from statistics students, but only one of those is what the statistics profession means by the Type I error rate. Likewise, an expression such as "the probability of rejecting a true null hypothesis" equivocates among three possible meanings (Cohen 1994; Benjamini and Hochberg 1995; Howson 1997a,b).

The probability of a Type I error can be construed as the number of Type I errors, 4 in this example, divided by the column total, row total, or grand total. First, some students and scientists think that the Type I error rate is the conditional probability of rejecting the null given that it is true, which is 4 divided by the column total of 95, or approximately 0.0421. Second, others think that the Type I error rate is the conditional probability of a true null given that it is rejected, which is 4 divided by the row total of 7, or 0.5714. Third, many students and scientists think that the Type I error rate is the probability that the null is rejected and it is true, which is 4 divided by the grand total of 100, or 0.040.

The formal definition of the probability of a Type I error rate, however, is the first of those interpretations – not the second, and not the third. Likewise, the proper definition of a Type II error rate is the conditional probability of accepting the null given that it is false, which is $2/5 = 0.4$ for this example – not $2/93$ and not $2/100$. The Type I probability is symbolized by α, the Type II probability by β, and the "power" of a statistical test by $1 - \beta$.

Let H be a generic label for a hypothesis, either that a null is true, or else false, and let D be a generic label for the data that prompt acceptance or rejection of the null hypothesis. So H represents the columns, and D the rows, in Figures 7.4 and 7.5. Then the first of the three interpretations of a Type I (or Type II) error rate is a probability of the form $P(D \mid H)$, the second $P(H \mid D)$, and the third $P(H \cap D)$. By definition, the Type I and II error rates are actually both of the first form, $P(D \mid H)$. In other words, the Type I error rate concerns sampling exclusively from cases where the null is true, and the Type II error rate concerns sampling exclusively from cases where the null is false. More specifically, the Type I error rate is $\alpha = $ P(reject null | null true), the Type II error rate is $\beta = $ P(accept null | null false), and the power of a statistical test is $1 - \beta = $ P(reject null | null false). Also, the meaning of a p-value, as a post-experimental Type I error rate, is of the form $P(D \mid H)$, exactly as one should gather from the way a p-value is calculated, as in Table 7.2.

(2) Type A and B Error Rates. Scientists, doctors, lawyers, and others are concerned about the probabilities of various kinds of errors as they go about the business of drawing inferences or diagnoses from experiments, tests, and data. *But just exactly which errors are these?* The error rates usually on offer are the Type I and II probabilities regarding the null hypothesis, P(reject | null true) and P(accept | null false), or α and β.

To introduce some new terminology, however, there also exist Type A and B error rates, P(null true | reject) and P(null false | accept), symbolized by the uppercase Greek letters A and B. Note that Type I and A probabilities involve the same error *event* (rejection of a true null), but different error *rates* (rejection given a true null, or else a true null given rejection); and a similar remark applies to Type II and B error events and rates.

Note that α and A (and β and B) always have different meanings that answer different questions. They also usually have different numerical values, sometimes wildly different. For instance, in this example, the rate of false positives among rejections (or significant results, 0.5714) is an order of magnitude larger than the rate of false positives among tests for which the null hypothesis is true (0.0421). Therefore, Type I and A error rates should not be confused, nor Type II and B error rates.

It is a frequent error for scientists to to think that they want Type I and II error rates when actually they need Type A and B error rates. This error amounts to a rather fundamental confusion and reversal about what is given and what is required in scientific inference, including medical diagnosis. What is given are the data, leading to accepting or rejecting the null hypothesis (say, of no disease), whereas what is required is an inference or diagnosis about the true state of nature (such as a disease being absent or present). Merely to understand what is given and what is required is to see that the relevant quantities are the Type A and B error rates (Cohen 1994; Benjamini and Hochberg 1995). So, even granted a paradigm featuring error rates of statistical tests, still a choice must be made regarding exactly which error rates are most relevant to research purposes.

(3) The "Bayesian" Misreading of Frequentist Hypothesis Tests. A Bayesian report for a hypothesis test is of the form $P(H \mid D)$, whereas a frequentist report is of the form $P(D \mid H)$. However, as many statisticians and educators have observed and lamented, scientists seem bound and determined to interpret all statistical reports with Bayesian meanings. No quality or quantity of statistical training seems capable of correcting this persistent misreading of frequentist reports, particularly p-values.

When scientists run an experiment, they measure certain differences between the treatments and also a certain variability among replicates, and they perform a statistical analysis of the data. Scientists understand quite correctly that there are two possibilities regarding the treatment differences: They could be real, caused by actual treatment effects, or they could be spurious, caused by chance and random sampling fluctuations. Scientists use statistics to quantify the probabilities of these two possibilities.

In other words, scientists want to know whether the null hypothesis or an alternative is true. Given this entirely natural and eminently practical question, scientists tend to impose a relevant meaning on any statistical tool that is placed conveniently at their disposal, including p-values. Accordingly, Schield (1996) represents the voice of experience, as his paper opens with these words: "When students obtain a statistically significant sample at a 5% level of significance, they may conclude they can be 95% confident that the alternate hypothesis is true." To add my own experience with students, when analysis of their experiments gives a p-value of 0.05, they typically take that to mean that there is only a 5% probability that the observed treatment differences are due to chance and a 95% probability that they are real. That is, the experimental data provide rather strong ("statistically significant") evidence against the null hypothesis of no treatment effects.

But that cannot possibly be the meaning of a p-value (Cohen 1994). Indeed, as the preceding Bayesian paradigm has shown, to reach a conclusion about the probability that a hypothesis is true, one must specify the hypothesis set and must specify the prior probabilities of the hypotheses. On both counts, the frequentist paradigm is ineligible to speak about the probabilities of hypotheses being true. In other words, the Bayesian paradigm is, and the frequentist paradigm is not, a story about the evidential bearing of data on hypotheses.

First, comparing Tables 7.1 and 7.2, there is a profound difference in that the Bayesian analysis has H_W compete against the specified alternative H_B, whereas the frequentist analysis has H_W as the null hypothesis assessed on its own, with no specification of its competitor. How can an experiment provide evidence for or against a null hypothesis when its competitor is unspecified? Well, it cannot.

For example, imagine that we are told that an urn contains marbles that can be either blue or white, but are told nothing else about how many blue and white marbles there are. Then the hypothesis set for the probability of a white draw P(white) is the whole range from 0 to 1, which includes the particular null hypothesis H_W that the urn contains 3 white marbles and 1 blue marble, so P(white) = 0.75. Now the data given in Table 7.2 are that 15 draws yielded 11 blue and 4 white draws, so the experimental outcome is P(white) = $4/15 \approx 0.2667$.

Is this outcome evidence against the null hypothesis? Well, if this null hypothesis goes against the alternative that P(white) = 0.25, then this experiment does give evidence *against* the null. But if this same null goes against the alternative that P(white) = 0.8, then this same experiment does give evidence *for* the null. And if the alternative hypothesis is not specified, then this experiment does not give evidence either for or against the null hypothesis. A horse that runs a race alone against no other competitors cannot beat anyone! And neither can a lone horse get beaten!

Second, even if a *p*-value could allow an experiment to provide evidence for or against the null, it still would be possible that the prior probabilities might provide even stronger evidence in the reverse direction, so the conclusion drawn from the experiment in isolation from other relevant knowledge would be likely to go contrary to the truth. This observation is particularly important in the context of medical diagnostic tests and treatment options.

So, if a *p*-value is not a measure of an experiment's evidence either for or against a null hypothesis, what is it? Well, to reiterate what has already been said, a *p*-value answers this question: Assuming that the null hypothesis is true (or, in other words, that we are sampling exclusively from cases for which the null is true), what is the probability of observing an outcome as extreme as or more extreme than that actually observed in the experiment?

Finally, another way to challenge misinterpretations of *p*-values is to realize that substituting a statistical procedure's performance in the long run for its performance in an individual experiment is an instance of the fallacy of division – of assuming that a property of a whole is also a property of a part. So it is helpful here to recall this fallacy from the earlier chapter on deductive logic.

(4) The Strangely Influential Stopping Rule. Every experiment must stop for some reason, or else incur infinite cost and run forever! But the Bayesian and frequentist paradigms have strikingly different views about the influence that the stopping rule should have on statistical inferences (Berger and Berry 1988; Kadane, Schervish, and Seidenfeld 1999:349–381). The frequentist paradigm endows the stopping rule with tremendous influence that most scientists find quite strange.

Consider again the frequentist analysis of the marble experiment. The stopping rule for this experiment given in the footnote to Table 7.1 is that "the experiment stops at 15 draws." The implications of that reason for stopping, however, have deliberately not been explored until now. Many scientists would be surprised to learn that changing this stopping rule would change p-values, even if the data remained exactly the same. So different experiments giving the same data could reach different conclusions, quite against science's ideal of objectivity!

For example, the stopping rule for this particular marble experiment with its 15 draws might have been to stop at 11 blue draws. Or it might have been to stop at 4 white draws. All three stopping rules result in exactly the same data for this marble experiment. These three rules differ, however, in the imaginary outcomes that would result as the frequentist envisions numerous repetitions of the experiment. In the frequentist paradigm in general, and in p-value calculations in particular, imaginary as well as actual outcomes are integral parts of the story. The rule to stop at 15 draws has exactly 16 possible outcomes, as listed with their probabilities in Table 7.2. The rule to stop at 4 white draws, however, has an infinite number of possible outcomes, with 0 to ∞ blue draws, and their probabilities are different from those listed in Table 7.2.

Incidentally, careful readers may have noticed a subtle shift between the analyses in Tables 7.1 and 7.2, the former experiment stopping when the margin between blue and white draws equals 7, but the latter stopping at 15 draws. But because both of these tables analyze exactly the same marble experiment with its 15 draws, one might suppose this shift would make no difference. In fact, the rule to stop at 15 draws was used in Table 7.2 to obtain a simple calculation that is easy to explain. The original rule involves a more complicated frequentist calculation, but the needed probability theorems can be derived from the preceding probability axioms, and the result is rejection of the null hypothesis H_W at a somewhat smaller and hence more significant p-value of 0.000096 (interested readers can consult any elementary probability text, such as Ross 1994).

Countless other stopping rules could be imagined that would all happen to stop this particular marble experiment at the same point, thus generating a panoply of different p-values for the same experimental data. Some of those rules could be quite vague and subjective, such as "Stop after collecting data for two minutes," and we might receive an experimenter's data with no explanation of why the experiment was stopped at a particular point. Berger and Berry (1988) cited a disturbing example in which frequentist analyses of a single experiment gave p-values of 0.021, 0.049, 0.085, and any other value up to 1 just by assuming different stopping rules.

So a p-value depends on the experimental data *and* the stopping rule. In other words, it depends on the actual experiment that did occur *and* an infinite number of other imaginary experiments that did not occur. Different stopping rules are generating different stories about just what those other imaginary experiments are, thereby changing p-values. But such reasoning seems bizarre and problematic, opening the door to unlimited subjectivity, quite in contradiction to the frequentists' grand quest for objectivity.

Although a p-value has a problematic meaning and value, as just explained, we must also understand how this problem disappears from view in an ordinary, standard statistical analysis. For a concrete and representative example, consider the data analysis for a doughnut experiment analyzed in a popular statistics text now in its eighth edition: "During cooking, doughnuts absorb fat in various amounts.... [We want] to learn if the amount absorbed depends on the type of fat used. For each of four fats, six batches of doughnuts were prepared" (Snedecor and Cochran 1989:217). The data are presented, and the standard frequentist calculations are performed. The conclusion is to reject the null hypothesis, of no differences between the four fats, at the 1% level of significance or more exactly a p-value of 0.0070, so "the fats have different capabilities for being absorbed by doughnuts" (p. 224). Yet the stopping rule is never specified. Was a fixed number of six batches planned, for whatever practical considerations, from the outset? Or was the experiment to be continued until it triggered a highly significant p-value of 0.01 or smaller? Or did it stop for some other reason? We do not know. So instead of telling us what we need to know to calculate a p-value properly, the calculations proceed in a manner that equates to assuming some stopping rule whose existence and influence are never revealed.

Generalizing from this doughnut example to ordinary statistical practice among scientists, canned software for frequentist tests magically provides mechanical, automatic, objective answers without ever asking the user to specify the stopping rule. But inside such software, many assumptions about the experiment are being incorporated without ever making the assumptions explicit or asking the user whether or not they are true. Consequently, the apparent objectivity is illusory.

The frequentist's emphasis on the stopping rule induces a related problem with experiments having periodic monitoring and analysis of the data, as is common for medical trials, in which one wants to stop giving patients inferior treatments as soon as the bad treatments are identified. Frequentist analyses penalize for interim looks at the data, which causes an experiment to run longer before reaching a conclusion. But Bayesian analyses, as in the marble example, impose no penalty for monitoring the results.

Another problem with p-values is that they can seriously overestimate or underestimate the strength of the evidence. For instance, the calculations in Table 7.2 overestimate significance, because they give so much attention to imaginary experiments more extreme than the actual experiment (Berger and Berry 1988). For example, if the marble experiment's p-value of 0.000096 from frequentist analysis is interpreted (or, better, misinterpreted) as a probability of 0.000096 that H_W is true, then comparison with the Bayesian probability of 0.000457 will show that the frequentist rejection of H_W is almost five times as strong as the Bayesian rejection.

Which assessment is right? Does the frequentist analysis overestimate the evidence, or does the Bayesian analysis underestimate the evidence? Without question, the physical reality here is that rejection of H_W on the basis of these marble data would prove incorrect at a frequency of 0.000457, or about one mistake per 2,200 experiments, just as the Bayesian analysis says.

What are the implications of p-values often overestimating the strength of the evidence? After all, historically the standard statistical tests have played a sustained and significant role in research to improve airplanes, bridges, cars, dog foods, doughnuts, medicines, and so on. The airplane you ride and the dog food you buy are what they are in part because of statistical decisions based on p-values during research and development of those products. Regrettably, many of those decisions have been based on weaker evidence than appears to be the case because of misleading frequentist analyses.

Finally, the suggestion may be offered that if scientists' understanding of frequentist statistics is to be improved in the future, the key may have more to do with statistical software than statistical instruction. For the most part, statistics texts and courses are entirely clear about the meaning of p-values. Admittedly, sometimes ambiguous wordings appear, such as "a Type I error is to reject a null hypothesis *when* it is true," which is unfortunate, given students' common misinterpretations. Also, sometimes more could be said about what p-values do *not* mean. Anyway, from the day students finish their statistics courses, the greatest influence on their day-to-day statistical practice probably is not the courses that they took years or decades ago, but rather the statistical software that they use routinely.

Accordingly, one great advance would be for statistical software performing frequentist hypothesis tests to query the user for the experiment's stopping rule and to refuse to perform an analysis until that rule had been specified precisely. The frequentist paradigm absolutely demands that the stopping rule be specified, so frequentist software absolutely should require that input. Scientists need to see that the data input is not the only thing that affects frequentist analyses and conclusions.

A second great advance would be for every appearance of a p-value in computer outputs to be accompanied by an explanation somewhat along the following lines: "This p-value is the probability of an outcome as extreme as or more extreme than the experiment under the assumption that the null hypothesis is true. It is *not* a measure of the strength of the experimental evidence either for or against the null hypothesis or any other hypothesis. For the data to have evidential bearing on the hypotheses, one must specify the prior probabilities of the hypotheses and perform Bayesian calculations."

PARADIGMS AND QUESTIONS

Scientists should understand how consulting statisticians see their role. The statistician's role is not to tell an agronomist what she should be asking about crops nor to tell an astronomer what he should be asking about stars. At most, the consulting statistician will merely prompt the scientist to express the questions with greater clarity and precision. Rather, given the scientist's research objectives, the statistician seeks to design efficient experiments for gathering the data and to formulate powerful inference procedures for analyzing the data. But the original and most essential job, of framing interesting questions, is the scientist's responsibility. So the message to scientists is, Know your own research questions, and then choose an appropriate statistical paradigm.

Obviously, there are debates within the Bayesian camp as well as within the frequentist camp, but nevertheless there is a dominant difference between the camps. For the sake of brevity, this section's analysis of the Bayesian–frequentist debate must focus on just the main issues, as represented by only a few champions on each side. The main representative of the Bayesian side is the book by Howson and Urbach (1993), and the frequentist side features the book by Mayo (1996). An advantage of these particular choices is that these proponents have argued their positions in papers by Howson (1997b) and Mayo (1997) that were published together and introduced by Giere (1997). Such direct exchange fosters unusual clarity, broad perspective, and real understanding.

The Bayesian and frequentist paradigms have the same ultimate objective: to learn from experiments and other experiences of the physical world. The setup includes three components: hypotheses, data, and inference procedures. But that is where the similarity ends. A Bayesian procedure and a frequentist procedure ask different questions, although at the most general level both statistical procedures concern learning from experience.

The Bayesian question is, What is the probability that a hypothesis is true, given the data and any prior knowledge? This prior knowledge refers to what is known beyond the one particular experiment at hand, which may be substantial in one case or negligible in another case.

The frequentist question is, How reliable is an inference procedure, by virtue of not rejecting a true hypothesis or accepting a false hypothesis? The basic idea is that a reliable procedure with low error rates is a suitable tool for sorting true from false hypotheses.

Note that the Bayesian paradigm weighs hypotheses, whereas the frequentist paradigm weighs procedures. But because statistical procedures are used to judge hypotheses, the Bayesian and frequentist questions might seem to be virtually equivalent because of their shared goal of helping scientists to learn from experience. However, that difference in questions does induce a series of further differences, including the following four differences.

(1) The Prior and Required Data. The best-known element of the Bayesian–frequentist debate is the prior. Bayesian analysis requires prior probabilities for the hypotheses, which are evaluated prior to (or apart from) receiving the new experiment's data. But frequentists object that prior probabilities are often difficult to specify accurately. So the Bayesian procedure imposes an extra burden on scientists to supply data for quantifying prior probabilities, and those potentially inaccurate and controversial priors introduce undesirable subjectivity that erodes science's presumed objectivity. On the other hand, Bayesians reply that it is absolutely impossible to eliminate the prior and that frequentist procedures always sneak in some functional equivalent of the prior. Also, if there is substantial previous knowledge, as is often the case, then efficiency is gained by using an informative prior, whereas if there is negligible previous knowledge, a noninformative prior can be used, and the robustness of the result checked by varying that prior.

(2) The Concept of Probability. Statistics uses probability at every turn, but Bayesians and frequentists differ in their concepts of probability. For Bayesians, probabilities are personal degrees of belief in propositions (or hypotheses), given a person's relevant knowledge. For frequentists, probabilities are long-run frequencies for an event (or experiment) that is repeated numerous times, with the repetitions being either actual or imaginary. As discussed in the earlier chapter on probability, however, this particular debate is rather academic. Bayesians and frequentists alike need to talk about both beliefs and events in order to connect with ordinary scientific usages of probability. Furthermore, both beliefs and events can be handled by the same probability axioms, so a broad and unified concept of probability is possible.

(3) Aesthetics. Some differences between Bayesians and frequentists are best described as differences in aesthetics. For example, the Bayesian paradigm emerges from merely several axioms and principles and thereby possesses a demonstrable coherence. For Bayesians, that is elegance and beauty. But frequentists think that such elegant coherence has been purchased at the price of not addressing the complicated, messy world of real experiments. Rather, science calls for a hodgepodge of statistical ideas and methods that make a virtue of the necessities imposed on us by a complicated world. For frequentists, that is realism and beauty.

(4) The Likelihood Principle and Imaginary Data. The likelihood principle, which is a foundational part of the Bayesian paradigm, says in essence that statistical procedures and inferences should depend only on the actual outcome of an experiment, not on possible but imaginary other outcomes (James Robins and Larry Wasserman, in Raftery et al. 2002:431–443). By contrast, frequentist statistics (such as p-values) depend on both actual and imaginary outcomes. Because of this difference, Bayesian procedures are indifferent to the reason or rule for stopping an experiment, but stopping rules matter for frequentist procedures. Consequently, two experiments could give the same actual data but could have been stopped for different reasons (that induced different sets of imaginary other possible outcomes), whereupon the Bayesian would reach the same conclusion for both experiments, whereas the frequentist would reach different conclusions (that could even disagree about whether or not the results were significant at a given level, even though the two experiments produced exactly the same data). Likewise, when an ongoing experiment is monitored and analyzed regularly, as often occurs in medical trials, a Bayesian will say that interim looks at the data will not affect the final conclusion, but a frequentist will say that they will make the final conclusion weaker, so that more data will be needed if the conclusion is required to reach a specified significance level.

Imagine a simple clinical trial in which pairs of subjects are randomly assigned, one getting drug A and the other drug B, and the measured response is which subject survives longer. Consider the following data and question:

> *Data.* For 10 pairs of subjects, drug A won 8 times, and drug B won 2 times.

> *Question.* Is this substantial evidence that drug A is better than drug B, or, in more quantitative terms, what is the probability that drug A is better?

Many scientists may be surprised to learn that one thing that Bayesian and frequentist methods have in common is that neither can answer that question. Given just the data, neither Bayesians nor frequentists can proceed with answering any questions about the implications of the data. Rather, both paradigms require additional information, beyond the experimental results as such, to be able to calculate an answer.

But the nature of the additional information is different for the two paradigms, which fuels an interminable debate. Bayesians dislike the kind

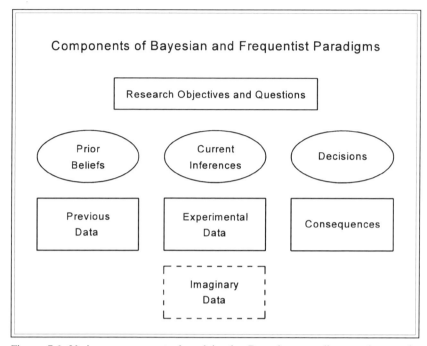

Figure 7.6. Various components found in the Bayesian paradigm or frequentist paradigm or both. Major differences are that only the Bayesian paradigm explicitly models prior beliefs, whereas only the frequentist paradigm considers imaginary data.

of additional information that frequentists require, and frequentists dislike the additions that Bayesians require. Ironically, each camp believes that the other is making moves that compromise science's objectivity. In some sense, scientists want a statistical paradigm that bases inferences and decisions on public experimental facts, not on personal subjective factors, so that everyone will get the same, objective answer. Science's objectivity and hence credibility are at stake in this significant statistical debate, so it is imperative that scientists distinguish reality from illusion.

Figure 7.6 shows several components of statistical paradigms. Some components count in the Bayesian paradigm, others count in the frequentist paradigm, and still others count in both, as will be explained momentarily. Consequently, a scientist's intuition or understanding of just what should or should not count in scientific investigation can be a powerful resource for deciding which paradigm is most sensible and even most objective.

Beginning at the top of the figure, the Bayesian and frequentist paradigms have in common the most important feature of scientific endeavor: research objectives and questions. For instance, medical researchers and their consulting ethicists and statisticians want to know how to make medical research efficient and how to improve clinical practice and patient care. Again, shared goals are exactly what make comparisons of Bayesian and frequentist analyses meaningful, with real potential for winners and losers. In the remainder of the figure, ovals represent beliefs and decisions residing in persons' minds, whereas rectangles represent data and events regarding the external world. The rectangle with a dashed border represents imaginary physical events.

Moving toward the middle of the figure, experimental data are shared by the Bayesian and frequentist paradigms. All statisticians want data with which to inform their conclusions. It is the empirical data, not the statistical analysis, that can bring science into contact with physical reality and provide opportunity for objective results.

But having acknowledged shared goals and shared data, that is where the similarity between the Bayesian and frequentists paradigms ends. In order to draw inferences or conclusions from the data, both Bayesians and frequentists need to add something else, but their additions differ.

What Bayesians need to add to the experimental data to reach their current (or posterior or post-experimental) inferences is information on prior beliefs reflecting the available previous (or pre-experimental) data. For example, recall the drug trial with 8 wins for A and 2 for B. To analyze those experimental results, a Bayesian must ask what was known about those drugs prior to running the experiment. And the Bayesian can operate in either of two modes, depending on whether or not previous data are available: If there are useful previous data, such as an earlier smaller trial with 2 wins for A and 1 for B, they can be incorporated explicitly. On the other hand, if there are no useful previous data, a noninformative prior can model the initial uncertainty or ignorance.

All scientists would agree that it makes good sense to use all available evidence to reach an informed conclusion. Also, using all of the available information can reduce the amount of data required for an experiment to reach a definitive conclusion. Furthermore, computer technology and communication are steadily making it easier to extract and use relevant previous data. But such advantages come at a cost. Whereas the data from the current experiment are specific and objective, there for everyone to see, data from other experiments and other sources are relatively diffuse and subjective, with different individuals having at their disposal different sets of data.

For example, imagine two doctors who read a report on a recent clinical trial that favors new treatment B over standard treatment A on the basis of moderate but not enormous evidence. One of those doctors may have given standard treatment A to 11 patients and observed good results in 9 cases, and hence may be reluctant to switch to new treatment B, whereas the other doctor may have given A to 14 patients with only 6 successes, and hence may be willing to try new treatment B. Consideration of other data from outside sources can introduce a subjective element into Bayesian conclusions, for not everyone possesses or considers exactly the same other data. Also, analyzing other data involves extra work. However, that is not a sufficient reason to ignore other data when such extra work often is handsomely rewarded by increased efficiency and more reliable conclusions.

On the other hand, what frequentists need to add to the experimental data to reach their conclusions are calculations about imaginary data that might have been observed but were not. As was explained in the preceding section, to enumerate those imaginary outcomes, the principal extra information that frequentists need is an explanation of why the researcher stopped the experiment (Berger and Berry 1988). Again recall the drug trial with 8 wins for A and 2 for B. Was it stopped upon reaching 10 comparisons? Or was it stopped upon reaching 8 wins for A? Or was it stopped upon reaching 2 wins for B? Or was it stopped upon reaching a difference of 6 wins between the two drugs? Or was it stopped by some other stopping rule? Each stopping rule will induce different sets of imaginary outcomes that might have occurred but did not, and each of those stories about imaginary data will lead the frequentist to reach different conclusions about the import of the actual data.

Furthermore, stopping rules not only affect the conclusions as such but also affect the amount of data required to reach a confident conclusion. Consequently, in the frequentist paradigm, subjective intentions affect research efficiency, such as the number of experimental animals or subjects that must be tested to achieve adequate results. One person's subjective choice of imaginary data could impose great suffering on an extra 20 dogs or an extra 50 humans, as compared with some other frequentist's imaginary data.

Understand that the context here is the analysis of a given set of data, whereas the design of an experiment before its actual execution is a different context. At the design stage, all one has to contemplate is a spectrum of imaginary outcomes and the opportunity to compare proposed statistical analyses over that range of possible outcomes. So analysis of imaginary data is legitimate and relevant at the design stage. One way to implement design is to investigate the frequentist properties of Bayes rules, which are often quite good (Berger 1985:521–558).

Focusing next on the center of Figure 7.6, after analyzing the current experimental data, both statistical paradigms yield current inferences. But the options of basing inferences on experimental data plus prior beliefs, as in the Bayesian paradigm, or else basing inferences on experimental data plus stopping rules and imaginary data, as in the frequentist paradigm, lead to profound differences in the meaning of research inferences or conclusions.

Those differences can be seen most clearly by emphasizing that the two paradigms answer two different questions. The Bayesian paradigm allows researchers to ask and answer questions such as, Given the data, what is the probability that treatment A is better than treatment B, or what is the probability that some particular treatment is best for a new patient? The frequentist paradigm asks and answers the question, Given the assumption that the null hypothesis, of no treatment differences, is true, what is the probability of an outcome as extreme as or more extreme than the observed data? Philosophically, this difference is nothing less than the distinction between inductive reasoning from data to model and deductive reasoning from model to data.

Lastly, moving to the right in Figure 7.6, an enormous distinction, caused by Bayesian and frequentist inferences having different meanings, is that Bayesian inferences lead readily to decision procedures, whereas frequentist inferences do not. As explained earlier in this chapter, Bayesian analysis of old and new data yields probabilities of the various states of nature. Those probabilities are multiplied by entries in a consequences matrix, specifying the benefit or loss of each action-and-state combination, to calculate each action's expected utility. Then a decision can be made using some decision criterion, such as maximizing the expected utility. But that natural connection between inferences and decisions works because Bayesian inferences provide probabilities of the states of nature, such as the probability that treatment A is superior to B.

On the other hand, frequentist inferences provide no satisfying connection to decision problems. Frequentist inferences answer some obscure question about imaginary outcomes given the null hypothesis, rather than the direct question about the probabilities of the various states of nature. So even if ample facts provide good estimates of the consequences of each action given each state, and even if frequentists and Bayesians can agree on a consequences matrix, still the frequentist has no estimate of the probabilities of the various states of nature.

What a tangled debate! Frequentists dislike the subjectivity introduced by the Bayesians' prior, whereas Bayesians dislike the subjectivity introduced by the frequentists' role for stopping rules. Most distressingly, the Bayesian gives a convoluted answer to the frequentist's question, and the frequentist gives a convoluted answer to the Bayesian's question. Nevertheless, with modest effort, scientists can understand the basic ideas of both paradigms and discern which suits their applications. After some misconceptions have been dispatched, the main necessity is merely that scientists know exactly what questions they want to ask.

The bottom line is simply that scientists must know their own questions. If they are asking about the probability of an outcome as extreme as or more extreme than that in the experiment, under the assumption that the null hypothesis is true, given the data and the stopping rule, then a frequentist analysis will give the answer. If they are asking about the probability that a hypothesis is true, given the data and the prior probabilities, then a Bayesian analysis will give the answer.

INDUCTION LOST

So far, this chapter's account of inductive logic has been, on the whole, rather confident and cheerful. Induction – or, in more common terminology, statistics – has been depicted as a great success. There are enough problems with induction to keep statisticians employed, no doubt, but not enough problems to make philosophers despair. Induction's success is essential for science's rationality and credibility because induction is such a critical component of scientific thinking.

However, many other accounts of induction have been rather dark, dreary, and skeptical. A tremendous philosophical battle has been fought over induction, from ancient Greek skeptics to the present, with David Hume's critique being especially well known. Without doubt, inductive logic has suffered considerably more numerous and more drastic criticisms than all of the other components of scientific reasoning combined. Dozens of books, mostly by philosophers, have been written on the so-called problem of induction.

Unfortunately, the verdict of history seems to be that "the salient feature of attempts to solve Hume's problem is that they have all failed" (Friedman 1990:28). Broad's oft-quoted aphorism says that induction is "the glory of science and the scandal of philosophy" (Broad 1952:143; Howson 2000:10), and Whitehead (1925:25) called induction "the despair of philosophy." Howson (2000:14–15, 2) has concluded that "Hume's argument is one of the most robust, if not the most robust, in the history of philosophy," and it simply is "actually correct."

Hume's critique of induction appeared in his anonymous, three-volume *A Treatise of Human Nature*, which was a commercial failure and drew heavy criticism from his fellow Scottish philosophers Thomas Reid and James Beattie. Subsequently, his admirably brief *An Enquiry Concerning Human Understanding* reformulated his critique, and that punchy book was a great success. Because Hume's advertisement in the latter work dismisses the former as a juvenile work, the discussion here follows the usual custom of examining just the *Enquiry*.

Although Hume's presentation in Chapters 4 and 5 of his *Enquiry* is quite brief and readable, scholars have produced insightful and succinct summaries of Hume's argument (Gustason 1994:178–180; Couvalis 1997:36–44; Tom L. Beauchamp, in Beauchamp 1999:26–32; Howson 2000:6–21). There are three key premises, followed by the conclusion: (1) Any verdict on the legitimacy of induction must result from deductive or inductive arguments, because those are the only kinds of reasoning. (2) A verdict on induction cannot be reached deductively. No inference from the observed to the unobserved is deductive, specifically because nothing in deductive logic can ensure that the course of nature will not change. (3) A verdict cannot be reached inductively. Any appeal to the past successes of inductive logic, such as that bread has continued to be nutritious and that the sun has continued to rise day after day, is but worthless circular reasoning when applied to induction's future fortunes. (4) Because deduction and induction are the only options, and because neither can reach a verdict on induction, there is no rational justification for induction.

Incidentally, whereas the second premise, that of no deductive link from the past to the future, had been well known since antiquity, the third premise, that of no (legitimate, non-circular) inductive link from the past to the future, was Hume's original and shocking innovation (Howson 2000:12). The bottom line circulates widely in formulations such as "There is no noncircular justification of induction," or "The future might not resemble the past," or "There is no sound inductive reasoning from empirical observation alone."

Induction suffered a second serious blow half a century ago, which was two centuries after Hume, when Goodman (1983) propounded his "new riddle" of induction – the first edition of that book having appeared in 1954, and a forerunner of the argument in an even earlier 1946 paper. "The new riddle of induction has become a well-known topic in contemporary analytic philosophy.... There are now something like twenty different approaches to the problem, or kinds of solutions, in the literature.... None of them has become the majority opinion, received answer, or textbook solution to the problem" (Douglas Stalker, in Stalker 1994:2).

Briefly, Goodman's argument ran as follows: Consider emeralds examined before time t, and suppose that all of them have been green (where t might be, say, tomorrow). The most simple and foundational inductive procedure, often called the "straight rule" of induction, says that if a certain property has been found for a given proportion of many observed objects, then the same proportion applies to all similar unobserved objects as well as to individual unobserved objects. For example, if numerous rolls of a die have given an outcome of 2 with a frequency of nearly $1/6$, then inductive logic leads us to the conclusion that the frequency of that outcome in all other rolls will also be $1/6$, and likewise that the probability of any particular future roll giving that outcome will be $1/6$. In this case, those observations before time t of many emeralds that are all green support the inductive conclusion that all emeralds are green, which also supports the prediction that if an emerald is examined after time t, it too will be green.

Then Goodman introduced a new property, "grue," with the definition that an object is grue if it is examined before time t and is green or if it is not examined before time t and is blue. Then scientists examining a sample of emeralds before time *t* will discover that they are all green, and yet they also are all grue. But of course that is a problem, because emeralds examined after time *t* will be green and hence will fail to be grue. This problem shows that not all properties are appropriate (projectable) for application of the straight rule of induction. So how can one decide in a nonarbitrary manner which properties are projectable?

All too predictably, Hume had complained that all received systems of philosophy were defective and impotent for justifying even the simple straight rule of induction (Beauchamp 1999:131–133). Goodman's complaint, however, was the exact opposite (Friedman 1990:37–42). His concern was not that induction proved too little, but rather that it proved too much. Induction can support contradictory predictions – it can prove anything, which means it proves nothing. Understand that Goodman, like his predecessor Hume, was not intending to wean us from common sense, such as causing us to worry that all of our emeralds would turn from green to blue tomorrow. Rather, he was deploying the new riddle to wake us to the challenge of producing a philosophically respectable account of induction.

Finally, and perhaps most importantly, the great generality of those old and new problems of induction must be appreciated. Hume and Goodman expressed their arguments in terms of time: past and future, or before and after time *t*. But thoughtful commentators have discerned their broader scope. Gustason (1994:205) has assimilated Hume's argument to a choice among various standard and nonstandard inductive logics. Accordingly, the resulting scope encompasses any and all inductive arguments, including those concerning exclusively past outcomes. And Howson (2000:30–32) has followed Goodman in interpreting Goodman's argument as a demonstration that substantial prior knowledge about the world enters into our (generally sensible) choices about when to apply induction and how much data to require.

Couvalis (1997:48) has cleverly said it all with a singularly apt example: "Having seen a large number of platypuses in zoos and none outside zoos, we do not infer that all platypuses live in zoos. However, having seen a small number of platypuses laying eggs, we might infer that all platypuses lay eggs." Similarly, Howson (2000:6, 197) has observed that scientists are disposed to draw a sweeping generalization about the electrical conductivity of copper from measuring current flow in a very few samples. But obviously, for many other scientific generalizations, scientists demand enormous sample sizes.

INDUCTION REGAINED

The preceding section told the sad story of induction lost, but this section tells the happy story of induction regained. Capable philosophers and scientists have chronicled many attempts to solve the problem of induction

(Broad 1968; Lakatos 1968; Stove 1973; Swinburne 1973, 1974; Burks 1977; Rescher 1980; Friedman 1990; Kornblith 1993; Stalker 1994; Howson 2000). The essence of the most promising responses follows.

Hume said that we need not fear that skeptical philosophical doubts about induction "should ever undermine the reasonings of common life," because "Nature will always maintain her rights, and prevail in the end over any abstract reasoning whatsoever," and "Custom . . . is the great guide of human life" (Beauchamp 1999:120, 122). Hume's conclusion is not that induction is shaky, but rather that induction is grounded in custom or habit or instinct, which we share with animals, rather than in philosophical reasoning (Howson 2000:20). So common sense must trump skeptical doubt.

Indeed, when philosophy's roots in common sense are not honored, a characteristic pathology ensues: Instead of natural philosophy happily installing science's presuppositions once, at the outset, by faith, in a trifling trinket of common-sense knowledge, a death struggle with skepticism gets repeated over and over again for each component of scientific method, including induction. The proper task, "to explain induction," swells to the impossible task, "to defeat skepticism and explain induction." If (Hume's) philosophy cannot speak in induction's favor, that is because it is a truncated version of philosophy that has exiled animal habit, rather than having accommodated our incarnate human nature as an integral component of philosophy's common-sense starting points, as Reid recommended.

Plainly, all of the action in Hume's attack on induction derives ultimately from the concern that the course of nature might change, but that is simply the entrance of skepticism. His own examples include such drastic matters as whether or not the sun will continue to rise daily and bread will continue to be nutritious. Such matters are nothing less than philosophy's ancient death-fight with skepticism! They are nothing less than the end of the world! In the apocalypse proposed by those examples, not only does induction hang in the balance, but also planetary orbits and biological life. As Himsworth (1986:87–88) has pointed out in his critique of Hume, if the course of nature did change, we would not be here to complain! So as long as we are here or we are talking about induction, deep worries about induction are unwarranted. Consequently, seeing that apocalypse as "the problem of induction" rather than "the end of the world" is like naming a play for an incidental character. The rhetoric trades in obsessive attention to one detail.

Turning next to Goodman's new riddle of induction, it shows that although the straight rule of induction is itself quite simple, judging whether or not to apply it to a given property for a given sample is rather complicated. These judgments, as in the example of platypuses, draw on general knowledge of the world and common sense. Such broad and diffuse knowledge resists tidy philosophical analysis.

SUMMARY

Induction reasons from actual data to an inferred model, whereas deduction reasons from a given model to expected data. Both are important for science, composing the logic or "L" portion of the PEL model. Probability is the deductive science of uncertainty, whereas statistics is the inductive science of uncertainty.

A particularly important theorem derivable from probability axioms is Bayes's theorem, saying that the posterior equals the likelihood times the prior. It has been applied to an inference problem regarding blue and white marbles drawn from an urn. Then Bayesian decision theory, which requires just one more axiom, has been illustrated with a simple agricultural example. The frequentist paradigm has also been explained, particularly the meaning of p-values. Because scientists' questions often concern the probabilities of hypotheses being true, which only the Bayesian paradigm can answer, Bayesian statistics merits greater attention.

Inductive logic (which includes modern statistics) has been under severe philosophical criticism, especially since Hume. More recently, Goodman has uncovered additional problems. But given common-sense presuppositions, induction can be defended effectively.

CHAPTER EIGHT

PARSIMONY AND EFFICIENCY

The principle of parsimony recommends that from among theories fitting the data equally well, scientists choose the simplest theory. Thus, the fit of the data is not the only criterion bearing on theory choice. Additional criteria include parsimony, predictive accuracy, explanatory power, testability, fruitfulness in generating new insights and knowledge, coherence with other scientific and philosophical beliefs, and repeatability of results. The principle of parsimony has four common names, also being called the principle of simplicity, the principle of economy, and Ockham's razor (with Ockham sometimes Latinized as Occam).

Parsimony is an important principle of the scientific method for two reasons. First and most fundamentally, parsimony is important because the entire scientific enterprise has never produced, and never will produce, a single conclusion without invoking parsimony. Parsimony is absolutely essential and pervasive.

Second and more practically, parsimonious models of scientific data can facilitate insight, improve accuracy, and increase efficiency. Remarkably, parsimonious models can be more accurate than their data. Or, in other terms, parsimonious models can be extremely efficient, requiring considerably less data collection than do more complicated models to achieve the same accuracy and results. These advantages of accuracy and efficiency are important because scientists want to find the truth, but they also want to spend no more time and money finding the truth than is necessary.

This chapter addresses science's evidence, the "E" portion of the PEL model. Chapter 4 introduced this model, saying that every scientific conclusion, if fully disclosed, involves three kinds of premises, regarding presuppositions, evidence, and logic. Chapter 4 also disclosed science's presuppositions, the "P" portion of the PEL model, and Chapters 5–7 described science's deductive and inductive logic, the "L" portion. Now this chapter will complete the tour of the PEL model.

Most aspects of evidence are rather obvious to scientists, and most aspects involve specialized techniques useful only within a given discipline. For those two reasons, most of what needs to be said about scientific evidence is in the domain of specialized disciplines, rather than general principles. But the one great exception is parsimony, which is not obvious to many scientists but does pervade all of the sciences. Accordingly, this book's account of science's evidence takes the form primarily of a detailed analysis of parsimony.

Parsimony is not an unusually difficult topic, compared with the ordinary topics routinely studied by chemists, geologists, and other scientists. Also, because parsimony pervades all of science, it is easy to find interesting examples and productive applications. Nevertheless, the implementation of parsimony has always faced serious obstacles. In the first place, many scientists seem inclined to think that only a few words, such as "Prefer simpler models," can exhaust the subject. Such complacency does not motivate further study and new insight. Second, parsimony has been so thoroughly neglected in scientists' training that most scientists have no idea that something is missing. Parsimony is like a delicious tropical fruit that is not missed because it has never been tasted. Third and finally, the literature on parsimony is scattered in philosophy, statistics, and science, but few scientists read widely in those areas. Yet each of those disciplines provides distinctive elements that must be combined to achieve a full picture.

In some areas of science and technology, such as in signal processing, the principle of parsimony has already been well understood to great advantage. But in most areas, a superficial understanding of parsimony has been a serious deficiency of scientific method, costing scientists billions of dollars annually in wasted resources. Frequently, a parsimonious model that costs a few seconds of computer time can provide insight and increase accuracy as much as would the collection of more data that would cost thousands or millions of dollars. If more scientists really understood parsimony, science and technology would gain considerable momentum.

HISTORICAL PERSPECTIVE ON PARSIMONY

Before beginning this history, one must understand that parsimony has been discussed with two distinct but related meanings. On the one hand, parsimony has been considered a feature of nature, that nature chooses the simplest course. On the other hand, parsimony has been deemed a feature of good theories, that the simplest theory that fits the facts is best. That is, humans choose the simplest theory or explanation. These are ontological and epistemological conceptions, respectively, concerning nature itself and humans' theories about nature. Of course, these two conceptions are related, because science's epistemological methods aim at ontological knowledge about nature.

The venerable law of parsimony, the *lex parsimoniae*, has a long history. Aristotle (384–322 B.C.) discussed parsimony in his *Posterior Analytics*: "We may assume the superiority *ceteris paribus* [other things being equal] of the demonstration which derives from fewer postulates or hypotheses" (McKeon 1941:150). "That is done in vain by many means which may equally well be done with fewer" (Nash 1963:173). "They [the principles] should, in fact, be as few as possible, consistently with proving what has to be proved" (Aristotle, quoted by Hoffmann, Minkin, and Carpenter 1996). Likewise, in his *Almagest*, Claudius Ptolemy (A.D. 100–178) used parsimony to help decide between theories about planetary motions.

Aristotle also used parsimony as an ontological principle. For example, Aristotle rejected Plato's Forms on the basis of a parsimony argument. Plato (c. 427–347 B.C.) believed that both the perfect Form of a dog and individual dogs existed, but Aristotle held the more parsimonious view that only individual dogs existed. Hence, even something as elemental as the tendency in Western thought to regard individual physical objects as being thoroughly real derives from an appeal to parsimony. Likewise, in his influential commentary on Aristotle's *Metaphysics*, Averroes (Ibn Rushd, A.D. 1126–1198) regarded parsimony as a real feature of nature.

Robert Grosseteste (c. 1168–1253), who greatly advanced the use of experimental methods in science, also emphasized parsimony, as here in commenting on Aristotle: "That is better and more valuable which requires fewer, other circumstances being equal, just as that demonstration is better, other circumstances being equal, which necessitates the answering of a smaller number of questions for a perfect demonstration or requires a smaller number of suppositions and premises from which the demonstration proceeds" (Crombie 1962:86). Grosseteste held parsimony not merely as a criterion of good explanations or theories, but more fundamentally as a real, objective principle of nature. Thomas Aquinas (c. 1225–1274) espoused a rather ontological version of parsimony, writing that "If a thing can be done adequately by means of one, it is superfluous to do it by means of several; for we observe that nature does not employ two instruments where one suffices" (Hoffmann et al. 1996).

William of Ockham (c. 1285–1347) probably is the medieval scholar best known to modern scientists, through the familiar principle of parsimony, often called Ockham's razor (Thornburn 1918; J. J. C. Smart, in Fetzer 1984: 118–128; Wright 1991:77–104; Jefferys and Berger 1992; Gauch 1993; Marilyn M. Adams, in Audi 1999:629; Hoffmann et al. 1996). The principle of parsimony had been valued by many philosophers and scientists from Aristotle to

Grosseteste, but Ockham advanced the discussion considerably: "It is quite often stated by Ockham in the form: 'Plurality is not to be posited without necessity.' (*Pluralitas non est ponenda sine necessitate*), and also, though seldom: 'What can be explained by the assumption of fewer things is vainly explained by the assumption of more things' (*Frustra fit per plura quod potest fieri per pauciora*). The form usually given, 'Entities must not be multiplied without necessity' (*Entia non sunt multiplicanda sine necessitate*), does not seem to have been used by Ockham" (Boehner 1957:xxi; also see Thornburn 1918 and Hoffmann et al. 1996).

Just what does this principle mean? "What Ockham demands in his maxim is that everyone who makes a statement must have a sufficient reason for its truth, 'sufficient reason' being defined as either the observation of a fact, or an immediate logical insight, or divine revelation, or a deduction from these" (Boehner 1957:xxi). The Latin original, "*Nulla pluralitas est ponenda nisi per rationem vel experientiam vel auctoritatem illius, qui non potest falli nec errare, potest convinci,*" is given by Hoffmann et al. (1996). They add the perceptive remark that "in the context of science, especially interesting" is the part of Ockham's razor "that experience ('*experientia*') can serve to justify plurality." However, Ockham's principle of sufficient reason tends to reach modern scientists in a somewhat thinner, more focused version of parsimony, saying something like: Prefer the simplest model that fits the data accurately (Jefferys and Berger 1992; Gauch 1993), or "one should not complicate explanations when simple ones will suffice" (Hoffmann et al. 1996).

Ockham insisted that parsimony was an epistemological principle for choosing the best theory, in contrast to his predecessor Robert Grosseteste and his teacher John Duns Scotus, who had interpreted parsimony as also an ontological principle for expecting nature to be simple. In Ockham's view, "This principle of 'sufficient reason' is epistemological or methodological, certainly not an ontological axiom" (Boehner 1957:xxi). Recall again Ockham's insistence that experience can require plurality, which can be restated in a more modern idiom by saying that experiment can justify complexity. "William of Ockham opposed this tendency [of Grosseteste and many other medieval writers] to read into nature human ideas about simplicity. He felt that to insist that nature always follows the simplest path is to limit God's power. God may very well choose to achieve effects in the most complicated of ways. For this reason, Ockham shifted emphasis on simplicity from the course of nature to theories which are formulated about it" (Losee 1993: 38–39).

A striking example of Ockham's application of parsimony was his rejection of the impetus theory of projectile motion advanced by Jean Buridan, based on earlier ideas from Aristotle. Ockham "defined motion as a concept, having no reality apart from moving bodies, that was used to describe the

fact that from instant to instant a moving body changed its spatial relationships with some other body without intermediate rest. There was no need to
postulate any external or internal efficient cause to explain such a sequence
of events" (Crombie 1962:176). That is, motion was neither a separate thing
nor a property of a thing, but rather a modification of existing things, namely,
a change in location from one time to another. Ockham's penetrating insight
about motion became the basis for the seventeenth-century theory of inertia,
replacing the earlier concept of impetus (Adams 1987:799–852).

Nicolaus Copernicus (1473–1543) inherited the geocentric cosmology of
Aristotle and Ptolemy. It was commensurate with the data (saved the appearances) within observational accuracy, accorded with the common-sense
feeling that the earth was unmoving, and enjoyed the authority of Aristotle;
but its one major flaw was lack of parsimony, what with its complicated cycles
and epicycles for each planet (Dampier 1961:109). Consequently, Copernicus
offered a new theory: that the earth revolved on its axis daily and journeyed
around the sun annually. The kinds of arguments advanced by Copernicus are
important for revealing his underlying conception of scientific reasoning. His
argument was not based on better fit, because the geocentric and heliocentric
models had the same accuracy, and both used eccentrics and epicycles.

Rather, his main argument featured parsimony: that the heliocentric model
was simpler, involving fewer epicycles, and the various motions were inter-
linked in a harmonious system: "I found at length by much and long observation, that if the motions of the other planets were added to the rotation
of the earth, and calculated as for the revolution of that planet, not only the
phenomena of the others followed from this, but that it so bound together
both the order and magnitudes of all the planets and the spheres and the
heaven itself that in no single part could one thing be altered without confusion among the other parts and in all the Universe. Hence for this reason . . . I
have followed this system" (Dampier 1961:110). For example, the Ptolemaic
system had a separate parameter for the time required for each planet's journey (supposedly around the earth), but it was curious and inexplicable that
those parameters were identical for Mercury, Venus, and the sun's seasonal
cycle, namely, one year (Williams and Steffens 1978:12). The Copernican system, with its ordering of the planets that placed the smaller orbits of Mercury
and Venus (as well as the sun's central location) inside the earth's larger orbit, provided an elegant and simple explanation that allowed one parameter
to replace three.

Another possible consideration was an experimental test, the obvious test
being parallax: "If the earth did move around the Sun then there ought to be
a perceptible angle difference when a fixed star is observed from the Earth
at opposite sides of the Earth's orbit" (Williams and Steffens 1978:19). But
failure to observe parallax would not necessarily prove that the earth was

stationary, because the stars might be too far away for the earth's movement to cause measurable parallax. Unfortunately, the experiment had a negative, inconclusive outcome until centuries later, when in 1838 Friedrich Bessel first achieved accuracy adequate to show parallax. Actually, the very first direct evidence for the motion of the earth around the sun had come a century earlier, in 1728, when James Bradley had discovered the aberration of starlight, an effect more than ten times as great as that of parallax and hence easier to detect.

Model accuracy was ambivalent and experimental tests were inconclusive for the Copernican theory choice, so the weight of the argument fell on parsimony. From the perspective of science, Copernicus was revolutionary for placing the sun in the center of the cosmos; but from the perspective of scientific method, Copernicus was revolutionary for elevating parsimony to a prominent position.

Although parsimony has an essential and salutary role in scientific method, in fairness this historical review must mention an instance of parsimony gone awry. Galileo Galilei (1564–1642) had inherited from Aristotle the view that there were two kinds of motion: Beyond the sphere of the moon, motions were circular; but below the lunar sphere, undisturbed motions were rectilinear (with heavy bodies moving straight down toward the earth's center, and lighter-than-air bodies moving straight up). In an ill-fated application of parsimony, Galileo unified and simplified that theory by claiming that all undisturbed motion was circular and that rectilinear motion was just an illusion. For example, a ball dropped from a tower appeared to fall in a straight line, but supposedly an observer off the earth would see the true curvilinear motion. "Circular inertia" was a key concept in Galileo's physics, but it has proved to be false. Nevertheless, as shown by Newton and refined by Einstein, the more basic element in Galileo's thinking has proved correct: A single parsimonious theory of motion does work, both beyond and below the moon's orbit.

Isaac Newton (1642–1727) further anchored parsimony's importance with the four rules of reasoning in his monumental and influential *Philosophiae Naturalis Principia Mathematica* (Cajori 1947:398–400). Parsimony was the first rule, expressed in a vigorously ontological version concerning nature that echoed words of Aristotle and Duns Scotus: "We are to admit no more causes of natural things than such as are both true and sufficient to explain their appearances." Newton explained: "To this purpose the philosophers say that Nature does nothing in vain, and more is in vain when less will serve; for Nature is pleased with simplicity, and affects not the pomp of superfluous causes." Again, parsimony, in a distinctively epistemological version concerning theories about causes, was the second of Newton's rules, corollary to the first: "Therefore to the same natural effects we must, as far as possible, assign the same causes," such as for "respiration in a man and in a beast" and "the reflection of light in the earth, and in the planets." And even his third and fourth rules about experiments and induction rested on the presupposition that nature "is wont to be simple." So in Newton's science, various aspects of parsimony were nothing less than his four fundamental rules of reasoning in science. Parsimony was the centerpiece of his scientific method.

Gottfried Leibniz (1646–1716) also gave parsimony a distinctively onto-logical interpretation. He used differential calculus to prove that the path of a light ray from one medium to another, described by Snell's law, minimized the path difficulty (the length of the path times the resistance of the medium). Leibniz took that success "as support for the metaphysical principle that God governs the universe in such a way that a maximum of 'simplicity' and 'perfection' be realized" (Losee 1993:104).

More recently, Albert Einstein (1879–1955) employed parsimony in his discovery of general relativity: "Perhaps the scientist who most clearly under-stood the necessity for an assumption about the simplicity of [scientific] laws was Albert Einstein. In an informal conversation he once told me about his thoughts in arriving at The General Theory of Relativity. He said that after years of research, he arrived at a particular equation which, on the one hand, explained all known facts and, on the other hand, was considerably simpler than any other equation that explained all these facts. When he reached this point he said to himself that God would not have passed up the opportunity to make nature this simple" (Kemeny 1959:63; also see Schilpp 1951:23, 137, 255–257). Likewise, Einstein spoke of "the grand aim of all science, which is to cover the greatest possible number of empirical facts by logical deduc-tions from the smallest possible number of hypotheses or axioms" (Nash 1963:173). He also remarked that "Everything should be made as simple as possible, but not simpler" (Hoffmann et al. 1996).

Similarly, Ernst Mach (1838–1916) said that "science itself can be consid-ered as a problem of minimum, which consists in telling the facts as perfectly as possible, with the least intellectual expense" (A. Sevin, in Hoffmann et al. 1996). Henri Poincaré (1854–1912) related parsimony to generalization: "Let us first observe that any generalization implies, to a certain extent, belief in the unity and simplicity of nature. Today, ideas have changed and, neverthe-less, those who do not believe that natural laws have to be simple, are obliged to behave as if it was so. They could not avoid this necessity without render-ing impossible all generalization, and consequently all science" (A. Sevin, in Hoffmann et al. 1996). Yet Poincaré also appreciated the subtlety of simplic-ity, bringing counterpoint with his view that "simplicity is a vague notion" and "everyone calls simple what he finds easy to understand, according to his habits" (A. Sevin, in Hoffmann et al. 1996).

Historically, philosophers and scientists have been the scholars who have written about parsimony. More recently, statisticians have also explored this subject, offering two new and important results. First, simple theories tend to make reliable predictions (Frank 1957:352–353; Jeffreys 1983:1–5; Friedman 1990; Gauch 1993). Second, Bayesian analysis automatically gives simple theories the highest prior probabilities of being true [Jeffreys 1973, 1983; Mary Hesse, in Edwards 1967(7): 445–448; Jefferys and Berger 1992; MacKay 1992; Berger and Pericchi 1996; Hoffmann et al. 1996; Elliott Sober, in Craig 1998(8):780–783]. Furthermore, even if the competing models have equal prior probabilities, the resulting posterior probabilities will favor simpler models.

In line with these statisticians, the philosopher Richard Swinburne also sees simplicity as evidence of truth and of reliable predictions: "I seek ... to show that – other things being equal – the simplest hypothesis proposed as an explanation of phenomena is more likely to be the true one than is any other available hypothesis, that its predictions are more likely to be true than are those of any other available hypothesis, and that it is an ultimate a priori epistemic principle that simplicity is evidence of truth" (Swinburne 1997:1).

From its inception with Aristotle, the criterion of parsimony has always had a *ceteris paribus* clause, and that remains so in its modern statistical formulation. Quoting from this chapter's first sentence, the statisticians' version of other things being equal is "among theories fitting the data equally well," after which "The principle of parsimony recommends that ... scientists choose the simplest theory." Hence, parsimony is actually a dual criterion, weighing both the theory's fit with the data and the simplicity of the theory.

But scientists never have the luxury of a theory combining perfect fit and perfect simplicity, so a delicate trade-off ensues between fit and simplicity. Often one theory has the best fit, whereas another has the greatest simplicity, so the best choice is not obvious. Statisticians have made tremendous contributions in understanding and optimizing this trade-off in quantitative detail. They have also shown that the optimal trade-off depends on how the scientific results are to be used, particularly whether the scientists' intention is to fit old observations or to predict new observations. And statisticians have demonstrated that parsimonious models often can yield considerable gains in accuracy and efficiency, which is an important practical benefit that had never emerged clearly in earlier philosophical discussions. Indeed, statisticians have discovered so much about parsimony that their work is beginning to attract attention from philosophers [Wright 1991:85–88; Forster and Sober 1994; Kukla 1995; DeVito 1997; Kieseppä 1997, 2001a,b; Swinburne 1997:15–19; Elliott Sober, in Craig 1998(8):780–783; Bandyopadhyay and Boik 1999; Malcolm R. Forster, in Audi 1999:197–198; Forster 1999; Elliott Sober, in Newton-Smith 1999; Mulaik 2001].

It may be suspected that this late blooming of the practical aspect of parsimony is attributable not to our forebears' lack of insight but rather to their lack of computers. Fitting a parsimonious model to data typically requires millions of arithmetic steps, which would have been inconceivable to Ockham or Copernicus. But since around 1960, computers have become increasingly available to scientists. Coincidentally, beginning with Stein (1955), conceptual advances in statistics have revealed the practical potential of parsimonious models to help scientists gain accuracy and efficiency. Although affordable computers and relevant theory have now been in place for some decades, most areas of science and technology still show little understanding of parsimony's important role in scientific method.

As a criterion in theory choice, parsimony can be regarded as one aspect of a more general criterion, beauty. McAllister (1996, 1998) has examined the question whether or not beauty is a sign of truth in scientific theories. He has also documented historical changes in scientists' aesthetic ideals.

This brief review has explained something of parsimony's intriguing past, but more engaging will be its exciting future. History shows that many notable paradigm shifts in science were precipitated by arguments concerning parsimony, rather than better fit to the data. Why? Evidently, false scientific theories often get into trouble with parsimony before they get into trouble with the issue of more extensive or more accurate data. For example, Copernicus realized that the geocentric theory was in trouble on grounds of parsimony centuries before Bessel proved that it was in trouble for not fitting with the data that showed parallax. The lesson here for contemporary scientists is that those scientists who also consider a theory's parsimony, rather than only its fit with the data, are often the ones on the cutting edge of science.

PREVIEW OF BASIC PRINCIPLES

This chapter's primary tools for exploring parsimony are the following four examples of parsimony at work in science. But simplicity is a complicated topic! Accordingly, this section first previews six basic principles.

Figure 8.1 characterizes the typical relationship between parsimony and accuracy that has been observed for countless data sets and diverse models (Gauch 1993). The abscissa depicts a sequence of increasingly parsimonious models moving toward the left (or increasingly complex models moving toward the right). For example, this could be a polynomial family, with its most simple model (the constant model) at the extreme left, then the more complex linear, quadratic, cubic, and higher models progressing toward the right, and finally its most complex model (the full model) at the extreme right. The full model has as many parameters as the data, and its estimates automatically

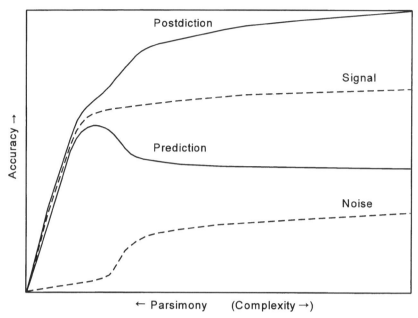

Figure 8.1. Predictive and postdictive accuracies of models differing in terms of parsimony. The abscissa represents more parsimonious models to the left (and more complex models to the right), and the ordinate shows model accuracy. Imperfect data are mixtures of real signal and spurious noise. Signal is recovered quickly at first as models become more complex, but thereafter signal is recovered slowly. By contrast, noise is recovered slowly at first while much signal is being recovered, then for a brief time noise is recovered more quickly, but thereafter slowly. Postdictive accuracy increases as the signal plus the noise, so it always increases for more and more complex models. But predictive accuracy increases as the signal minus the noise, so it rises to a maximum for some relatively parsimonious model, and thereafter declines.

equal the data exactly (such as a quadratic equation with its three parameters automatically going through three data points exactly). The ordinate shows model accuracy or goodness of fit.

(1) Signal and Noise. To understand this figure, recognize, first of all, that data are imperfect, mixtures of real signal and spurious noise. These terms, "signal" and "noise," originated in the context of radio communication, where a receiver picks up the signal from a transmitter plus noise from various natural and human sources. But in statistics, these terms are used more generally to refer to treatment effects and random errors. The signal consists in responses to imposed treatment factors, such as five different diets fed to different groups of mice, whereas the noise is uncontrolled variation, such as individual differences among the mice, slight variations between batches of a theoretically standard food mixture, and measurement errors.

The fundamental concept is that the data equal the signal plus the noise. Hence, the noise in the data is the discrepancy between the true signal and the imperfect data. Likewise, the noise in a model is the discrepancy between the true signal and the model's estimates.

Increasingly complex models capture more and more of the total variation in a data set, but the signal and noise are captured differentially, as depicted by the dashed lines in Figure 8.1. Signal is recovered quickly at first, but slowly thereafter. However, noise recovery has a very different pattern. At first noise is recovered slowly, then briefly somewhat more quickly, and finally slowly thereafter.

The reason for those differences is that the signal ordinarily is relatively simple, caused by only a few major treatment differences or causal factors, whereas the noise typically is relatively complex, caused by numerous small uncontrolled factors. Because the signal is parsimonious, it can be captured by an appropriate parsimonious model, but complex noise inevitably requires a complex model. Ordinarily, the signal is more powerful than the noise, but in any case the initial focus on signal suppresses recovery of noise. After most of the signal has been captured, the focus then shifts to the noise. At that point, chance correlations in the noise can be exploited briefly by statistical analyses to accelerate the recovery of noise, but after that opportunity has been largely exhausted, noise is recovered slowly.

This selective recovery of signal and noise, with signal recovered first and noise last, has profound implications for model choice and accuracy. Clearly, if one knew just where to quit, one could have a model that would capture most of the data's signal but discard most of the data's noise. This separation of signal from noise offers a strategy for gaining accuracy, which equates to gaining efficiency. Consequently, parsimonious models can achieve accurate results with fewer data, which is valuable because scientific data are costly.

(2) Populations and Samples. Another fundamental concept is that the data usually are limited, comprising only a sample from a larger population of interest. For example, five different diets could be fed in a replicated experiment to groups of eight mice for each diet, for a total of 40 mice. But ordinarily the conclusion of the research, such as "diet C yields better growth than the other four diets," is intended to apply to mice in general, or at least to mice of some particular kind or species, rather than to only the 40 mice actually in the study. Thus scientific experiments usually concern two distinguishable entities: the set of actual individuals included in an experiment, called the "sample," and the set of all individuals addressed by the research's questions and conclusions, called the "population." Ideally, the data comprise a large and representative sample from the population of inference.

(3) Prediction and Postdiction. The fundamental distinction between a sample and a population leads to a further distinction between predictive accuracy and postdictive accuracy, as depicted by the solid lines in Figure 8.1. Here "prediction" and "postdiction" have technical statistical meanings that in general do not equate to their literal meanings of speaking about the future and the past. Rather, the distinction here concerns whether or not the data are used to make ampliative inferences beyond the data themselves. Predictive accuracy concerns a model's fit to the population, whereas postdictive accuracy concerns a model's fit to a sample. Hence, in their technical statistical meanings, prediction might be called "population-diction" because it is speaking about a population, and postdiction might be called "sample-diction" because it is speaking about a sample. Although those two terms are more obvious, they are also quite clumsy, so it would be burdensome to recommend that they replace the conventional terms. Nevertheless, they merit mention just once to facilitate a clear and simple grasp of the essential meanings of prediction and postdiction.

Prediction and postdiction are two different functions entailing different principles, different model choices, and different relevances and accuracies for different applications. The crucial feature of statistical prediction is that a model not only must fit the sample data but also must fit other data representing more of the entire population. Such other data might be collected in the future, though not necessarily. The goal in postdiction is to fit with the sample data at hand, whereas the goal in prediction is to fit with the entire population from which the sample was drawn.

Consider the line for prediction in Figure 8.1. Envision collecting data on two individuals sampled from the same population, so they are replicate observations. The paradigm of noisy data can then be represented as $Data_1 = Signal + Noise_1$ for the one individual, and $Data_2 = Signal + Noise_2$ for the second individual. The signal, due to treatment factors such as diet, is held in common among all individuals or replicates in a given treatment group. But the noise, due to uncontrolled factors such as experimental errors, is idiosyncratic. Consequently, replicated data have the same signal but different noise and hence different values (which is why Data and Noise carry subscripts, but not Signal). So knowing the signal for one individual helps when predicting the response for another replicate given the same treatment. But taking seriously the noise for one individual when making a prediction for another replicate individual is a mistake, because noise is idiosyncratic, of no predictive value. Consequently, predictive accuracy is helped by capturing signal but is hindered by capturing noise. Accordingly, the line for predictive accuracy in Figure 8.1 is depicted as the signal line minus the noise line.

Consider next the line for postdiction in Figure 8.1. The goal in postdiction is to model the sample data at hand, with no serious concern about a larger population or about the distinction between signal and noise. More complex models continue to fit the data more accurately until finally a full model automatically fits the data perfectly. Recovery of signal and recovery of noise are rewarded equally. Accordingly, the line for postdictive accuracy in Figure 8.1 is depicted as the signal line plus the noise line.

The lines for prediction and postdiction are different because of noise. The pattern minus the noise line for prediction and the pattern plus the noise line for postdiction would be identical if the noise were negligible.

Quite importantly, these two lines have different shapes, reaching their peaks of maximum accuracy at different places. Postdictive accuracy is automatically maximized by the most complex model at the extreme right, the full model that equals the data. But predictive accuracy is maximized for some relatively parsimonious model closer to the left, rather than at the extreme right where a full model equals its data. This means that parsimonious models can be more predictively accurate than their data! That is the central message of this chapter, offering scientists a tremendous opportunity to gain accuracy and efficiency.

Also notice that the model that optimizes predictive accuracy for the population frequently is different from and is more parsimonious than the model that optimizes postdictive accuracy for a sample. Prediction and postdiction are different processes that lead, in general, to different model choices.

Given the distinction between predictive accuracy and postdictive accuracy, which is more relevant and important for science? Almost always, the objectives of scientific research are thoroughly predictive, not postdictive, and accordingly the relevant criterion is predictive accuracy. That is, ordinarily the data comprise only a sample, not the entire population, yet the statistical inferences pertain to the whole population from which the sample was taken. To give just one representative example, the primary objective motivating the funding of cancer research is to identify superior treatments to benefit a large population of cancer patients, not merely to describe what has happened to a small sample of persons included in the clinical trial. Rescher (1998) has provided a nice overview of prediction.

(4) The Curve-fitting Problem. In the literature on parsimony, the archetypical example is the curve-fitting problem. The abscissa and ordinate represent the input and output of some process, dots show the data, and various curves show the relationships between inputs and outputs expected on the basis of various theories or models, as in Figure 8.2. The scientist's task is then to select the best theory by comparing theoretical predictions with actual data. That immediate task must be preceded or accompanied, however, by the deeper philosophical and scientific task of specifying sensible and workable criteria defining exactly what is meant by the "best" theory, including specifying acceptable trade-offs when multiple criteria are in conflict. The story varies, however, depending on whether the data are relatively accurate or noisy, so Figure 8.2 has two panels.

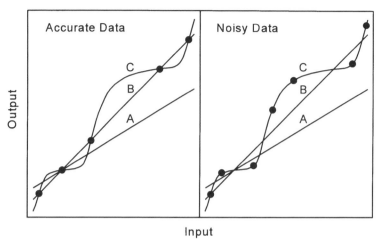

Figure 8.2. Choice among three models, A or B or C, for the relationship between the input and output of some process. Dots show the data, which can be accurate data (left panel) or noisy data (right panel). Goodness of fit and parsimony are relevant criteria for a model choice.

Consider first the simpler case of extremely accurate data, shown on the left in Figure 8.2. There are five data points, and there are three different theories, generating straight line A, straight line B, and curved line C. Given these data and a choice among these three theories, which theory do you pick? Why? And what is wrong with the other two lines that you rejected?

Obviously, virtually every scientist would pick straight line B that goes right through all five data points. But understand that the rejected alternatives, A and C, are rejected for different reasons. What is wrong with line A? Obviously it is rejected for failing to fit the data. Line A fits just one datum, whereas the better line B fits all five data. However, the other line, C, is not rejected for that same reason of lack of fit. Indeed, curved line C fits the data just as well as the preferred line B, going right through all five data points. So what is wrong with curve C? It is rejected on grounds of parsimony. Among theories that fit the data equally well, we prefer the simplest, and a straight line can be specified by a simpler equation with fewer terms than can a curved line with several inflections.

One reason why scientists value parsimony is that it tends to help predictions, that is, parsimony promotes the future or further success of a model. For example, the straight line B seems more likely than the curved line C to predict sensible values for future data from new levels of the input. For an overly complex model, like curve C, "including too many terms will *lose* accuracy in prediction instead of gaining it" (Jeffreys 1983:46). Indeed, assuming that the current data are representative of the future or other data that might be collected, model B has already provided some evidence of predictive success, because its mere two parameters fit five data points.

> It is axiomatic that any function with a number of adjustable parameters equal to or greater than the number of observations can be made to pass exactly through all of the ... [data] points on the graph. However, it is not necessarily true that a function with fewer adjustable parameters than the number of observations will pass through all of the points. If it turns out that it does, then the function ... has already had some success in describing [or predicting] one or more events that we have measured experimentally.... Models that predict are valued.... [We] are naturally more inclined to choose the model that has already had the greater success in "predicting" the measurements we have made so far. This will be the model with ... the smaller number of adjustable parameters – ie, the simpler model. (Hoffmann et al. 1996)

Besides their superior properties for prediction, parsimonious models are valued for three additional virtues (Hoffmann et al. 1996). First, "The simpler model is likely to be more *vulnerable* to future falsification, because with fewer adjustable parameters it will have less flexibility." Karl Popper rightly insisted that a legitimate scientific hypothesis risks falsification (Mulaik 2001). Second, one could say that "the simpler model provides a clearer and more readily *comprehensible* description." It better suits the limited capacities of human understanding. Lastly, "The number of equally satisfactory models in a given class is generally related to the complexity of the class," so simpler models provide a smaller class and hence an easier choice. For example, in the left panel of Figure 8.2, among simple straight lines, only one line is obviously best, model B, not model A. But among complex polynomial models with four or more parameters, such as model C, an infinite number of candidates could be offered that would be equally acceptable, so the current data would provide no basis for selecting one specific model over its competitors.

Anyway, both goodness of fit and parsimony are required to account for the choice of theory B over rejected theories A and C. Goodness of fit alone cannot justify the choice of theory B.

Consider next the harder case of rather inaccurate data, depicted on the right in Figure 8.2. There are seven data points and exactly the same three theories as before, A, B, and C. Assume that the noise in these data can amount to scattering data points up or down by as much as the distance from the label "A" to the label "B" or occasionally even to the label "C." Again, which theory do you pick? Why?

Now the choice is tougher, but most scientists would still pick theory B. Why? Well, line A is rejected fairly easily because it lies farthest from the data, at least for the four data points on the right. However, exactly because these data are noisy, there is some possibility that a true theory A could have generated the observed data. Consequently, the presence of noise makes the rejection of A a weaker matter of inductive probability, in contrast to the original absence of noise that made the rejection of A a strong matter of deductive certainty. Therefore, knowledge of a data set's accuracy (from replication or whatever) is an important consideration in theory choice.

The remaining competition between B and C depends substantially on whether curve C was constructed from other knowledge apart from these particular seven observations or was constructed *ad hoc* specifically to fit these observations. Assuming that the latter is the case, and knowing that these data are rather noisy, a smart scientist will pick B over C because of its advantage of parsimony. That argument could be strengthened if known physical laws or other similar experiments were to support a simple relationship between the input and output in this figure. Also, a more nearly definitive model choice could result if more data could be collected.

Finally, actual scientific research, unlike the fiction in Figure 8.2, often has experiential, experimental reasons for expecting simple relationships: "With the important and fascinating exception of systems on the threshold of chaotic behavior, or those near phase transitions [or other causes of abrupt change], our experience suggests to us that the universe is much more a system of smooth curves than jagged edges. It is not often that small changes in some control factor cause wild and unpredictable swings in the response of the system under study" (Hoffmann et al. 1996). To invoke metaphysics to comment about nature exhibiting smooth changes and simple relationships might seem contrary to Ockham's sentiments about the razor that now bears his name, but remember Ockham's own respect for experience. Perhaps an acceptable way to express or legitimate the ontological aspect of Ockham's razor is to say that scientists have experienced far more instances of smooth changes and simple relationships than instances of abrupt changes and chaotic behavior, so when encountering some new phenomenon, they are inclined to put most of the prior probability on a parsimonious hypothesis, but then to let experimental outcomes determine the final verdict. Perhaps another way to express scientists' thinking is to say that parsimony arguments operate on two levels: a subject-matter preference for parsimonious models, motivated by a higher-order expectation for simple phenomena that has emerged from extensive general experience of the physical world.

(5) Related Data. A significant feature of ordinary data sets is that the data are related, concerning some particular phenomenon or topic, rather than being an odd collection of assorted facts. Related data partially illuminate one another. Hence, knowing one outcome provides clues about related outcomes.

Table 8.1. Equivalent conductivities
(Ω^{-1} cm^2 equiv^{-1}) of aqueous solutions of
hydrochloric acid at various concentrations and
temperatures

Concentration	Temperature		
(mol l^{-1})	25°C	30°C	35°C
4.5	183.1	196.6	209.5
5.0	167.4	—	191.9
5.5	152.9	165.0	175.6

Source: These data, from Lide (1995:5-87), are reproduced with kind permission from CRC Press. © 1995
CRC Press, Boca Raton, Florida.

For example, consider the following specific chemistry question, and attempt to answer it from memory without recourse to books. What is the
equivalent conductivity of an aqueous solution of hydrochloric acid at a
concentration of 5.0 mol l^{-1} and a temperature of 30°C? Doubtless few
chemists could give a reasonably accurate answer offhand, and even fewer
other scientists.

But now consult Table 8.1, which does *not* answer this question. However,
it does give values for eight experiments related to our question, namely, for
eight surrounding combinations of concentrations and temperatures (Lide
1995:5-87). Equipped with these data, which do *not* answer the question that
you were asked, can you now give a respectable estimate for the desired
quantity?

Well, now the task is easy! Everyone can supply a good approximation,
that the answer is about 180 Ω^{-1} cm^2 equiv^{-1}. Among various sensible approaches, one easy solution is just to average the surrounding eight values,
which gives 180.2. This answer is just an approximation, but it happens to
agree exactly with the measurement published (Lide 1995).

This estimate (or any other similar estimate) is motivated by a parsimony
argument, saying that responses typically change in a simple and smooth
manner with small changes in inputs. Hence, we expect something like 180
in the middle of this table, rather than some wild number like 15 or 572.
These data are related by a simple theory about smooth changes, so these
values partially illuminate one another, even to the extent of making possible
a decent estimate of a missing value.

How do parsimonious models increase accuracy? The key means is extraction of additional information from related data. The estimates provided
by a parsimonious model fitted to numerous observations make use of more

data than do estimates merely reflecting individual observations (or individual averages over replicates). The use of more data means more accuracy. This claim will be documented momentarily, but it is hoped that this basic intuition already seems plausible.

(6) **Statistical Terminology.** At this point, several statistical terms require definition. The root-mean-square (rms) of a collection of numbers is the square root of the arithmetic mean of the squares of the numbers. For example, consider the data 7, 8, and 9. Their rms equals $[(7^2 + 8^2 + 9^2)/3]^{0.5} \approx 8.04156$. For comparison, their average is 8. The rms is similar to the average, but places somewhat greater weight on the larger values in the data.

The sum of squares (SS) is the sum of squared values, usually calculated after first subtracting the grand mean. The total number of degrees of freedom (df) for a data set is the number of observations minus one. And the mean square (MS) is SS divided by df. For example, consider the three observations 118, 94, and 88. The grand mean is 100, which is subtracted from the data to obtain 18, −6, and −12. Accordingly, $SS = 18^2 + (-6)^2 + (-12)^2 = 504$, with $3 - 1 = 2$ df, and hence $MS = 504/2 = 252$.

Models also have degrees of freedom. In the simplest case, the model's df equals the model's number of parameters. For example, the quadratic equation $y = a + bx + cx^2$ has three parameters (the constants a, b, and c), so it has 3 df. Particularly for similar models from a given model family, a model's df is a convenient measure of the model's complexity. Or, to say the same thing in reverse, models with fewer degrees of freedom are more parsimonious.

An important feature of an experiment is the relative magnitude of signal and noise, the signal-to-noise (S/N) ratio. The specific definition used here is that the S/N ratio equals the signal sum of squares divided by the noise sum of squares. A large S/N ratio means that an experiment is accurate, and thus it can detect relatively small treatment differences.

This chapter's examples use two model families, the polynomial family and the Additive Main effects and Multiplicative Interaction (AMMI) family. Polynomial equations are too familiar to require explanation here.

AMMI uses a combination of analysis of variance and principal-components analysis to model data arranged as a two-way data table (Gauch 1992). For present purposes, it suffices to explain that the AMMI family is like the polynomial family in that it generates a sequence of increasingly complex models. For a data set with R rows and C columns, the most simple and most complex extremes are denoted AMMI-0, with no principal-components axes, equivalent to the additive analysis-of-variance model, and the full model AMMI-F, with F equal to the minimum of $R - 1$ and $C - 1$, and this full model equals the raw data. Hence, AMMI analysis provides a sequence of models from the additive model to intermediates to the raw data. For example, a data matrix with 19 rows and 13 columns, such as will be encountered momentarily, has an AMMI family with 13 members. The most parsimonious model is AMMI-0, then AMMI-1, AMMI-2, and so on, up to the most complex model AMMI-12, which is the full model, AMMI-F. Hence, the situation for AMMI analysis of a 19 × 13 matrix is analogous to that for the polynomial family for a data set with 13 data points, having 13 polynomial models of order 0, 1, 2, and so on, up to 12, which is the full model.

This chapter's first graph showed, by way of preview, the relationship between model complexity and accuracy. The ordinate of Figure 8.1 was marked "accuracy," with no numerical scale given. Although a vague concept of "accuracy" was sufficient for the preview, as this chapter progresses, an exact and quantitative measure of accuracy will be needed. The convenient measure used here is statistical efficiency.

Statistical efficiency is defined as the variance of the data around the true values divided by the variance of the model around the true values. Equivalently, it is the SS of the data around the true values divided by the SS of the model around the true values, because this change merely multiplies both the numerator and denominator of this ratio by the same value, the number of observations. As a hypothetical example, assume that for three treatments the true values are 10, 20, and 30, the data are 7, 15, and 32, and a model's estimates are 14, 22, and 29. Then the SS of the differences between the data and the true values is $(-3)^2 + (-5)^2 + 2^2 = 38$, and the SS of the differences between the model and the true values is $4^2 + 2^2 + (-1)^2 = 21$. Hence, the statistical efficiency is $38/21 \approx 1.81$. The larger the statistical efficiency, the better the model estimates the true values.

Statistical efficiency also has a convenient interpretation or equivalent definition in terms of replications. The variance of the average of N replications equals the variance of individual observations divided by N. Accordingly, the statistical efficiency of a parsimonious model equals the number of replications that would be needed by the full model to achieve the same predictive accuracy as the parsimonious model, divided by the number of replications supplied to the parsimonious model. For example, if the full model needs 3.62 replications to match the accuracy of a parsimonious model supplied only 2 replications, then the statistical efficiency of the parsimonious model is $3.62/2 = 1.81$. The larger the statistical efficiency, the better the model extracts information from its data.

These two aspects of statistical efficiency, in terms of accuracy and efficiency, are related conceptually and unified mathematically. By "accuracy" we mean model estimates being close to the true values, and by "efficiency" we mean getting close with as little data or as few replications as possible. Integrating these concepts, by "statistical efficiency" we mean getting good accuracy from limited data – getting the most out of the data. An efficient model requires fewer data to accomplish a given job. Consequently, the modest effort and cost invested in vigorous statistical modeling can often repay scientists with much greater savings in the cost of collecting data. Indeed, scientists need to be reminded more frequently that it is costly to collect valuable data and then extract only a small fraction of the information therein. That is to be data-rich but information-poor.

To understand some of the graphs later in this chapter that concern statistical efficiency, not all of these technicalities need be remembered, but it is essential to retain the basic meaning. Statistical efficiency measures a model's accuracy and efficiency, which are the two sides of a single coin, the coin of getting more out of a given quantity of data. A statistical efficiency of 1 means that a model yields the same accuracy as its data provide. A statistical efficiency of 2 or 3 means that a model gives the some accuracy as would averages over replications based on twice or thrice as much data. And a statistical efficiency of 0.5 means that a model offers less accuracy than its data.

EXAMPLE 1: MENDEL'S PEAS

This section begins this chapter's four examples of parsimony at work in science. These examples become progressively more difficult. They illustrate parsimony in the mathematical, physical, and biological sciences.

For a simple but beautiful first example of parsimony and efficiency, consider Gregor Mendel's classic experiments regarding the genetics of garden peas (Figure 8.3) (Mendel 1865; Bennett 1965; Gauch 1993). Mendel crossed different pea varieties to obtain hybrid progeny and then self-pollinated the progeny for one or more generations. The original two parental lines in a given cross differed for certain traits, such as tall or short plants and yellow or green seed endosperms. In regard to any such characteristic, the hybrid progeny all expressed the trait of just one parent, and Mendel termed that the dominant trait; but the other trait reappeared again intact in later generations, and he termed that the recessive trait. For example, a cross between tall and short plants gave only tall progeny, but the next generation included both tall and short plants, so tall was dominant and short was recessive. Mendel conducted experiments using seven different traits.

For the progeny from hybrids between tall and short plants, Mendel reported that "Out of 1064 plants, in 787 cases the stem was long, and in 277 short. Hence a mutual ratio of 2.84 to 1" (Bennett 1965:18). But as Mendel's experiments and thinking continued, that estimate of the tall:short ratio was

Figure 8.3. Garden pea, *Pisum sativum* L., the main species that Gregor Mendel used in his genetics experiments. (From Vilmorin-Andrieux, 1885:391.)

revised in interesting ways that illustrate the workings of related data and parsimonious models.

Mendel's seven experiments with progeny from hybrids yielded the following: 5,474 round and 1,850 wrinkled ripe seeds, for a 2.96:1 ratio; 6,022 yellow and 2,001 green seed endosperms, for a 3.01:1 ratio; 705 dark and 224 white seed coats, for a 3.15:1 ratio; 882 inflated and 299 constricted ripe pods, for a 2.95:1 ratio; 428 green and 152 yellow unripe pods, for a 2.82:1 ratio; 651 axial and 207 terminal flower positions, for a 3.14:1 ratio; and 787 tall and 277 short plants, for a 2.84:1 ratio. His theoretical insight was that all seven experimental ratios were related instances of a single phenomenon – that round:wrinkled, yellow:green, tall:short, and so on, are all instances of dominant:recessive. Accordingly, he observed that "If now the results of the whole of the experiments be brought together, there is found, as between the number of forms with the dominant and recessive characters, an average ratio of 2.98 to 1" (Bennett 1965:19). Seven separate ratios were combined, by the theory of dominant and recessive traits, into a parsimonious model with one ratio. Thus the empirical result for tall:short of 2.84:1 was revised to a more accurate estimate of 2.98:1 by combining information across related experiments using 19,959 observations instead of only 1,064. That strategy represented a tremendous gain in efficiency, because all of the data were used to estimate each of the seven ratios – each datum giving seven times as much benefit.

But to complete the earlier quotation, Mendel actually wrote of "an average ratio of 2.98 to 1, or 3 to 1." Those last few words are significant. Even his final empirical ratio, based on a substantial sample of 19,959 plants or seeds, was calculated from a finite sample. Notice the clear tendency for Mendel's larger experiments to come closer to 3:1 ratios than did his smaller experiments. So more data helped, but even the combined data were less than perfect. Accordingly, Mendel surmised that the true ratio was 3:1. That conjecture can be interpreted as an appeal to a higher-level parsimony argument, to an implicit theory that the current phenomenon was like numerous other physical phenomena known to have simple principles and ratios.

In review, the estimate of the tall:short ratio progressed from 2.84:1 to 2.98:1 to 3:1 as more data and more theory were added to the initial experimental result. Some other biologist, less insightful than Mendel, could have had exactly the same data and yet never progressed beyond the 2.84:1 experimental result for tall:short. Incidentally, additional experiments with second and subsequent generations from hybrids, with crosses involving multiple traits, and with hybrids of other plant species all confirmed his 3:1 ratio.

The ultimate test for the 3:1 ratio of tall:short progeny would be to discover the physical phenomenon behind the ratio. What is in pea plants that causes this outcome? How is the tall or short trait inherited by progeny from their parents?

Of course, in the middle of the nineteenth century, Mendel was not in a position to explain the physical basis for the 3:1 ratio. He could cite a few known, rudimentary facts about pollen and egg cells. Beyond that, in a wise and tempered fashion, he called his further reflections mere hypotheses.

But current genetic knowledge about DNA, genes, chromosomes, mitosis, meiosis, and fertilization provides a lucid explanation. Briefly, the hybrid from a tall × short cross has one chromosome with the gene for tall plants and a paired chromosome with the gene for short; after meiosis, pollen and egg cells carry the gene for one or the other of those traits in equal proportions; and after fertilization with random assortment, $3/4$ of the progeny receive one or two copies of the dominant gene and express the dominant trait, whereas the other $1/4$ of the progeny receive two copies of the recessive gene and express the recessive trait, leading to a 3:1 ratio for tall:short plants.

Also, current statistical knowledge about the binomial distribution provides a quantitative understanding of the inaccuracies and fluctuations in finite samples, explaining Mendel's experimental ratios not being exactly 3:1.

Thus Mendel's 3:1 ratio is explained and confirmed. It has borne well the test of time. Indeed, his 3:1 ratio accords with the underlying physical process more accurately than do the 2.98:1 or 2.84:1 imperfect experimental results. Likewise, the 3:1 ratio is more accurate for predicting the outcome of new experiments than would be those other ratios. Good data plus good theory, including two levels of parsimony arguments, got at the truth.

The genes for all of Mendel's traits have been mapped precisely on the pea genome. Remarkably, two traits have been characterized in full molecular detail. Wrinkled seeds are caused by an insertion in a gene encoding a starch-branching enzyme (Bhattacharyya et al. 1990). Short plants are caused by a single mutation in the DNA that changes an alanine to a threonine near the active site of gibberellin 3β-hydroxylase, resulting in altered gibberellin with reduced activity as a growth hormone (Lester et al. 1997).

Mendel's use of parsimony was a charmingly simple application of simplicity! He merely averaged seven ratios and rounded the results to small integers. Nevertheless, to perceive that parsimony is at work even in such rudimentary scientific reasoning is to begin to understand that parsimony arguments pervade scientific thinking, even though the arguments are so familiar and frequent that they are seldom stated explicitly.

The remaining three examples of parsimony in this chapter illustrate increasingly complicated uses of simplicity. Calculation using the following parsimonious models requires millions of arithmetic steps, which is easily and inexpensively done by us with computers, but would have been unimaginable for Mendel and his contemporaries.

EXAMPLE 2: CUBIC EQUATION

The salient features of this example are that the true model is already known exactly, and the noise is also known exactly. Knowing both the signal and noise exactly allows for an unusually penetrating analysis, elucidating principles that subsequently can be recognized in more complex and realistic settings.

Obviously, to get an example with signal and noise known exactly requires that we scientists place ourselves in a very unusual position. Such an example must be constructed by us, not offered to us by nature. Accordingly, it must come from mathematics, not from the empirical sciences. Only by constructing a little mathematical reality can we know it completely.

The advantage of a mathematical example is that we can trace the recovery of signal and noise exactly, thereby gaining deep and incontrovertible insights into the workings of parsimony. But the disadvantage, which must

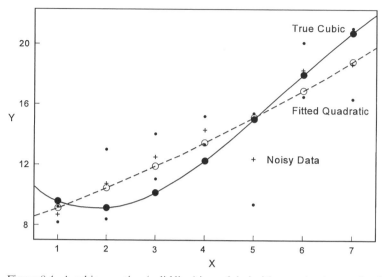

Figure 8.4. A cubic equation (solid line) is modeled with a quadratic equation (dashed line). Values of the true cubic equation are shown at seven levels of X (●). Noisy data are generated at each level for two replicates (•) and their average (+), with these averages having an S/N ratio of 5. Also shown are values for the quadratic equation fitted to the noisy data (○). Note that at every level except the sixth, the fitted quadratic's values are closer to the true values than are the data, the averages over replications. Remarkably, this parsimonious model is more accurate than its noisy data. It achieves a substantial statistical efficiency of 2.10, meaning that on average the quadratic model based on only two replications is slightly more accurate than averages based on twice as much data, four replications. So modeling helps predictive accuracy as much as would collecting twice as much data, but modeling is far more cost-effective when the data are expensive. (Adapted from Gauch 1993, and reproduced with kind permission from *American Scientist*.)

be overcome by further examples, is that some simplifying assumptions and easy calculations allowed here do not transfer to real-world research. So this mathematical example is both a good start and a bad finish in this business of comprehending parsimony. Accordingly, subsequent real-world examples from chemistry and agriculture will add required elaborations.

Figure 8.4 shows a cubic equation, $y = 12.00 - 3.50x + 1.17x^2 - 0.07x^3$, and its values at seven levels, $x = 1, 2, \ldots, 7$. By construction, this cubic equation is the true model or signal, known exactly. To mimic imperfect experimental data, random noise is added that is also known exactly. This noise has a normal distribution adjusted to have a variance of 0.2 times that of the cubic equation's data, resulting in an S/N ratio of 5. Frequently, experiments are replicated, which is represented here by showing these noisy data as averages of two replicates (that have twice as much variance as do their averages).

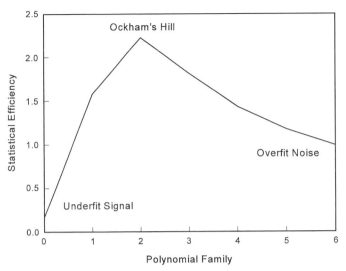

Figure 8.5. Ockham's hill for the noisy cubic data having an S/N ratio of 5, using the polynomial family encompassing the constant model up to the to sixth-order model. The quadratic model is at the peak of Ockham's hill, achieving the greatest statistical efficiency of 2.22. To the left of the peak, excessively simple models underfit real signal; to the right, excessively complex models overfit spurious noise.

Finally, this figure also shows the least-squares quadratic equation fitted to these noisy data, $y = 7.95 + 1.13x + 0.06x^2$.

Note that at every level except the sixth, the fitted quadratic's values are closer to the true values than are the data, the averages over replications. This model is more accurate than its data! In this case, the SS of differences between the data and true values is 24.67, and the SS of differences between the quadratic model and true values is 11.75, so the statistical efficiency is $24.67/11.75 = 2.10$. This experiment has two replications, but the quadratic model is as accurate as would be averages based on $2 \times 2.10 = 4.20$ replications. So modeling increases accuracy as much as would collecting twice as much data!

The case shown in Figure 8.4 invites three generalizations. Instead of measuring performance with just one set of random noise values, what would be the performance averaged over numerous repetitions with different noise? Instead of presenting results for just the quadratic model, what would be the results for the entire polynomial family? And what would happen at various S/N levels?

Figure 8.5 shows statistical efficiencies for the entire polynomial family for noisy cubic data with an S/N ratio of 5. There are seven data points, so the polynomial family encompasses the constant model (marked 0 on the abscissa), linear model (1), quadratic model (2), and so on, up to the sixth-order model (6). The statistical efficiency of the quadratic model for the single case analyzed in Figure 8.4 was 2.10, but this figure shows that the average over numerous repetitions with different noise is slightly different, 2.22. Figure 8.5 shows a typical response, as previewed earlier in the line for prediction in Figure 8.1. Using an apt phrase from MacKay (1992), the shape of this curve is called "Ockham's hill" in honor of William of Ockham, the fourteenth-century English philosopher who emphasized parsimony (Copleston 1953:43–121; Hyman and Walsh 1973:649–702).

The most predictively accurate member of the polynomial family for these noisy cubic data is the quadratic model, achieving a substantial statistical efficiency of 2.22. In either direction away from the peak, efficiency declines, but for different reasons. To the left of the peak, excessively simple models are inaccurate because they underfit real signal. To the right of the peak, excessively complex models are inaccurate because they overfit spurious noise. Optimal accuracy requires a balance between these opposing problems. These results emerge from the more fundamental phenomenon, previewed in Figure 8.1, that relatively simple signal is selectively recovered in early model parameters, whereas relatively complex noise is selectively recovered in late model parameters ordinarily relegated to a discarded residual.

Figure 8.6 generalizes the results for a wide range of noise levels, S/N ratios of 0.1 to 100. Beginning with familiar material from Figure 8.5, note the same results for an S/N ratio of 5 (located about seven-tenths of the way from 1 to 10 on this logarithmic abscissa), with the quadratic model most accurate. This figure shows that for rather accurate data having S/N ratios above 16.6, a cubic model is most predictively accurate, achieving a statistical efficiency of 1.82. But as noise increases, progressively simpler models are best. However, the fourth-order and higher models never win, including the sixth-order model which is the full model. So which model will be most predictively accurate will depend on the noise level. It makes sense that as noise increases, fewer of the true model's parameters can be estimated accurately enough to be helpful, until finally only the grand mean, which is the parameter used by the constant model, can resist the onslaught of noise. On the other hand, cleaner data can support more parameters.

Note the different behaviors of models of various complexities relative to the true model. Models of the same order as or higher orders than the true cubic will have flat responses, unaffected by the noise level. But models more parsimonious than the true model will be responsive to noise, achieving greater efficiency as noise increases. Moving to the left beyond the range included in Figure 8.6, as noise increases infinitely, the quadratic will reach an asymptote of about 2.46, the linear will reach about 3.67, and the constant will reach exactly 7. Going to the right, as noise decreases to nothing, these

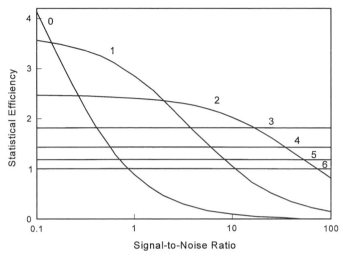

Figure 8.6. Statistical efficiency of the polynomial family over a range of S/N ratios, with the constant model (0), linear model (1), and others up to the final sixth-order polynomial (6). The constant model is most predictively accurate for extremely noisy data, with S/N below 0.15; the linear model is best for S/N from 0.15 to 2.0; the quadratic model is superior for S/N from 2.0 to 16.6; and the cubic model wins for relatively accurate data, with S/N above 16.6. With seven data points, the full model is the sixth-order equation. This full model is always most postdictively accurate, but never most predictively accurate. Notice that for accurate data with an S/N ratio above 16.6, the true cubic model is most predictively accurate; but for noisier data, progressively simpler models are most accurate. Consequently, diagnosing the most predictively accurate member of a model family and determining the true model are distinguishable goals, sometimes having different answers. (Adapted from Gauch, 1993, and reproduced with kind permission from *American Scientist*.)

three curves will all go to 0. It makes sense that modeling helps more when there are noisier data, whereas perfect data need no help.

Finally, if one steps back momentarily from the details to focus on the big picture, this example portends two momentous stories: a comforting scientific story and a scary philosophical story. The wonderfully comforting story is that even without having the true model, modeling can gain accuracy! For example, a quadratic fit can gain considerable accuracy and efficiency even when the true model is something else, a cubic equation in this instance. Hence, as this chapter's next two examples leave behind a tidy mathematical world of our own making and venture forth into the real world of chemical and agricultural data, not having the true model in hand will add some computational complexities, but we need not anticipate any fundamental or debilitating hindrances. Indeed, without ever knowing the true model, one can readily achieve sizable statistical efficiencies that can help science and technology dramatically. Parsimony can help even imperfect scientists with fragmentary knowledge and noisy data!

This scientific benefit, however, comes at a philosophical cost. Predictive accuracy traditionally has been regarded as a helpful guide to truth: More predictively accurate models are more likely to be true. But the results here are disquieting, showing, for example, that a quadratic can be most predictively accurate when a cubic is the true model. Consequently, diagnosing the most predictively accurate member of a model family and determining the true model are distinguishable goals, sometimes having different answers. Noise complicates the relationship between predictive success and truth.

Later in this chapter we shall return to this philosophical quandary, assessing just what it does or does not mean for science. But additional scientific aspects of parsimony must be mastered first before philosophical aspects can be tackled with profit. Accordingly, for now we press on with the happy business of using parsimony to help scientists get more benefit from their hard-won data.

EXAMPLE 3: EQUIVALENT CONDUCTIVITY

Equivalent conductivity is a measure of a solution's ability to carry an electric current. It is measured by inserting two electrodes into the solution, applying a rapidly alternating voltage, and measuring the current (see Gauch 1993 for a simplified explanation, or a physical chemistry text for a rigorous explanation). The example considered here is the equivalent conductivity of aqueous solutions of hydrochloric acid measured in a two-way factorial design with 19 concentrations and 13 temperatures (Lide 1995:5-87). The 19 concentrations are 0.5 to 9.5 mol l^{-1} and the 13 temperatures are 0 to 65°C in increments of 5°C, except that 60°C is missing. Hence, there are $19 \times 13 = 247$ measurements (of which eight were given earlier in Table 8.1). These equivalent conductivities range from 52.3 to 552.3 Ω^{-1} cm^2 equiv^{-1}. These data are rather accurate, carrying about 4 significant digits.

For current purposes, however, it is useful to generate noisy versions of these data to mimic the effect of using less accurate measuring instruments. Accordingly, the original data have been artificially degraded to seven levels of noise by rounding the data to the nearest 1, 3, 5, 10, 20, 30, and 50. This equates to reducing the original 4 significant digits to only 3–1 significant digits. The corresponding S/N ratios range from 129,191 to 56.50, or log S/N ratios from 5.11 to 1.75. Rounding to the nearest 10, 30, and 50 will here be termed "low," "medium," and "high" noise, respectively.

The salient feature of the earlier mathematical example was that the signal and the noise were both known exactly. By contrast, to take one step in the direction of hard and realistic scientific problems, this conductivity example is designed to have only the noise known accurately, and the signal known poorly.

The noise in these edited data sets, due primarily to rounding, is known nearly exactly. For example, from Table 8.1, the equivalent conductivity of a 4.5 mol l^{-1} hydrochloric acid solution at 30°C is 196.6 Ω^{-1} cm^2 equiv^{-1}. For the lowest of the seven noise levels, involving rounding to the nearest 1, this original 196.6 is degraded to 197, introducing a known error of 0.4. Likewise, the most severe rounding, to the nearest 50, results in 200, introducing a known error of 3.4. So all 247 errors are known at all seven noise levels. Furthermore, the original measurement errors are negligible compared with these substantially larger introduced rounding errors, so the noise is known nearly exactly.

But the signal is known poorly, in the sense that the model used here is AMMI, which exploits no knowledge about conductivity from physical chemistry. There is no pretense whatsoever that AMMI's parameters reflect specific physical entities and processes. Of course, for different purposes, known chemical principles could be incorporated into a conductivity model. However, the purpose here is to show how parsimonious models that are not the true model, when given noisy data, can still gain accuracy. Although AMMI models are applied here to noisy versions of the data, the original data are still available at the end for rather exact assessment of each model's accuracy.

Figure 8.7 shows statistical efficiencies for the conductivity data with medium noise using the AMMI family. The AMMI-1 model is at the peak of Ockham's hill. Rounding these 247 values to the nearest 30 introduces a noise SS of 18,450. But the AMMI-1 model's SS around the true values is only 2,763. Thus, AMMI-1 achieves a statistical efficiency of 6.68. The AMMI-1 model is considerably more accurate than its data because most of the signal goes into this parsimonious, signal-rich model, whereas most of the rounding noise goes into the discarded, noise-rich residual (in line with the principles previewed in Figure 8.1). To the left of the peak, an excessively simple model underfits real signal; to the right, excessively complex models overfit spurious noise. (Not shown are the AMMI-8 to AMMI-12 models, which decline further in statistical efficiency, to exactly 1 for the last model, the full model that equals the supplied data.)

This chemical example is quite different from the earlier mathematical example. Nevertheless, the broad features are the same, as can be seen by comparing Figure 8.7 with Figure 8.5 (and with the line for prediction in Figure 8.1). Predictive accuracy is maximized by a model of intermediate complexity, complex enough to fit most of the signal, and yet parsimonious enough to avoid much of the noise.

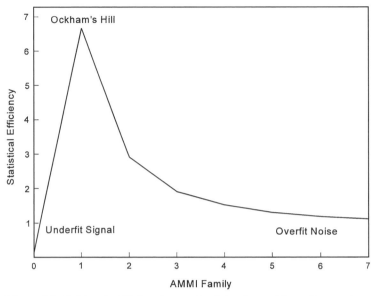

Figure 8.7. Ockham's hill for the conductivity data with medium noise, using the AMMI family. The AMMI-1 model is at the peak of Ockham's hill, achieving the greatest statistical efficiency of 6.68. To the left of the peak, an excessively simple model underfits real signal; to the right, excessively complex models overfit spurious noise.

Significantly, parsimonious models help scientists to achieve robust results. As noise increases, the model's estimates rattle around the true values less than do the raw data (which is exactly what a statistical efficiency above 1 means). Only a fraction of the noise gets into the model, with much or most of it captured instead by the discarded residual. So the model is more robust than its data. Furthermore, although careful model diagnosis is desired to reach the very top of Ockham's hill, modeling is also robust in the additional sense that ordinarily a slight misdiagnosis will give better results than no modeling at all. For example, in Figure 8.7, the AMMI-1 model is optimal, but AMMI-2 and even AMMI-3 are still markedly better than the raw data. Hence, parsimonious models are doubly robust in that neither the data nor the model need be perfect in order to gain substantial efficiency and accuracy, thereby increasing the rate of scientific and technological progress.

Figure 8.8 generalizes the results for a wide range of noise levels, from log S/N ratios of 1.75 to 5.11 (due to rounding to the nearest 50 to 1). Figure 8.7 gave results for just one noise level, a log S/N ratio of 2.19 (due to rounding to the nearest 30), for which AMMI-1 is the most predictively accurate member of the AMMI family, achieving a statistical efficiency of 6.68.

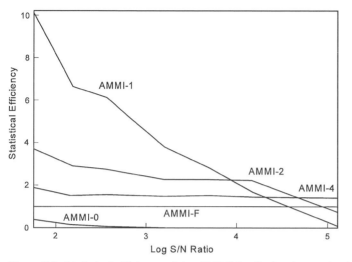

Figure 8.8. Statistical efficiency of the AMMI family for the conductivity data with added noise over a range of S/N ratios. The full model, AMMI-F, equaling the supplied data, has a statistical efficiency of 1 by definition. Progressively noisier versions of the data support progressively simpler AMMI models. The greater the noise, the greater the accuracy gain from AMMI modeling.

Were there no noise, prediction and postdiction would coincide, and there would be no Ockham's hill. So noise is a pivotal element in the story about parsimony. Furthermore, the exact magnitude of the noise, quantified by the S/N ratio, affects model performance and choice. The full model, AMMI-F, equaling the supplied data, has a statistical efficiency of 1 by definition. It is never the most predictively accurate model. AMMI-4 is most accurate for the least noisy version of the data, rounded to the nearest 1. Then AMMI-2 is best for somewhat noisier data, rounded to the nearest 3. Finally, AMMI-1 is best for the five still noisier versions of the data, reaching a statistical efficiency of 10.15 for the noisiest data rounded to the nearest 50.

Two trends are apparent. First, noisier data support simpler models. This makes sense because the worse the noise, the fewer the model parameters that can be estimated with sufficient accuracy to be useful. Second, the greater the noise, the larger the accuracy gain from modeling. For the noisiest data at the left, AMMI achieves an impressive statistical efficiency of 10.15, whereas for the cleanest data at the right, it achieves only 1.42. This also makes sense because the worse the data, the more room there is for modeling to gain accuracy. Note that Figures 8.6 and 8.8 show these same two trends, with noisier data supporting simpler models achieving greater statistical efficiencies.

What would a figure like Figure 8.7 look like if it showed results for high, medium, and low noise, not just medium noise? Bearing these two trends in mind, Ockham's hill for high noise would be farthest left and highest, for medium noise it would be intermediate, and for low noise it would be farthest right and lowest. More noise shifts the peak of Ockham's hill toward simpler models and toward higher statistical efficiencies, which is left and upward in figures such as Figures 8.5 and 8.7.

How can parsimonious models be more accurate than their data? Why does Ockham's hill occur? So far, one explanation has been given regarding signal–noise selectivity, as in Figure 8.1. Signal is captured mostly in early model parameters, whereas noise is captured mostly in late model parameters relegated to a discarded residual. So by stopping at the right point, a parsimonious model gains accuracy by capturing most of the signal but discarding most of the noise. This chemistry example lends itself to giving a second, complementary explanation in terms of direct and indirect information.

As this explanation begins, recall the earlier exercise of being asked to give the equivalent conductivity of 5.0 mol l^{-1} hydrochloric acid at 30°C. Table 8.1 did not answer this question directly. But it did supply related information that allowed for a decent answer indirectly. This simple insight, that indirect information can help, is the key to the following explanation of how parsimony increases accuracy.

The conductance data involve 13 temperatures and 19 concentrations, giving 247 related measurements. From the perspective of any single measurement, there exist 1 direct datum and 246 indirect data. For these conductances, consider the AMMI family with AMMI-0 to AMMI-12 (AMMI-F), but also consider the simplest possible model, merely the grand mean. The extremes of this extended family are the most parsimonious grand-mean model, with 1 parameter, and the most complex full model, with 247 parameters. The parsimonious extreme simply averages all the data to calculate the grand mean, so for its estimate of each matrix entry, the direct datum receives a weight of 1/247 or 0.00405, and the indirect data receive the complementary weight of 0.99595. On the other hand, the complex extreme simply offers each datum as its own estimate, so the direct datum receives a weight of 1, and the indirect data 0. Between these opposite extremes, the AMMI-0 to AMMI-11 models offer a sequence of intermediate blends, with higher models placing more weight on the direct data and less weight on the indirect data.

There is a fundamental difference between a parsimonious (or reduced) model and the full model. For each estimate, the full model uses just the datum for that one entry – the 1 direct datum. But a parsimonious model uses all the data – the 1 direct datum *and* the 246 indirect data. Were one entry changed (say, because a typographical error is found and corrected), then the full model would change that one entry, whereas in general a parsimonious model would change all of its 247 estimates.

This difference is expressed most forcefully by presenting it as a question. For estimating the value of a single matrix entry, what are the relevant data? The full model responds that only the data for that one matrix cell are relevant. A parsimonious model responds that all of the data from the entire experiment are relevant.

Clearly, a model sequence of increasingly complex models (such as appears in Figures 8.1, 8.5, and 8.7) can be interpreted as a shifting of weight increasingly toward the direct information. The complexity of AMMI calculations, however, tends to hide this interpretation.

Accordingly, for the purpose of cultivating an important insight, the direct and indirect information can be artificially separated, subsequently blending back all the information in a transparent manner. By temporarily withholding the data from each matrix cell in turn, throughout the whole matrix, one cell at a time, the indirect information can be extracted separately by using a missing-data version of AMMI to impute the maximum-likelihood value for the matrix cell with no data (Gauch 1992:153–162). For each of the 247 conductances, two estimates are formed: One estimate uses the 1 direct datum, and a separate estimate uses the other 246 indirect data. Finally, the two kinds of information are blended in a single combined estimate by giving the direct estimate some weight from 0 to 1 and the indirect estimate its complementary weight between 1 to 0. This blending is done for the entire spectrum of direct weights from 0 to 1. This process mimics the behavior of a model family, from parsimonious to complex models; but now the shift in weight between the indirect and direct information is transparent.

Figure 8.9 presents the results for AMMI-1 analyses of three noisy versions of the conductance data, with high noise generated by truncating the data to the nearest 50, medium noise to 30, and low noise to 10. The outcome is Ockham's hill, with some blend of direct and indirect information being more accurate than either kind of information alone. For the high-noise data, the optimal blend gives the direct information a weight of 0.30233 and the indirect information a weight of 0.69767, achieving a statistical efficiency of 4.91. At the left side of the high-noise curve, the 246 indirect observations give a statistical efficiency of 2.83, so, on average, each $246/2.83 = 87$ indirect observations are as informative as 1 direct observation. The indirect information is rather dilute (by a factor of 87), but it is also quite abundant (by a factor of 246), and therefore it is well worth extracting.

Turning next to the other two lines in this figure, as the noise decreases from high to medium to low noise, the optimal blend shifts to the right, placing more weight on the direct data. Also, the attainable statistical efficiency decreases. These trends make sense because better data leave less room for

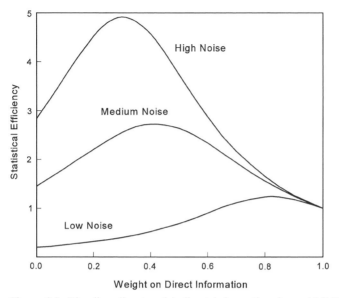

Figure 8.9. Blending direct and indirect information from AMMI analyses of the conductivity data with added noise at three noise levels. The remarkable statistical efficiency of parsimonious models arises from exploiting both the direct and indirect information. As noise increases, optimally accurate models place greater weight on the indirect information. Also, with greater noise, more accuracy gain is attainable.

modeling to borrow strength from the indirect data and thereby to increase accuracy.

Again, the purpose of this exercise has been to gain insight into Ockham's hill, not to construct an optimal model. Two clarifications reinforce this point. First, for the high-noise data, in Figure 8.9 the optimal model with separated information achieves a statistical efficiency of 4.91, whereas in Figure 8.8 the ordinary AMMI-1 model using all the data together achieves 10.15. Separating the direct and indirect information helps our intuition but does not help modeling. Second, AMMI serves here as a generic, off-the-shelf model for these conductance data. A vigorous Bayesian approach could offer a superior model, however, especially if it incorporated known physical laws about ions in solution carrying an electric current. Indeed, the interesting lesson here is that even a rather minimal modeling effort suffices to gain considerable accuracy and efficiency.

Another example of optimal blending of direct and indirect information is provided by Azzalini and Diggle (1994). A parsimonious Bayesian model is used to gain accuracy by combining direct data on a soil sample's respiration rate with indirect information about the soil's temperature, moisture content, and soil type. Still another example, concerning the three-dimensional shapes of complex molecules, will be one of the case studies in the following chapter.

EXAMPLE 4: CROP YIELDS

The fourth and final example of parsimony at work is the case most familiar to me from my own research, agricultural yield trials. Plant breeders use yield trials to select superior genotypes, and agronomists use them to recommend varieties, fertilizers, and pesticides to farmers. Worldwide, several billion dollars are spent annually on yield trials. These experiments help plant breeders to continue to increase crop yields, typically by about 1% to 2% per year for open-pollinated crops such as corn and 0.5% to 1% per year for self-pollinated crops such as soybeans.

The most common type of yield trial tests a number of genotypes in a number of environments that are location-year combinations, often with replication. The example used here is a New York State soybean trial. The data are for 7 genotypes in 10 environments with 4 replications (Gauch 1992:56). Figure 1.6 showed a soybean yield trial.

The salient feature of this agricultural example is that neither the signal nor the noise is known exactly, quite in contrast to the easy earlier example with a known cubic equation with known added noise. We receive the soybean data from nature, with the signal and noise already mixed together.

The soybean yield trial's grand mean is 2,678 kilograms/hectare (kg/ha), and the error rms is 318.058 kg/ha. Hence, the standard error of a mean over 4 replicates is $318.058/4^{0.5} \approx 159$ kg/ha. Assuming that the errors are distributed approximately normally, an interval of plus or minus 1.96 standard errors gives 95% coverage, which is ±312 kg/ha here. Thus, even with 4 replications, these yield observations carry only 1 significant digit. This is not a tidy physics experiment, but rather a messy agricultural experiment producing noisy data.

Furthermore, modeling here uses the AMMI family as a generic model. There is no pretense that the AMMI model parameters are specified in advance to refer to specific physical entities, such as soybean genes or environmental features. Nevertheless, in retrospect some broad patterns may be discernible and interpretable.

Figure 8.10 shows statistical efficiencies for the soybean data using the AMMI family. The AMMI-2 model is at the peak of Ockham's hill, achieving a statistical efficiency of 1.44, with AMMI-1 a close second. To the left of the peak, excessively simple models underfit real signal; to the right, excessively complex models overfit spurious noise. Clearly, Figure 8.10 for the soybean data is reminiscent of Figure 8.7 for the conductivity data and Figure 8.5 for the cubic data. Ockham's hill is the parsimony–accuracy trade-off that pervades science and technology. However, because the true signal is now

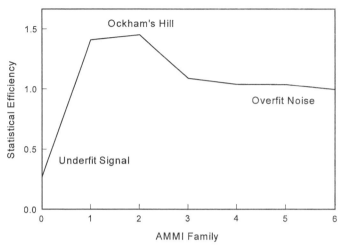

Figure 8.10. Ockham's hill for the soybean data using the AMMI family. The most predictively accurate member of the AMMI family is AMMI-2, achieving a statistical efficiency of 1.45, but AMMI-1 is almost as accurate.

unknown, the method for calculating statistical efficiencies in Figure 8.10 is rather different and more complicated than for the earlier figures, as will be explained next.

Figure 8.11 shows model validation for the soybean experiment based on replicated noisy data rather than on access to the true signal (as in Figure 8.5) or to quite accurate data (as in Figure 8.7). The calculations have been explained in detail elsewhere (Gauch 1992:134–153), so a somewhat informal account is sufficient here. Results are shown for the entire AMMI family from AMMI-0 to AMMI-6, which is the full model in this case. The top line in Figure 8.11 shows the rms prediction difference for each model in the AMMI family. The prediction difference is the difference between an AMMI model's yield predictions and the validation observations' actual yields. It has been calculated here by a data-splitting procedure that splits the data into modeling data and validating data. This experiment has 4 replications, and given its factorial design, with 7 genotypes tested in 10 environments, there are 70 yields. For each yield individually, 3 replicates are chosen at random for fitting the AMMI family. This produces 7 yield predictions, one for each of the 7 models, AMMI-0 to AMMI-6. Then data splitting uses the remaining 1 replicate as a validation observation. Yield predictions from the 7 members of the AMMI family are each compared with the validation data by calculating the rms prediction difference, which equals the sum of squared differences, finally dividing by the number of validations and taking the square root. This calculation occupies an ordinary computer for only a fraction of a second, so the entire procedure is repeated thousands of times

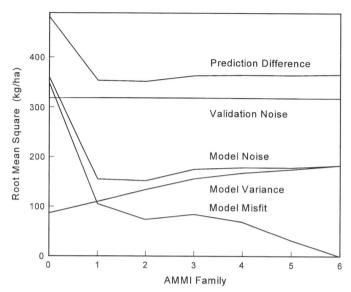

Figure 8.11. Prediction errors of the AMMI family for the soybean data. Prediction differences between validation data and model estimates are partitioned into two components, validation noise and model noise. Then model noise is further partitioned into components for variance and misfit (or bias). The most predictively accurate model, having the least noise, is AMMI-2, with AMMI-1 a close second. Predictive accuracy results from a variance–bias trade-off, with moderate amounts of both problems being better than an excessive amount of either problem. (Adapted from Gauch, 1992:143, and reproduced with kind permission from Elsevier Science.)

with different randomizations, and the results are averaged to achieve good accuracy. The model with the smallest rms prediction difference is most predictively accurate, lying closest to the validation data. In this case, AMMI-2 is the winner, with an rms prediction difference of 352.734 kg/ha.

Understand that this quantity, the rms prediction difference, is a thoroughly empirical quantity calculated from concrete data. This quantity is of interest because the replicates have essentially independent errors, so by using different data for constructing and validating a model, the rms prediction difference rewards a model's recovery of signal but penalizes recovery of noise. Hence, this data-splitting procedure discriminates between signal and noise – even without having access to a known signal (or to known noise).

This top line for rms prediction differences has a remarkable feature. Recall that AMMI-6 is the full model, so its yield estimates automatically equal the data exactly, which are the averages over the 3 replicates used for modeling in this instance. The rms prediction difference for AMMI-6 is 366.834 kg/ha, which is higher and worse than that for AMMI-2, with 352.734. So the AMMI-2 model is more predictively accurate than its data! Although second best, AMMI-1 is also more accurate than its data, and even AMMI-3 through AMMI-5 exhibit slight accuracy gains. Evidently, the AMMI-2 model is capturing much signal while relegating much noise to a discarded residual, thereby gaining accuracy.

Because rms prediction differences are calculated from differences between model predictions and validation observations, these values are not perfect (zero) for two reasons: The model predictions are imperfect, and the validation data are imperfect. Both fail to equal the true means exactly. Hence, although rms prediction differences have the practical advantage of being empirical quantities computable from the data, they have the conceptual disadvantage of mixing two different problems. What scientists would really like to know is how close a model is to the truth, rather than how close it is to some imperfect data.

Accordingly, the next two lines in Figure 8.11 show how rms prediction differences are partitioned into two components, validation noise and model noise, reflecting imperfections in the validation data and in the model predictions separately. The validation data are simply individual yield measurements, so their noise is just the error MS, or 101,161, and its square root is 318.058 kg/ha. This value is the same regardless which AMMI model is considered, as shown by the flat line for validation noise in this figure. Then the model noise is the difference between the prediction difference and validation noise, doing this calculation in terms of mean squares, but presenting the results in this figure by square roots of those quantities in order to get back to units of yield (kg/ha).

For example, for AMMI-2, its MS prediction difference is $352.734^2 = 124,421$, and the MS validation noise is $318.058^2 = 101,161$, so the MS model noise is $124,421 - 101,161 = 23,260$, and its square root is 152.512 kg/ha, as shown in Figure 8.11. This means that the rms difference between AMMI-2 yield estimates and the true values is 152.512 kg/ha. It is remarkable to get such a number even though the true values are not available. Of course, without knowing the true yields (or the exact errors), it is impossible to know just how far each individual yield measurement is from its true value. One might suppose that because we cannot determine individual errors for single measurements, we also cannot determine the average errors for numerous measurements, but that supposition is not true. In fact, it is possible to determine the typical errors for a collection of numerous yield measurements. And those average performance values are quite sufficient for determining which model within a model family is most predictively accurate.

Incidentally, because this soybean trial's grand mean is 2,678 kg/ha, this rms prediction error of 152.512 kg/ha for the AMMI-2 parsimonious model using 3 replicates amounts to a typical error of about 5.7%, which is rather good for this sort of agricultural experiment. For comparison, yield estimates based the AMMI-6 full model that merely averages the 3 replicates have an rms prediction error of 182.773 kg/ha, or an error of about 6.8%, which is worse. Again, the AMMI-2 model is more accurate than its data. Remarkably, this conclusion can be reached even without ever knowing the exact true values.

The line for model noise, near the middle of Figure 8.11, is the most important feature of this graph. Note that because the rms model noise is a measure of inaccuracy, this line shows Ockham's hill drawn upside down. It has exactly the same general shape, inverted, as does Figure 8.10. Furthermore, a simple calculation relates these two figures. The statistical efficiency of each AMMI model is simply the noise of the full model divided by the noise of the AMMI model, expressing noise in terms of MS values.

For example, recalling figures from preceding paragraphs, the MS noise for the full model is $182.773^2 = 33,406$, and the MS noise for the AMMI-2 model is $152.512^2 = 23,260$. So the statistical efficiency of the AMMI-2 model, as shown earlier in Figure 8.10, is $33,406/23,260 \approx 1.44$. This means that the AMMI-2 model using 3 replications is as accurate as mere averages over replications based on $3 \times 1.44 = 4.32$ replications. Because there are 70 yields and this analysis provides $4.32 - 3 = 1.32$ free replications, AMMI-2 analysis improves accuracy as much as would collecting data for $70 \times 1.32 \approx 92$ more yield measurements. Effective statistical analysis allows for greater accuracy and for more information to be extracted from available data.

The bottom two lines in Figure 8.11 partition the model noise into two components: model variance and model misfit (or bias). Model variance is caused by imperfect data having experimental errors, whereas model bias is caused by imperfect models whose predictions would not equal the true values even if the model were supplied perfect data (such as incredibly accurate data from averaging an enormous number of replications). The details have been published elsewhere (Gauch 1992:142), but the model variance can be estimated fairly easily, and then the model misfit can be obtained by subtracting the model variance from the model noise. For example, for AMMI-2, its noise is $152.512^2 = 23,260$, which is partitioned into $133.686^2 = 17,872$ from variance and $73.403^2 = 5,388$ from misfit.

Note that as more complex AMMI models are used, the variance increases, and the misfit decreases. Model variance increases with model complexity because more model parameters capture more model noise. And model misfit decreases with model complexity because the full model's predictions equal the averages over replications, which are the best linear unbiased estimators with no misfit.

This variance–bias trade-off provides another insight for understanding how parsimonious models gain accuracy. It is better to have moderate problems with both variance and bias than to have an extreme problem with either.

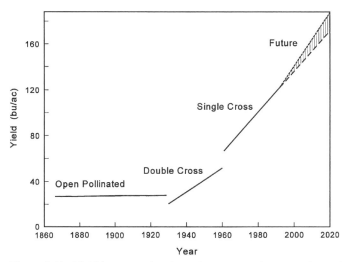

Figure 8.12. Yield increases for the U.S. corn crop have accelerated with new technologies. From 1866 to 1929, open-pollinated varieties dominated, and the average
gain in yield was only 0.02 bu/ac per year. From 1930 to 1960, double-cross hybrids
gained 1.04 bu/ac per year. From 1961 to 1992, single-cross hybrids gained 1.79 bu/ac
per year. The yield in 1992 was about 122 bu/ac, so the contemporary rate is 1.5% per
year. If this same gain of 1.79 bu/ac continues until 2020, future hybrids will then reach
172 bu/ac (dashed line). But if aggressive statistical analysis increases the rate by 0.4%
per year to 1.9% per year, that larger gain of 2.32 bu/ac per year will reach 187 bu/ac
(dotted line). The shaded area shows the extra yield attributable to aggressive statistical analysis, amounting to an increment of 9% by 2020. If similar yield increments
were to occur for most major crops, the additional food would feed hundreds of millions of people. (The data are from A. Forrest Troyer and DEKALB Genetics, and
this figure is adapted from A. Forrest Troyer, in Wilkinson 1991:165–177, and Gauch
1993.)

A relatively parsimonious model, here AMMI-2, has the optimal trade-off,
thereby maximizing accuracy. In the hypothetical situation with virtually infinite replication, the complex full model is automatically most accurate; but
in the practical situation with limited replication, a relatively parsimonious
model is ordinarily most accurate.

Emerging now from the technicalities to the big picture, Figure 8.12 shows
the implications of better accuracy for future crop yields, specifically U.S.
corn yields (personal communication, A. Forrest Troyer; also see A. Forrest
Troyer, in Wilkinson 1991:165–177; Gauch 1993; Troyer 1996, 1999). During the past century, yield increases have accelerated as new technologies
have improved corn varieties. This figure shows linear regression lines for
the annual data from three historical periods dominated by open-pollinated
varieties, double-cross hybrids, and finally single-cross hybrids. From 1866 to

1929, open pollinated varieties gained only 0.02 bushel/acre (bu/ac) per year. Subsequently, the introduction of double-cross hybrids and then single-cross hybrids accelerated yield gains, in 1992 reaching a yield of 122 bu/ac and a gain of 1.79 bu/ac per year. If this rate continues until 2020, future corn hybrids will then reach 172 bu/ac.

It is fortunate that U.S. corn yields have quadrupled over the past century, because the U.S. population has also quadrupled. Otherwise, an economically impossible and ecologically disastrous amount of land would be needed to grow this crop. Nevertheless, with all due respect for this great achievement by agronomists and breeders, some remarks must be offered regarding the statistical methods used by those scientists to analyze their data.

In his presidential address to the American Society of Agronomy, Nielsen (1992) said that "We still rely on statistical methods developed more than one-half century ago before we had anything but a slide rule, a primitive calculator, or a paper and pencil." That assessment is certainly valid as regards the analysis of yield data. The most common method for analyzing yield trials has been linear regression of yields on environment means, a technique invented by Mooers (1921) and Yates and Cochran (1938) and later popularized by Finlay and Wilkinson (1963). More generally, my observations (Gauch 1992:251) agree with those of Nielsen: "An extensive survey would be quite interesting, but I suspect that of the statistical analyses published in recent issues of agricultural journals, over 90% were invented before 1940 and over 99% have no distinctively predictive criteria or optimization. But many factors, especially the increased availability of computer hardware and software, presage rapid change at this time." Articles in agricultural journals evince little awareness of parsimony or Ockham's hill.

Consequently, the yield increase of 1.79 bu/ac per year during the past three decades should be interpreted as the progress achieved while relatively primitive statistical analyses extracted only a portion of the potentially useful information residing in the yield data. My argument (Gauch 1992:263–264) is that modern AMMI analysis of yield data combined with refinements in experimental design can allow agricultural researchers to get more out of their data and hence to achieve faster yield increases.

Chapter 1 mentioned that parsimonious models of yield trials typically can boost the rate of yield increases by 0.4% per year, and it mentioned the practical importance of that claim, leaving the technical documentation for that claim to this chapter. That claim originated from simulation results (Gauch 1992:178–185). The simulation concerned a yield-trial experiment with only 70 yield measurements, which is smaller than typical experiments and hence less favorable for exploiting parsimony. And it assumed a statistical efficiency from a parsimonious AMMI model of only 1.78, which again is less favorable than typical experiments. So that simulation was intended to produce a conservative, realistic estimate of the advantages to be expected in routine applications. The finding was that exploiting parsimony could increase yield by 0.4% per year, which is dramatic given that breeding programs typically gain about 1% per year.

Several years have passed since that claim was made in 1992. What has subsequent research shown? The most direct test of that claim would be to run two agronomy or breeding programs in parallel, basing selections on parsimonious models in one program while using ordinary procedures in the other. After a few years, the varieties produced by those two programs would then be compared to see if parsimony really pays. At present, several such direct tests are beginning, so results can be expected in a few years. Meanwhile, genetic and statistical theory has identified the components needed to accelerate yield gains, and empirical results are already available for all of these components. This indirect evidence strongly supports the claim of a 0.4% advantage.

First, even as early as 1996, Richard W. Zobel and I (in Kang and Gauch 1996:85–122) had cited 31 reports of accuracy gains using AMMI for a great diversity of crops and agroecosystems, and many more reports have appeared subsequently. Statistical efficiencies of 2 to 4 are common. The highest value reported so far is 5.6 (Ebdon and Gauch 2002). Furthermore, besides AMMI, a broad class of multivariate analyses can also gain accuracy, so this is a very general phenomenon, as expected from statistical theory (Moreno-Gonzalez and Crossa 1998; Cornelius and Crossa 1999). So the primary assumption in the 1992 simulation, of statistical efficiencies of 1.78 or better from widespread use of parsimonious models, has been supported by definitive evidence.

Second, an implicit assumption in the claim of a 0.4% advantage is that results from yield experiments in general, and AMMI parameters in particular, are reasonably repeatable. Occasional studies have reported poor repeatability (Sneller and Dombek 1995), but the overwhelming majority have shown substantial repeatability (Annicchiarico, Bertolini, and Mazzinelli 1995; Annicchiarico 1997, 1998, 2000; Sneller, Kilgore-Norquest, and Dombek 1997). Furthermore, this documented repeatability is often explainable by correlating model parameters with environmental and genetic factors that are known to be stable causal factors (Annicchiarico 1992, 1997; Nachit et al. 1992a,b; Romagosa et al. 1993, 1996; Kang and Gauch 1996:85–122; Brancourt-Hulmel, Denis, and Lecomte 2000).

Third, the simulation that predicted the 0.4% advantage presumed only one stage of selection. But most breeding programs have two or three stages running concurrently, so actually there are several opportunities to put parsimony to work. So the findings of routine statistical efficiencies of 2 or better and good repeatability are all that is needed to support that claim of a 0.4%

advantage, and yet the prevalence of two or three stages implies that only a half or a third of that 0.4% at each stage would suffice.

Finally, parsimony pays in additional ways besides gaining accuracy. Another role for parsimonious models in yield trials is to provide a simplified account of which genotypes perform best in which environments, including grouping the environments into so-called mega-environments in which a given genotype (or set of similar genotypes) will perform best (Gauch and Zobel 1997). That will allow breeders and farmers to take advantage of not only broad adaptations to the entire growing region but also narrow adaptations to particular environmental conditions, thereby increasing the opportunities to improve yields. Where subdivision into a few well-defined mega-environments has not yet been done, this refinement will confer a one-time benefit. Preliminary findings suggest that typically it may offer a yield gain of about 5% to 10% (Gauch and Zobel 1997; Annicchiarico 1998, 2000). That will equate to the claimed 0.4% gain accumulated over a decade or two. And perhaps more importantly, there will also be an ongoing benefit from mega-environments of increasing heritability, which will speed up breeding progress (Annicchiarico and Perenzin 1994; Moreno-Gonzalez and Crossa 1998; Annicchiarico 2000). Parsimonious models can also open up new possibilities for improving experimental designs, although that matter is better left to the technical literature (such as Gauch 1992).

The bottom line is that abundant empirical evidence now supports all of the components that together legitimate the 1992 claim that incorporating parsimony into yield-trial research typically can increase yield gains from 1.0% to 1.4% per year. Further confirmation from direct tests is expected in a few years.

Returning now to the big picture, again consider Figure 8.12. From 1961 to 1992, corn yields gained 1.79 bu/ac per year, and in 1992 the yield was 122 bu/ac, so the yield gain was about 1.79/122 or 1.5% per year. By helping breeders to estimate yields more accurately, to select better genotypes, to understand complicated genotype–environment interactions, and to design more efficient experiments, it is plausible that extensive use of AMMI and related modern statistical tools could boost yield increases by another 0.4% per year to achieve 1.9% per year. If that larger gain of 2.32 bu/ac per year continued until 2020, future corn hybrids would then reach 187 bu/ac, rather than the current projection of 172 bu/ac. Hence, the extra yield attributable to aggressive statistical analysis would be an increment of 9% by 2020. If similar yield increments were to occur for most major crops, the additional food could feed hundreds of millions of people.

There is, however, no time to waste. How long does it take for the world's population to grow by a million persons? About four days! Yet plant breeding is an incremental process of improving breeding stocks over a span of years,

so efficient research is needed now to achieve substantial yield gains a decade
or two or three from now. Globally the amount of suitable agricultural land
is steadily decreasing (because of soil erosion, expanding cities, and similar
factors), so the only way to grow more food is to increase crop yields.

EXPLANATION OF ACCURACY GAIN

Even if not one person on earth had the slightest clue as to how parsimonious
models can be more accurate than their data, that accuracy gain would still
stand as an established fact. For example, to recall just one number from
just one of the earlier examples, at an S/N ratio of five, the parsimonious
quadratic model achieves an average statistical efficiency of 2.22, meaning
that this model is substantially more accurate than its data. Because the signal
and the noise are both known exactly by construction, this fact is absolutely
incontrovertible.

However, a fact without an explanation is unsatisfying and sometimes
even unconvincing. Accordingly, this section collects and reviews insights
mentioned earlier in this chapter to explain exactly how this accuracy gain
works.

But first the context in which accuracy gain is possible (and indeed frequent) must be recalled. There are five assumptions about the data. The data
must be noisy, or else there is no room for improvement; the data must contain some signal that is not desperately small, or else modeling is hopeless;
the objective must be estimation of a population mean from sample data, or
equivalently prediction of new cases based on other cases, rather than mere
postdiction of a sample mean; the data must involve related experiments;
and the structure of the signal must be relatively simple compared with the
noise, ordinarily because the signal has few major causes but the noise has
numerous little causes. There is also one assumption about the model that
reflects the last of the five assumptions. The model must be suitable, capable
of recovering much signal in relatively few parameters compared with the
larger number required for noise. On the other hand, it is not required that
the true model be known, but rather a generic model family often will suffice.

There are three interrelated explanations for accuracy gain by parsimonious models. They concern signal–noise selectivity, variance–bias trade-off,
and direct–indirect information.

(1) Signal–Noise Selectivity. This chapter's preview explained accuracy
gain by parsimonious models in terms of signal–noise selectivity. Early model
parameters capture mostly the relatively simple signal, whereas late model
parameters capture mostly the relatively complex noise, as in Figure 8.1.
By selecting the most predictively accurate model at the peak of Ockham's
hill, a signal-rich model is separated from a discarded noise-rich residual.

An earlier study (Gauch 1992:125) goes into greater detail, examining noise recovery by individual model parameters.

(2) Variance–Bias Trade-off. Recall that Figure 8.10, for the soybean data, showed that progressively more complex models have higher variance but less bias. The most accurate model, sitting at the peak of Ockham's hill, had moderate amounts of variance and of bias, rather than an excessive amount of either problem.

That soybean example is representative of a common statistical phenomenon called the Stein effect (Berger 1985:168, 359–361). Stein (1955) discovered, quite surprisingly initially, that a particular model for related data was more accurate than its data. The explanation was that the model had a little more bias than its data but considerably less variance, and the substantial variance reduction helped more than the small bias increase hurt, so overall there was a gain in accuracy.

Decades later, however, the Stein effect has been observed for countless models applied to countless data sets, and it has come to be understood theoretically. What was a surprise has now become a commonplace. Hence, the accuracy gains reported for the preceding examples exemplify a very common and well-understood phenomenon. Naturally, the Stein effect is more likely and sizable for noisier data, because noise feeds the opportunity to reduce variance by a relatively parsimonious model, and reducing variance is the beneficial half of the variance–bias trade-off.

(3) Direct–Indirect Information. Recall that Table 8.1, which supplied eight conductivity measurements that were not the one you needed, showed quite simply that indirect data from related experiments can be informative. Then Figure 8.9 analyzed the conductivity data in greater detail, showing that accuracy is optimized by exploiting both the direct and the indirect information, placing increasing weight on the indirect data as noise increases.

Why are parsimonious models more accurate than their data? These models exploit more data – the direct and the indirect information in a set of related experiments that generated noisy data. For example, for each conductivity estimate, the full model uses just the 1 direct datum for that entry; but a parsimonious model uses all the data, the 1 direct datum *and* the 246 indirect data. The explanation of the accuracy gain concerns no magic and no mystery – just more data! Of these three complementary explanations, this third explanation is most powerful for achieving clear and simple insight regarding how models can surpass their data's accuracy. Using more of the data gives more accuracy.

Because parsimonious models gain accuracy by more extensive utilization of available data, a larger experiment with more data allows for greater accuracy gain. However, these words, "more data," are ambiguous, having two possible meanings. "More data" could mean more treatments, or it could

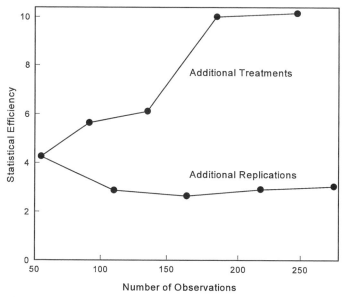

Figure 8.13. Statistical efficiency of AMMI analyses for the high-noise version of the conductivity data as a function of more data, supplied in the form of additional treatments or else additional replications. The top line shows results for data subsets of size 5×11, 7×13, 9×15, 11×17, and finally the complete data set of size 13×19, having from 55 to 247 observations. Additional treatments increase the statistical efficiency. The bottom line shows results for the 5×11 subset with 1 to 5 replications, having from 55 to 275 observations. Additional replications generally decrease the statistical efficiency.

mean more replications. Figure 8.13 shows the implications of this distinction for the conductivity data.

To obtain data sets with different numbers of treatments, several subsets of the conductivity data were formed. The original 13×19 matrix had 247 observations or treatments. The highest and lowest temperatures and the highest and lowest concentrations were trimmed from this matrix to form an 11×17 subset with 187 treatments. Repeating that process also formed 9×15, 7×13, and 5×11 subsets with 135, 91, and 55 treatments. To obtain data sets with different numbers of replications, five versions of the 5×11 subset were formed by dividing the high-noise rounding errors by the square roots of 1 to 5 to mimic having 1 to 5 replications and hence 55 to 275 observations.

Figure 8.13 shows that if "more data" means more treatments, then statistical efficiency increases. As the number of treatments increases from 55 to 247, the statistical efficiency rises from 4.28 to 10.15. As expected, more indirect data mean more accuracy gain. Incidentally, the AMMI-0 model is most accurate for the 5×11 subset with 1 and 2 replications (the leftmost two points on the bottom line of this figure), and AMMI-1 is best for all other cases.

However, Figure 8.13 also shows that if "more data" means more replications, then statistical efficiency generally decreases. The accuracy of averages over replications increases with the number of replications, which leaves less room for accuracy gain from modeling, as shown previously in Figures 8.6 and 8.8.

In general, more data in the form of more treatments increase statistical efficiency, but more data in the form of more replications decrease statistical efficiency. This basic story has, however, some elaborations and exceptions. For example, the bottom line in Figure 8.13 is only generally decreasing, not strictly decreasing. Likewise, additional treatments with tremendously different conditions and responses might reduce rather than increase the statistical efficiency (at least unless one switched to a different and more comprehensive model). Nevertheless, the usual story is that more data imply greater accuracy gain when "more data" means more treatments (but not more replications).

Actually, this is a very favorable outcome, because scientists are interested in comparing various treatments, whereas replication has no inherent interest, but rather is just a necessity for controlling errors. Because parsimonious models allow more treatments to substitute partially for more replications, they allow an experiment's resources to focus on the experiment's real action, its treatments. For example, a parsimonious AMMI model may allow a plant breeder to test 500 genotypes using 2 replications with the same accuracy as provided by raw data for 250 genotypes using 4 replications, but the greater focus on genotypes instead of mere replications allows for faster progress.

The preceding three explanations for accuracy gain can be integrated. Parsimonious models gain accuracy by more extensive utilization of available data, combining direct and indirect information. Indirect information helps more than it hurts, because parsimonious models capture most of the relatively parsimonious signal, while relegating much of the relatively complex noise to a discarded residual. Furthermore, even if a parsimonious model is a generic statistical tool not based on true scientific principles, still its reduction of variance may help accuracy more than its introduction of bias hurts accuracy. Imperfect models of imperfect data can use lots of data to separate simple and systematic signals from complex and idiosyncratic noise, reducing variance more than increasing bias, and thereby gaining accuracy.

Finally, benefits from parsimony abound in many contexts even when the word "parsimony" never appears. For instance, Brodley, Lane, and Stough (1999) have discussed data mining and knowledge discovery in databases (KDD), noting that the discovery of patterns in data is one of scientists' most important jobs. But modern scientific instruments can generate far more data than humans can inspect and grasp directly, so the objective of KDD is to train computers to search for patterns in large and often noisy data sets. That article never mentions parsimony explicitly, yet parsimony is a pervasive and critical part of the story. Specifically, models fitted to training data are validated by independent data in order to optimize predictive accuracy, which avoids overly complex models that attach too much significance to the errors or noise in the data. "The result is" a model "that may have less than 100 percent accuracy on the training set, but that will make fewer mistakes when it is applied to data outside of the training set" (Brodley et al. 1999). Any mention of data splitting, training sets, model validation, predictive accuracy, or noise reduction is a clue that parsimony probably is at work. When parsimony appears veiled, often a more explicit and vigorous understanding of parsimony can yield even greater advantages.

EFFICIENCY AND ECONOMICS

So far, this chapter has emphasized scientific principles and statistical tools, but this section places greater emphasis on economic implications. Parsimonious models can help scientists to get their money's worth out of their data. This is important because science is expensive, sometimes very expensive.

The basic story about economic advantages from modeling is obvious enough: Parsimonious models offer a cost-effective means for gaining accuracy and efficiency. Aggressive statistical analysis often can increase accuracy and improve decisions as much as would collecting several times as much data. When data are costly but calculations are cheap, as is often the case, this is a good deal. A researcher can collect the same amount of data as usual and apply modeling to gain accuracy, or alternatively can collect fewer data and use modeling to maintain the same accuracy as before. Either choice yields greater returns on research investments.

The first step in assessing modeling's economic worth is to connect dollars to some particular statistic measuring accuracy. Statistical efficiency is a sensible statistic for two reasons. First, these values also reflect effective replications, so a statistical efficiency of 2.5 equates to collecting 2.5 times as much data, which typically would cost about 2.5 times as much. Second, these values tend to give a conservative assessment, as compared with a detailed economic analysis, because even a small gain in productivity (resulting from greater accuracy and research efficiency) typically translates to a substantial gain in profitability. Also, competitive advantages tend to compound over time.

Figure 8.14 echoes Ockham's hill, but now transposing it into monetary terms. Beginning on the right, the most complex model is the full model that estimates treatment effects on the basis of the raw data, the averages

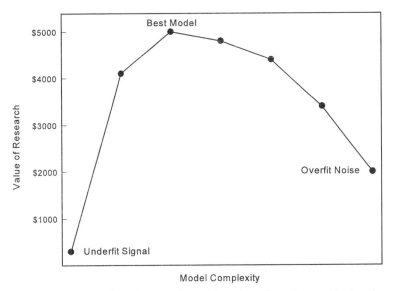

Figure 8.14. The value of research as a function of model complexity. Because of the trade-off between accuracy and complexity, as shown by Ockham's hill, a relatively parsimonious model is ordinarily most accurate. Parsimonious models increase accuracy and thereby increase the value of scientific research.

over replications. It overfits the noise because it captures not only all of the signal but also all of the noise in the raw data. This figure supposes that the economic value of having these data is $2,000. For comparison, assuming a middling statistical efficiency of 2.5 from modeling, the best model, at the peak of Ockham's hill, is worth $5,000. Finally, on the left, a severely underfit model is worth only $300.

Further suppose that collecting the experimental data costs $500, and calculating the best model costs $50. Then the full, overfit model costs $500 and is worth $2,000, so the experimental work returns $4 for each dollar invested. But the statistical work costs $50 and is worth an extra $3,000, so it returns $60 for each dollar invested. Consequently, an inexcusable research inefficiency and a sizable opportunity cost would result from failing to spend the $50 for modeling to make the data more valuable. In summary, $550 for wisely used data is a good deal, but $500 for poorly used data is a bad deal.

This example is representative in saying that modeling typically gives a return on investment that is an order of magnitude greater than that given by experimenting. Of course, these two research phases are not competitive because modeling requires data, but rather they are inherently complementary because modeling enables the data to do more for research objectives. This example is also representative in that statistical analysis ordinarily costs far less than data collection. Consequently, a research department should take care not to allow the avoidance of a little computation to mean that half of the information in the data is wasted. Data worth collecting are also worth analyzing.

PHILOSOPHICAL PERSPECTIVE ON PARSIMONY

This analysis of parsimony has been primarily from a scientific and technological perspective, with special interest in gaining accuracy and efficiency. But as this chapter draws toward a close, greater understanding of parsimony can emerge from adding some philosophical perspective, with special interest in truth. This section addresses three topics: the parsimony–accuracy trade-off, prediction and truth, and parsimony and nature.

(1) The Parsimony–Accuracy Trade-off. It has been considered a tangled philosophical problem that parsimony is inherently a dual criterion, combining simplicity with a *ceteris paribus* clause about an equally accurate fit to the data. Again, "The principle of parsimony recommends that . . . scientists choose the simplest theory," with the *ceteris paribus* clause, "among theories fitting the data equally well." But the difficulty with *two* criteria is that "we do not know what respective weight should be attributed to these two," simplicity and fit (Philipp G. Frank, in Frank 1956:3). Exactly how should scientists balance simplicity and fit, or, to use other synonyms, parsimony and accuracy?

Fortunately, recent advances have greatly clarified and quantified the parsimony–accuracy trade-off. The most important clarification has been a lucid distinction between the goals of prediction and postdiction. Are the data from only a sample of a larger population of interest, implicating prediction? Or are the data from the entire population, implicating postdiction? Almost always, the former is the case. In any event, this distinction is crucial, because the parsimony–accuracy trade-off for prediction and the trade-off for postdiction are strikingly different.

Figure 8.1 shows postdictive accuracy increasing continuously with model complexity for an entire sequence of more complex models. Hence, parsimony and accuracy are always in conflict, from start to finish. From a postdictive perspective, there is an awkward situation, with less parsimony promoting better fit until the bitter end, when the full model is reached, which automatically fits the data exactly. Of course, some arbitrary criterion can be imposed regarding accuracy, such as taking the first model that captures 75% or 90% or 95% of the variation in the data, or including those model terms that are statistically significant at the 5% or 1% or 0.1% level. But honestly, no objective and satisfying trade-off suggests itself.

Given this tortuous situation, it is fortunate indeed that scientific applications are rarely postdictive. Scientists rarely have the luxury of studying an entire population, but rather almost always have data from only a limited sample of a larger population. Consequently, this tortuous trade-off is usually irrelevant and avoidable.

Figure 8.1 also shows predictive accuracy as a function of parsimony, with a qualitatively different response than postdictive accuracy. Predictive accuracy increases until the peak of Ockham's hill is reached, after which neither accuracy nor parsimony is promoted. Hence, the tension between parsimony and accuracy does not last to the end, but rather lasts only until this peak is reached. Consequently, the ordinary solution to the parsimony–accuracy trade-off for a prediction problem is simply to pick the model at the peak of Ockham's hill. This solution is admirably objective, wonderfully easy, and clearly meaningful in that it optimizes a specific and important model property, predictive accuracy. In some cases, if the top of this hill is rather flat, rather than picking the exact peak one may prefer to sacrifice a little accuracy to gain considerable parsimony. But even then, model choice is still much easier than for the tortuous postdiction problem. So the primary resource for resolving the parsimony–accuracy trade-off is to distinguish prediction from postdiction.

(2) Prediction and Truth. Predictive success is often taken as evidence of truth. To cite one famous example, using Newton's theory of gravity, Edmond Halley (1656–1742) calculated the orbit of the impressive comet of 1682, which now bears his name, identifying it as the one that had appeared previously in 1531 and 1607, and predicting the time and place of its return in 1759. He did not live to see that return, but it did happen just as he had predicted. His striking predictive success was accepted universally as proof that his theory of comets' orbits was true, or at least very nearly true.

Generalizing from that familiar and yet representative example, predictive success is taken generally as evidence of truth, especially when numerous and diverse predictions are all correct, so that mere luck is an implausible explanation. Indeed, among theories that have attained strong and lasting acceptance among scientists, doubtless one of the most significant and consistent categories of supporting evidence is predictive success.

Nevertheless, this venerable formula, that predictive success implies truth, can be unsettled by interpreting or applying it too simplistically. Indeed, more specifically, a little reflection on Figure 8.6 should be disturbing, or at least thought-provoking. By construction, a cubic equation is known to be the true model. It is sampled at seven points, with addition of random noise (at an S/N ratio of five) to mimic measurement errors, and least-squares fits are calculated for the polynomial family. Although a cubic equation is the true model, the cubic model is less predictively accurate than another member of the polynomial family, the quadratic model. So even with the true model entered in the competition, the criterion of predictive accuracy gives the win to a false model! If such problems occur for easy cases with constructed and known models, what happens in the tough world of real scientific research? Does predictive accuracy have no reliable bearing on truth?

That a generic model, known not to be the true model, can still gain accuracy and efficiency is comforting to scientists because they often need to gain accuracy in situations where the true model is not available. But this outcome is troubling to philosophers, because predictive accuracy has often been taken as an indicator of closeness to the truth.

This problem, of predictive accuracy selecting a false model, cannot possibly be fixed by any philosophical or statistical repair in the definition or calculation of predictive accuracy. Rather, resolution must be in the form of precise claims about what predictive accuracy does (or does not) optimize. More specifically, careful distinctions must be made among a model's predictions, form, and parameters.

For a concrete example, recall the true cubic equation of Figure 8.4. It has the form $y = a + bx + cx^2 + dx^3$ and the parameters $a = 12.00$, $b = -3.50$, $c = 1.17$, and $d = -0.07$. At $x = 1$, the true value is $y = 9.60$. At an S/N ratio of 5, the most predictively accurate member of the polynomial family for this figure's data is the quadratic model. It has the form $y = a + bx + cx^2$; the least-squares-fitted parameters $a = 7.95$, $b = 1.13$, $c = 0.06$, and $d = 0$, as the quadratic equation has no cubic term; and the predictions $\hat{y} = 7.95 + 1.13x + 0.06x^2$. For example, at $x = 1$, this model's predicted value is $y = 9.14$. Comparing the true model and the fitted model, the fitted model's form is not the truth, and its parameter values are not even close to the truth. Yet its prediction of 9.14 is rather close to the true 9.60. Indeed, this prediction is substantially closer to 9.60 than is the noisy datum of 8.70.

Accordingly, for this example it is fine to claim that the quadratic model's predictions are more accurate (at an S/N ratio of 5) than those of any other member of the polynomial family, including the cubic model. Indeed, this is a fact! But it is confusing and false to interpret such a claim to mean that this model's form and parameters are accurate. Rather, a model's predictions, form, and parameters are three different things, so optimizing one is not necessarily optimizing the others.

Hence, it is somewhat vague or misleading to say that "the data can underwrite an inference concerning the curve's form based on an estimate of how predictively accurate it will be" (Forster and Sober 1994). Predictive accuracy *is* tremendously important, but it must be used to underwrite precise and legitimate claims.

On balance, it must be added that the most predictively accurate model may have a form or parameters that do reflect underlying physical processes or causes. For example, the parameters of an AMMI model of a yield trial are often interpretable in terms of known environmental factors and genetic pedigrees (Hugh G. Gauch and Richard W. Zobel, in Kang and Gauch 1996:85–122). This can be very useful. Furthermore, there is no mystery regarding why parameters of even generic models are often interpretable. Those parameters that capture sizable portions of the data's variation ordinarily reflect major causal factors that are already well known to scientists (Gauch 1992:231–236). However, the conclusion that a model's form or parameters can have a physical interpretation or can provide a causal explanation requires additional arguments and evidence besides that the model's predictions are accurate.

Even if a theory is in fact true, building a case for its truth using predictive success and other evidence can still be arduous. On the other hand, if a theory is false, falsification using predictive failure usually is relatively easy, although care still must be exercised to assure that the problem resides in the theory itself, rather than in some ancillary hypothesis or faulty instrumentation. In any case, whether predictive successes confirm one hypothesis or predictive failures disconfirm another hypothesis, the greatest necessity is that the true hypothesis be among those under consideration. When the truth is not even considered, so that all of the hypotheses are false, evidence has a misleading role.

Often scientists have no earthly prospects of ever knowing the exact true model, especially in fields like agriculture and medicine. This chapter's encouraging message is that even in those difficult cases, modest effort with generic models frequently will suffice to gain considerable accuracy and efficiency.

On the other hand, sometimes scientists do have an exact or nearly exact model based on well-established physical principles, especially in physics and chemistry. That luxury leads to a different situation. When the true equation is known completely, including both its form and its parameter values, then the true model does automatically make the most accurate predictions, regardless of the S/N ratio. For example, in Figure 8.4, the fitted cubic with parameters estimated from noisy data loses to the predictively superior fitted quadratic, but the true cubic with parameters known by construction would defeat all comers. Obviously, knowing the truth from the outset is quite different from receiving noisy data and starting to make sense of it.

Scientific research often has multiple steps and multiple purposes, however, which can defy simplistic accounts of parsimony's role. Indeed, a generic black-box use of parsimony in one step without truth claims (for the model's form or parameters) can help to support the search for causal explanations in another step with truth claims and testable predictions. For instance, in searching for beneficial genes to increase crop yields, pre-processing the yield data with a generic statistical model to reduce noise can subsequently help to locate genes on chromosomes more accurately, and then further yield trials with material carrying those genes can test whether or not those genes truly cause a yield increase.

In conclusion, predictive success generally indicates truth, in agreement with every scientist's intuition and experience. However, truth claims should focus carefully on precisely those specific aspects of a theory that do receive evidential support from the successful predictions. In the course of scientific development, predictive failures may have eliminated some hypotheses already, leaving several other contenders as multiple working hypotheses needing further research. Even if a theory has impressive predictive successes, thought should be given to the possibility that even better or more parsimonious theories that have not yet been considered might receive still stronger support from the data. Therefore, multiple working hypotheses are to be encouraged (Platt 1964; Hoffmann et al. 1996).

(3) Parsimony and Nature. Parsimony has two facets, as mentioned earlier in the historical review: an epistemological principle, preferring the simplest theory that fits the data, and an ontological principle, expecting nature to be simple. The epistemological aspect of parsimony has been emphasized here because it is part of scientific method. But the ontological aspect also merits attention.

Clearly, if parsimony is a sound epistemological principle, that must be because more fundamentally parsimony is also a realistic ontological principle. Simplicity of theories gets its value from simplicity of nature; otherwise, parsimony is ajar with reality and hence senseless. "If we accept a particular regulative principle," such as parsimony, and build it into our scientific method, then "we do so only because we also accept a certain conception of the object of knowledge," the physical world (Nash 1963:170). Note that this view follows Newton, whose first regulative principle concerned simplicity in nature, whereas the second, corollary to the first, concerned simplicity in theories.

So, is nature simple? Or more precisely, is nature simple at least in some significant sense that underwrites the methodological principle of parsimony? The answer to that question is, and must be, yes. But this answer is surpassingly difficult to explain satisfactorily. The problem is that virtually every basic presupposition about the world is immersed in parsimony, but in such a subtle way that usually there is no explicit mention of "parsimony" or "simplicity." Indeed, many large books on the philosophy and method of science lack such words in their index. Parsimony gets about as much attention as the air we breathe.

Nevertheless, simplicity does pervade nature. For starters, understand that the reality check is itself a *simple* theory about a *simple* world. It declares that "Moving cars are hazardous to pedestrians." This is simple precisely because it applies a single dictum to all persons in all places at all times. The quintessential simplicity of this theory and its world, otherwise easily unnoticed, can be placed in bold relief by giving variants that are not so simple. For example, were nature more complex than it actually is, more complicated variants could emerge, such as "Moving cars are hazardous to pedestrians, except for women in France on Saturday mornings and wealthy men in India and Colorado when it is raining." Although there is just one simple and sensible formulation of the reality check, obviously there are innumerable complex and ridiculous variants. Regarding cars and pedestrians, a simple world begets a simple theory. Or, to put it the other way around, a simple theory befits a simple world.

Capitalizing on this little example, meager thought and imagination suffice to see parsimony everywhere in the world – in iron atoms that are all iron, in stars that are all stars, in dogs that are all dogs, and so on. Parsimony touches our every thought. But to really understand parsimony, one must move beyond examples to principles.

Induction, uniformity, causality, intelligibility, and other scientific principles all implicate parsimony. Applications of induction to the physical world presuppose parsimony, specifically in the ontological sense expressed strongly by the law of the limited variety of nature. The uniformity of nature expresses the simplicity of nature, for a changing or capricious nature would be more complicated. Likewise, the law of causality, that similar causes produce similar effects, is an aspect of simplicity. Also, the ordinary scientific conviction that nature perdures is simpler than the contrary belief that the universe repeatedly flickers in and out of existence. That nature is intelligible to our feeble human reason shows that some significant features of reality are moderately simple. In a word, were nature not simple, science would lose all of its foundational principles at once. So one begins to understand parsimony properly upon recognizing that parsimony is not just one among several basic scientific principles, but rather parsimony also pervades all of the other scientific principles, such as induction and causality.

Yet the greatest influence of parsimony in scientific method is in the simplicity of the questions asked. Any hypothesis set that expresses a scientific question could in principle always be expanded to include more possibilities, and that action would make sense were the world more complex than it is. Were inductive logic bankrupt, were nature not uniform, were causes not followed by predictable effects, were it unlikely that nature perdures, and were nature barely comprehensible, then enormously more hypotheses would merit consideration, and science would languish with hopelessly complicated questions that would impose impossible burdens for sufficient evidence. The beginning of science's simplicity is its simple questions.

Having argued that nature is simple, this verdict should not be interpreted simplistically! Indeed, "there is complexity to the whole idea of simplicity" (Nash 1963:182). Also, one pebble or one leaf has more complexity than the world's scientists will ever fully master. Indeed, a mere one electron will keep many secrets from us mortals! Nature is enormously intricate relative to human understanding, and we ourselves are wonderfully made.

Simon (1962) offered remarkably keen insights regarding just which aspects of nature scientists expect to be simple and just which aspects they expect to be complex. In essence, the rich complexity of life and ecosystems emerges from the frugal simplicity of basic physical and chemical laws. From general experience, scientists and engineers ordinarily have a fairly reliable general sense of how simple or complex a given system or problem is. But speculations may be offered here that the next few decades will bring two big surprises that will confirm and extend the general perception that physics is simple and biology is complex.

First, fundamental physics may gain parsimony beyond the wildest imaginings of physicists during the twentieth century. One of the case studies in the following chapter mentions this exciting possibility.

Second, life may come to be understood to be vastly more complex than biologists currently realize. Only about a century ago, before some famous experiments were conducted by Louis Pasteur, biologists believed that life was simple enough for the spontaneous generation of mold, maggots, and even frogs in dirty rags and such. Contemporary biologists know that life is far too complex for spontaneous generation to occur, especially because information theory has quantified the tremendous complexity of DNA molecules. But as biological knowledge increases and new philosophical concepts and mathematical tools emerge to perceive complexity, it may turn out that thus far biology has taken only a few baby steps away from a pre-Pasteur conception of life and toward a realistic appreciation of life's true complexity.

Clearly, the verdict on parsimony is that "Ockham's Razor must indubitably be counted among the tried and useful principles of thinking about the facts of this beautiful and terrible world and their underlying causative links" (Hoffmann et al. 1996). Nevertheless, those authors also note the sensible reaction that "the very idea that Ockham's Razor is part of the scientific method seems *strange* ... because ... science is not about simplicity, but about complexity." Indeed, enthusiasts of parsimony, which should include all scientists, would do well to take a moment to ponder that sentiment. The plausible resolution that those authors offer, however, is that simple minds comprehend complex nature by means of ornate models made of simple pieces. The balance between a model's simplicity and the extent to which it approaches completeness requires a really, really delicate and skillful wielding of Ockham's razor. The comments on that paper by A. Sevin concur: "Our discovery of complexity increases every day.... This good old Ockham's razor remains an indispensable tool for exploring complexity." It may seem paradoxical, but simplicity is the gateway to complexity.

Finally, this chapter's treatment of parsimony is only introductory. It has substantial limitations. A scientist's choice from among competing theories can be much more difficult than has been illustrated by any of this chapter's examples. The competing models may be of diverse forms (such as an exponential model and a trigonometric model), rather than members of a single family. More fundamentally, competing models may posit different physical entities and processes. Also, theory choice may involve criteria besides goodness of fit and parsimony, such as explanatory power, fruitfulness in generating new insights or results, and coherence with other well-accepted theories.

Despite its limitations, perhaps this chapter will stimulate interesting new questions, even if it does not provide pat detailed answers. Merely to recognize that prediction is different from postdiction, or that parsimony can increase research efficiency, may suffice to move many scientists forward. It may be suggested that gaining an elementary understanding of parsimony, combined with ordinary statistical tools, can provide 90% of the practical benefit that would emerge from exemplary understanding and impeccable statistics.

SUMMARY

The principle of parsimony recommends that from among theories fitting the data equally well, scientists choose the simplest theory. Thus, fitting the data is not the only criterion bearing on theory choice. Building on earlier thinking by Aristotle, Grosseteste, and others, Ockham advanced our understanding of parsimony so significantly that the principle often bears his name, as Ockham's razor. In a famous paradigm shift, Copernicus chose the heliocentric theory over the geocentric theory, not on grounds of a better fit with the data, but rather on grounds of greater simplicity. As that case illustrates, sometimes false theories get into trouble with parsimony before they get into trouble with more extensive or more accurate data. Accordingly, considerations of parsimony can help to place scientists on the cutting edge of their specialties.

The workings of parsimony have been illustrated with four examples: Mendel's peas, a constructed cubic equation with added random noise, equivalent conductivities of hydrochloric acid solutions, and crop yields. Accuracy gain by the use of parsimonious models has been explained in terms of signal and noise selectivity, variance and bias trade-off, and direct and indirect information. Basically, parsimonious models gain accuracy by utilizing more of the information in the data.

In many areas of science and technology, a better understanding of the role of parsimony in the scientific method could yield considerable gains in accuracy and efficiency. Often a parsimonious model can increase accuracy as much as would the collection of several times as much data. Parsimony can help scientists achieve good returns on research investments. Scientists, statisticians, and philosophers all have had valuable insights into the meaning and usefulness of parsimony.

CHAPTER NINE

CASE STUDIES

The main thesis of this book is that the winning combination for scientists is mastery of the general principles of scientific method together with mastery of the specific research techniques of a chosen specialty. Neither can substitute for the other. Ordinarily, a scientist's training in specialized knowledge and techniques is strong. But all too often, training in science's history, philosophy, logic, presuppositions, and fundamental methods is weak.

The most compelling way to show that the general principles of scientific method matter is to give several examples of scientific research that are striking precisely because they incorporate an exceptional grasp of these principles. Accordingly, this chapter presents several case studies.

Recall that this book has two objectives, to increase productivity and enhance perspective. The first case study, concerning the elementary physics of motion, pursues a balanced perspective on science by defending science's credibility against the various philosophical attacks reviewed in Chapter 3. The remaining case studies, concerning advanced research in diverse pure and applied sciences, reflect remarkable productivity energized by an exceptional understanding of scientific method. It is hoped that reflection on these examples will suggest possibilities for some new advances using analogous applications in the reader's own specialty.

INTUITIVE PHYSICS

Recall from Chapter 3 that Sir Karl Popper, Thomas Kuhn, and other prominent philosophers have challenged the very foundations of science, citing four deadly woes: (1) Science cannot prove any theory either true or false. (2) Observations are theory-laden, and theory is underdetermined by data. (3) Successive paradigms are incommensurable. (4) What makes a belief scientific is merely that it is what scientists say. The upshot of these woes is that science is alleged to be irrational or arational, unable to claim any truths about the physical world. In a nutshell, the central problem is that nature does not significantly constrain theory choice.

Are these problems fact or fiction? Is science in big trouble or not? Most fundamentally, does nature constrain theory? This section approaches these questions in a practical way by looking at an area of physics accessible to every reader through the performance of some simple experiments involving the motions of familiar objects at ordinary speeds, far below that of light, such as a dropped or thrown ball. Accordingly, this topic will afford readers a suitable arena for testing the thesis that nature constrains theory and the antithesis that it does not.

McCloskey (1983) surveyed high-school and college students regarding their intuitive beliefs about motion. He also recounted both the medieval theory of impetus and the modern theory of inertia. Most importantly, he reported experiments on the actual motions of objects.

Briefly, impetus theory said that when a mover set an object in motion, that imparted an internal force or impetus that would act continuously to keep the object moving until the impetus had dissipated. By contrast, inertia theory says that an object will continue in its state of rest or its state of uniform motion in a straight line unless acted on by an external force. These incompatible theories predict different motions or paths in various settings. From informal experiences with moving objects, people tend to form ideas that are much like the medieval impetus theory.

As a specific instance of motion, imagine a running person who drops a ball from waist height. What path will the ball follow as it drops? Figure 9.1 shows the three possibilities: the ball will fall forward with the runner (path A), drop straight down (B), or fall backward (C). Which path do you expect? This analysis of the ball's path will progress through five stages. First, the subjective dimension of persons' beliefs and actions will be surveyed. Second, the objective physics of a ball's actual path will be explained by Newtonian mechanics. Third, the historical dimension of questions and discoveries about motion will be reviewed, from Aristotle to Newton. Fourth, educational challenges and reforms will be reviewed. Fifth and finally, there will be some philosophical reflection.

First, regarding persons' intuitive beliefs about the motion of a dropped ball, McCloskey (1983) surveyed numerous college students. The percentages espousing the three hypotheses in Figure 9.1 were as follows:

H_A: The ball will travel forward, landing ahead of the point of release (45%).

H_B: The ball will fall straight down, landing under the point of release (49%).

H_C: The ball will travel backward, landing behind the point of release (6%).

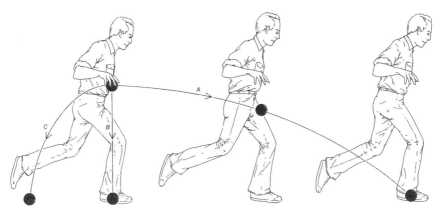

Figure 9.1. A ball is dropped by a running person. Students were asked if the ball would fall forward with the runner (path A), drop straight down (B), or fall backward (C). Most college students surveyed gave incorrect answers. (This drawing by Michael Goodman is reproduced with his kind permission.)

Note that hypothesis H_B was the most popular answer, the expectation that the ball would fall straight down. McCloskey was also interested in the relationship between belief and action. Accordingly, besides surveying subjects' verbal responses to physics questions, he monitored subjects' actual actions with physical objects. College students were given the task of dropping a ball while moving forward so that it would hit a target marked on the floor. Performance of that simple task should reveal each subject's belief about the path that the ball would take. There were three possible actions, corresponding to the three hypotheses. The percentages of students performing those three actions were as follows:

A_A: Drop the ball before reaching the target (13%).
A_B: Drop the ball directly over the target (80%).
A_C: Drop the ball after reaching the target (7%).

Note that these survey results, based on actions, generally were similar to the verbal survey. Again, A_B was most popular, and A_C least popular. Which action would you perform to make a dropped ball hit the target marked on the floor?

Second, let us now turn from the subjective dimension of persons' beliefs about motion to the objective dimension of a ball's actual path. The correct answer is hypothesis H_A, corresponding to action A_A: The ball travels forward with the runner as it falls, so it must be dropped before reaching the target marked on the floor. Was your intuitive belief true? Was your recommended action correct?

Using Newton's laws of motion, McCloskey (1983) explained the ball's path: When the ball is dropped by the runner, its inertia causes it to continue moving forward at the same speed as the runner, because (ignoring air resistance) no force is acting to change its horizontal velocity. As the ball moves forward, it also moves down at a steadily increasing velocity because of acceleration due to gravity. The forward and downward motions combine in a path that closely approximates a parabola, path A in Figure 9.1. This parabolic path was discovered by Galileo in 1604 and published in 1638 (Drake and MacLachlan 1975; McCloskey 1983; Jefferys and Berger 1992).

Regrettably, the correct answer was a minority position. Only 45% of the students surveyed held the correct belief, and worse yet, only 13% performed the correct action. Hence, for even such a simple instance of motion, subjective beliefs and objective facts are not in agreement as commonly as one might suppose: "One might expect that as a result of everyday experience people would have reasonably accurate ideas about the motion of objects in familiar situations. . . . It seems that such is not the case" (McCloskey 1983).

Third, the historical dimension of beliefs about motion is also surprising. Aristotle had two theories of motion, one for celestial objects and another for terrestrial objects, but both were incorrect on many counts. Medieval impetus theory was better, but still rather problematic. "What path do moving objects take?" was an archetypical science question for millennia, along with "What are things made of?" Both questions have proved amazingly difficult. The erosion of the medieval impetus theory by the work of William of Ockham and the advances in experimental and mathematical methods wrought by Robert Grosseteste positioned Galileo to first construct the theory of inertia and to discover the parabolic trajectory of a falling object with an initial horizontal motion. Subsequently, Newton subsumed those findings in his grand, unified mechanics, including his first law of motion regarding inertia (Cajori 1947:13). So, from the time that Aristotle posed simple questions about motion until a satisfactory solution emerged from Galileo and Newton, two millennia had passed. Contributions from many of the world's great intellects, advances by many successive generations, and two millennia were required to answer the seemingly simple question posed in Figure 9.1. And even now, centuries after Newton, most people still have erroneous beliefs about motion that reflect the medieval impetus theory more than Newtonian mechanics. Of course, no one is now taught that fallacious medieval theory as fact in school, yet from informal experiences with moving objects, apparently most people formulate an intuitive theory that generally agrees with impetus theory.

Fourth, this contrast between subjective beliefs and objective truths presents interesting educational challenges. Besides the motion of a dropped ball, McCloskey studied several other cases. In all cases, misconceptions were common.

Given the commonness of errors about motion, science educators must develop effective methods for correcting these errors. How can misconceptions be dispelled? How can students be convinced of the correct laws of motion? The obvious answer, as McCloskey notes, is instruction in Newtonian mechanics. However, intuitive beliefs are often difficult to modify. For example, before taking a physics course, only 13% of college students took the correct action of dropping the ball before reaching the target; but after the physics course, the figure increased by 60% to 73%, which still left erroneous preconceptions intact in 27% of the students. In another survey, 93% of high-school students believed in impetus, and a physics course reduced those erroneous assumptions only slightly, to 80%. Neither standard lectures nor laboratory demonstrations seem to be effective for overcoming false preconceptions (Halloun and Hestenes 1985a,b).

The problem is that new experiences are routinely interpreted within an existing paradigm: "It appears that students often rely on the intuitive impetus theory as a framework for interpreting new course material. As a result the material can be misinterpreted and distorted to fit the intuitive preconceptions" (McCloskey 1983). Consequently, educators must seek better instruction methods.

Better progress seems to result from having students articulate their beliefs, after which the differences between their initial beliefs and Newtonian mechanics can be sorted out. Implicit presuppositions may resist instruction and experiments, but explicit beliefs frequently respond. Often it is not enough just to be told the truth; rather, one must also recognize one's current beliefs and how they differ from new beliefs being offered by the instructor. In the confrontation between hypotheses and data, the hypotheses must be recognized explicitly. Explicit hypotheses plus experimental evidence can equal realistic learning.

Although most people have many delusions about motion, and though the historical path to correct knowledge was long and tortuous, the facts are now readily available. Indeed, the laws of motion are empirical and testable matters, requiring little time and resources to perform simple experiments, such as that in Figure 9.1. The actual motions are thus matters of public knowledge, because the required experiments are trivial and because Newton's laws of motion have been well known for centuries. Repetition of the experiments by various persons in various places dropping various objects can give results as certain as one desires. In short, true knowledge of motion is empirical, testable, public, objective, and certain.

Furthermore, knowledge about motion is important. Moving objects impact human welfare in a thousand arenas, from sports to warfare. Correct ideas work, whereas erroneous preconceptions fail. Ideas have consequences. Truth matters. For example, strategies other than releasing the ball before reaching the target invariably result in failure to hit the target. That is just the way the world is.

Astonishingly, in his interview in *Scientific American*, the noted philosopher Thomas Kuhn fondly recounted the great "Eureka!" moment in his career, when he concluded that "Aristotle's physics 'wasn't just bad Newton,'" but rather, "it was just different" (Horgan 1991). Well, what does it mean to say that Aristotle and Newton are "just different" when a projectile launched by Aristotle's theory would miss its target, whereas a projectile launched by Newton's theory would hit its target?

Recall the four deadly woes that allegedly challenge science. What is their meaning or credibility as applied to this specific case of moving objects? Supposedly, these four woes pervade all of science, which necessarily includes the current case. So, do these woes unsettle our presumed scientific knowledge of motion, or does our valid scientific knowledge of motion unsettle these woes?

(1) The first challenge is that science cannot prove any theory either true or false. Consider the three theories about a dropped ball's motion (Figure 9.1). It is within the reader's grasp, given a ball and a few minutes, to prove one of these theories true and two of them false. It is simply ludicrous to claim that science never finds any objective, final truth. More generally, millions of experiments have proved inertia theory true and impetus theory false.

(2) The second challenge is that observations are theory-laden and that theory is underdetermined by data. The intent of that challenge is to say that theory and observation are so profoundly intertangled that observation cannot constrain or guide theory choice. But manifestly, the three theories in Figure 9.1 are separable from observations of a ball's actual path, so observations can constrain theory. Furthermore, the only way to make the data underdetermine this theory choice is to throw off the common sense enjoined earlier by the reality check and thereby enlarge the original hypothesis set (H_A, H_B, and H_C) with other wild hypotheses. For example, add hypothesis H_D, that "The ball appears to follow path A, but actually we are just butterflies dreaming about physics experiments," or add hypothesis H_E, that "The universe thus far has followed inertia theory, but in ten minutes and six seconds it will switch to impetus theory." Then the data could not dictate theory choice. But as long as science presumes a scrap of common sense, such outlandish moves are but philosophical games.

(3) The third allegation is that successive paradigms are incommensurable, so competing theories are just different. A person who began with the majority view that a ball dropped by a runner would fall straight down (path B in Figure 9.1) but then considered the alternative possibilities and performed experiments revealing that a dropped ball would travel forward with the runner (path A) is a person who has undergone a paradigm shift. But after

shifting to H_A, there is no inability to understand one's former beliefs in H_B nor other persons' current beliefs in H_B. Indeed, anyone should be capable of understanding all three competing hypotheses, H_A, H_B, and H_C. The difference between them is not that adherents of any given hypothesis cannot understand what adherents of others are saying. Rather, the difference between them is that H_A is supported by empirical evidence and enables its knowers to accomplish certain tasks, whereas H_B and H_C are false theories commending misguided actions that will lead to failure. Of course different theories or paradigms are commensurable and comparable!

(4) The fourth and final charge is that what makes a belief scientific is merely that it is what scientists say. Is the only difference between H_A and the other hypotheses that H_A is what physicists say, whereas laypersons often espouse other hypotheses? Of course not. The greater difference is that H_A describes the actual path that a ball will travel, whereas the other hypotheses do not. At issue here are not only subjective beliefs but also objective facts, not only persons' beliefs but also the paths of falling objects. What makes H_A scientific is that it presents knowledge gained by the scientific method that accurately describes the true state of nature regarding a ball's motion.

The thrust of this case study is that science's truths emerge when hypotheses are confronted by data. Incidentally, even Galileo endorsed the impetus theory in his early writings, around 1590, before his experiments with physical objects forced a revision of his beliefs. The ultimate instruction about moving objects is actual experimentation with moving objects, particularly when the student recognizes explicitly the various possible hypotheses. Furthermore, scientists have learned millions of other things about the natural world that feature just as much objectivity and certainty as in this section's example of motion.

Finally, it might be objected that this section's reply to the four challenges underlying anti-realist or instrumentalist interpretations of science misses the mark because the real action concerns the status of unobservables, such as electrons, rather than observables, such as balls. However, that objection cannot be sustained.

It is simply false that this realist/anti-realist debate concerns only unobservables. Inexorably, the logic involved in answering these four woes can reach any claim about an external physical reality, regardless how ordinary or exotic the object might be. Indeed, philosophers and educators who promulgate skeptical and anti-realist positions have been quite forthcoming about including ordinary objects, such as a glass of water or a table, as discussed in Chapters 4 and 11.

Consequently, the realist/anti-realist debate can be discussed just as productively and just as fairly in the context of balls as in that of electrons. The real issues are far more fundamental than the trifling distinction between observables and unobservables. Nor is that distinction even very meaning-ful, for everything is observed with some degree of indirectness. Furthermore, a given object can change status because of perspectival differences that are without ontological interest, such as a runner approaching from a distance who is first unobservable, then a moving speck, then a human being, and finally a recognized friend.

PARSIMONY AND PHYSICS

by Millard Baublitz

Parsimony is especially important in physics. In Chapter 8 it was stated that parsimony has been viewed by some philosophers as a methodolog-ical axiom and by others as an ontological principle. One reason for the importance of parsimony in physics is that physicists pragmatically employ simplifying assumptions in their theoretical calculations and in the design of their experiments on a daily basis. On the other hand, some physicists have been among the boldest of scientists to espouse parsimony as an onto-logical principle. In addition to the two philosophical points of view about parsimony, it can be stated, broadly speaking, that parsimony in physics has taken two distinct forms: physical parsimony and mathematical parsimony. In this section, the two philosophical points of view about parsimony will be discussed, and then physical parsimony and mathematical parsimony will be compared.

An appropriate definition, consistent with the point of view of parsi-mony as a methodological assumption, states that parsimony is "economy of explanation in conformity with Occam's razor" (*Webster's Ninth New Collegiate Dictionary*, 1986:857). Furthermore, as noted in Chapter 8, William of Ockham (or Occam) suggested that "What can be explained by the as-sumption of fewer things is vainly explained by the assumption of more things." Parsimony, according to that dictum, is essentially a utilitarian as-sumption or perhaps even a labor-saving expediency. The prudent scientist considers the explanations that fit the data equally well and chooses the simplest explanation.

It is instructive to consider a pendulum, an instrument that has been important to natural philosophers and physicists since the time of Galileo (Matthews 1994:109–135; 2000). If the local acceleration due to gravity is g and a pendulum bob of mass m has been displaced from the vertical by a small angle θ, perhaps five degrees, then the restoring force that acts on the pendulum bob can be represented by the linear function $F_L(\theta)$:

$$F_L(\theta) = -mg\theta \qquad\qquad (1)$$

However, if the angle of displacement is not small, then the restoring force is represented more accurately by equation (2) for the angular values $0° \leq \theta \leq 90°$:

$$F(\theta) = -mg \sin \theta \tag{2}$$

Equation (1) is an approximation of the restoring force in equation (2) when the maximum angular displacement is small. The forces described by equations (1) and (2) both permit oscillatory motion, but the period of oscillation is independent of the angular displacement if equation (1) describes the exact restoring force, whereas the period of oscillation explicitly depends on the maximum angular displacement if equation (2) describes the restoring force. The restoring force in equation (2) is more generally valid; that is, it is more accurate for larger-angle oscillations of a pendulum, but the resultant description of the pendulum's motion with equation (2) is also somewhat more complicated.

Further thought reveals that many other physical variables might also be included to give a more thorough characterization of the motion of a pendulum. For example, if the pendulum bob is not a perfect uniform sphere or a point mass, then it may be important to use the moment of inertia of the bob instead of its mass. The mass of the string or chain that holds the pendulum bob may need to be considered. A physically real pendulum may experience a frictional force due to the surrounding air, and that damping force will lead to a decrease in the amplitude of the pendulum's swing. Of course, this brief recitation of additional physical variables that are needed for a more extensive description of a pendulum's motion is not exhaustive.

It may be helpful to contrast the simplifying approximation of small-angle oscillation for the pendulum with the use of parsimony as a methodological axiom in another discipline, agricultural science. An agricultural researcher interested in wheat yield per hectare must consider a multiplicity of often interrelated input variables, including precipitation, soil conditions, and resistance of the plants to disease. Furthermore, the uncertainties in the measurements, or the "statistical noise," may be relatively large. In the pendulum problem there are only a few relevant physical variables in the simplest descriptions, and the position of the pendulum as a function of time can be measured very precisely, so that uncertainties in the measurements are relatively small. Not all problems in physics are as amenable to analysis as the motion of the pendulum, but physicists perhaps have made their greatest progress when they have chosen and studied such problems.

It has been noted that from among the explanations that fit the data equally well, Ockham's razor instructs us to choose the simplest explanation. A physicist usually chooses the simplest model that "saves the appearances" of the salient features of the phenomena. The choice of the simplest model is usually straightforward if the data are fit *equally* well by the explanations, but the example of the pendulum indicates that models or explanations with varying degrees of sophistication can be chosen, depending on the level of agreement that the physicist seeks between the model and data. The linear restoring force in equation (1) might be adequate to describe a pendulum with a small-angle oscillation, and similar linear approximations are ubiquitous in physics. On the other hand, if a physicist is interested in the decrease of large-amplitude oscillations of a pendulum over an extended time, then equation (1) is inadequate as a description of the restoring force, and viscous forces also must be included explicitly in the calculations.

It is the nature of the physicist's trade and training to choose problems that are tractable, to observe the phenomena carefully, to identify the most notable features of the data, and to approximate the data with the simplest appropriate mathematical model. In practice, physicists often model phenomena using a method of successive approximations. Sober (1975:74–75) stated that "there is a tension between goodness of fit and simplicity. The former pulls toward higher order equations; the latter draws us toward lower order ones." This is as true in physics as in other disciplines.

Having considered parsimony as a methodological assumption in physics, next consider parsimony as an ontological principle. In other words, the proposition will be considered that there are a few, relatively simple fundamental laws that govern nature. Grosseteste was cited in Chapter 8 as one medieval philosopher who expected nature to exhibit simplicity, but the metaphor of natural laws was not developed fully until the 1600s. Robert S. Cohen (in Frank 1956:218–232) has suggested that development of the idea of natural law is attributable to the confluence of two trends: the influence of Judeo-Christian religion and the rise of centralized legal authority in the new nation-monarchies. He has described one element in the development of the metaphor of natural law: "In modern science, the natural law as metaphor became explicitly so by the seventeenth century (and distinguished from human and moral laws of behavior of man), the first separation into metaphor occurring about the time of Kepler. Descartes speaks of the laws which God put into nature.... Note that God, in earlier medieval time, was the divine will that brought the exceptions to the world, the *irregular* occurrences, for example comets and monsters. Now, after Descartes he is the lawgiver for the *regular* occurrences" (p. 226).

Cohen rightly acknowledges Kepler (Figure 9.2) as one of the early, major proponents of the concept of natural law, and the following excerpt from a letter by Kepler, of April 1599, written ten years before the publication of his *Astronomia Nova*, further explains his motivations in seeking "material laws" for nature: "To God there are, in the whole material world, material

Figure 9.2. Portrait of Johannes Kepler. (This frontispiece from Gunther, 1896, is reproduced with kind permission from the Division of Rare and Manuscript Collections, Carl A. Kroch Library, Cornell University.)

laws, figures and relations of special excellency and of the most appropriate order. . . . Those laws are within the grasp of the human mind; God wanted us to recognize them by creating us after his own image so that we could share in his own thoughts. . . . Only fools fear that we make man godlike in doing so; for the divine counsels are impenetrable, but not his material creation" (Baumgardt 1951:50).

Kepler's astronomical research was a landmark in the history of science. For two thousand years, astronomers had confidently assumed that the motions of planets should be described by circles or by combinations of circles, such as the epicycle and deferent shown in Figure 9.3. First, Kepler's *Astronomia Nova* of 1609 swept away the elaborate mechanisms of epicycles and deferents and showed convincing evidence that each planet moves in an ellipse, with the sun at one focus of the ellipse. Second, a major result in *Astronomia Nova* was the law of equal areas for the motion of a planet: If a line is drawn between the sun and a planet, the line sweeps out equal areas in equal times. Third, nearly a decade later, Kepler added to his successes and discovered a mathematical relation between the period of revolution of

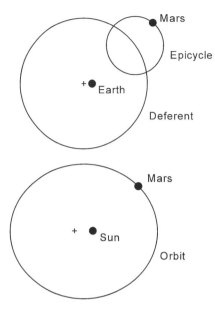

Figure 9.3. Simplified sketches of the path of Mars according to Claudius Ptolemy (top) and Johannes Kepler (bottom). According to Ptolemy's model, Mars revolved around an epicycle (a small circle), while the center of the epicycle revolved about the larger circle, which was called a deferent. That combination was intended to account for changes in the apparent speed of Mars, including occasionally slowing down and even reversing its motion. In Kepler's model, the orbit of Mars is an ellipse, with the sun at one focus. Here the shape of the ellipse is exaggerated to render it visibly different from a circle.

a planet, denoted here by T, and the mean radius of the planet's orbit, R. This relation is expressed in equation (3), where K is a constant with the same value for all of the planets.

$$R^3 = K\,T^2 \tag{3}$$

Kepler is remembered today by astronomers and physicists chiefly for those three findings, which have become known as Kepler's laws of planetary motion. But certain other attributes of his work will be emphasized here. Kepler recognized that the motions of the planets required explanation in terms of a physical theory, or as Stephenson (1987:2) has stated, Kepler was "the first actually to envision astronomy as a part of physics." Kepler also required that his model of planetary motion agree precisely with the voluminous, quantitative astronomical observations of Tycho Brahe. Brahe's data for the position of Mars had an uncertainty of about four minutes of arc, and Kepler discarded an earlier model, after years of intensive labor, because of an angular discrepancy of eight minutes between his earlier model and Brahe's observations. That insistence on precise agreement between theory and data established a new standard for the testing and verification of fundamental theories of physics.

Although Kepler attempted to understand astronomical motions in terms of physical ideas, his physics was fatally flawed because of the lack of understanding of inertia. The development of a consistent physics, which would include a unified understanding of both the motions of the heavenly bodies and motions on earth, had to await the genius of Isaac Newton.

Newton's *Philosophiae Naturalis Principia Mathematica*, first published in 1687, begins with an introduction and some definitions, and then states the following three axioms or laws of motion:

I. Every body perseveres in its state of rest, or of uniform motion in a right line, unless it is compelled to change that state by forces impressed thereon.

II. The alteration of motion is ever proportional to the motive force impressed; and is made in the direction of the right line in which that force is impressed.

III. To every action there is always opposed an equal reaction. (Newton 1964:23)

Newton ardently believed in the concept of simple natural laws and pronounced that those laws had an ontological basis, as shown by his statement that "Nature is pleased with simplicity, and affects not the pomp of superfluous causes" (Newton 1964:324). Because of the influence of Newton and his *Principia*, the concept of simple natural laws had become so firmly entrenched by the late eighteenth century, both in natural philosophy and in the contemporary popular culture, that even poets and hymn writers celebrated the concept of natural law.

During the nineteenth century, physicists made remarkable experimental discoveries in electricity, magnetism, optics, and other areas of study, and a search for additional natural laws ensued. That search for natural laws culminated in James Clerk Maxwell's unified theory of electromagnetism and light. Maxwell's synthesis is expressed today in terms of only four rather simple equations. The Maxwell equations, which are the fundamental equations of the classical theory of electromagnetism, are given here in modern notation (Feynman, Leighton, and Sands 1964:18-2). Electric and magnetic fields are represented, respectively, by the variables $\mathbf{E}$ and $\mathbf{B}$. The electric-charge density and electric-current density, respectively, are given by the variables ρ and $\mathbf{j}$. The speed of light is c, t is time, and ε_0 is a constant called the permittivity of free space.

$$\nabla \cdot \mathbf{E} = \frac{\rho}{\varepsilon_0} \tag{4}$$

$$\nabla \times \mathbf{E} = -\frac{\partial \mathbf{B}}{\partial t} \tag{5}$$

$$\nabla \cdot \mathbf{B} = 0 \tag{6}$$

$$c^2 \nabla \times \mathbf{B} = \frac{\mathbf{j}}{\varepsilon_0} + \frac{\partial \mathbf{E}}{\partial t} \tag{7}$$

Although the Maxwell equations are expressed in the language of mathematics, it must be emphasized that the equations are based on experimental results and the physical concepts of electric and magnetic fields. For example, equation (5) is derived from the empirical Faraday induction, shown in Figure 9.4.

It is also noteworthy that equations (5) and (7) include both electric and magnetic fields. Before the nineteenth century, electricity and magnetism were regarded as distinct topics within physics. But experiments by Ørsted, Faraday, and others, and the Maxwell equations (5) and (7), demonstrated that electric and magnetic phenomena are intimately related. Furthermore, mathematical solutions of the Maxwell equations suggested that light is an electromagnetic wave and that the classical results of optics can be derived from equations (4)–(7). Eventually, that understanding of electromagnetic waves led to the invention of communication by radio transmitters and receivers. Much earlier Newton had recognized that planetary motions could be understood as part of mechanics, but Maxwell had united the previously unconnected fields of electricity, magnetism, and optics into a single theory of great power and elegance.

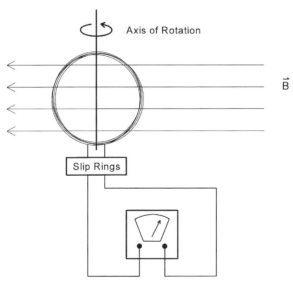

Figure 9.4. In this simple alternating-current generator, the coil of wire is rotated in a uniform magnetic field ($\vec{B}$), a voltage is induced, and electric current flows through the wire. Slip-ring electrical contacts connect the spinning coil of wire to the current meter. One of Maxwell's equations, equation (5), expresses the relationship between the induced electric field and the rate at which magnetic-field lines are cut by the rotating wire coil.

Maxwell had a profound belief in the interconnectedness of physical phenomena and the intrinsic simplicity of nature, as he stated in his inaugural address at the University of Aberdeen: "But as Physical Science advances we see more and more that the laws of nature are not mere arbitrary and unconnected decisions of Omnipotence, but that they are essential parts of one universal system in which infinite Power serves only to reveal unsearchable Wisdom and eternal Truth.... Is it not wonderful that man's reason should be made a judge over God's works, and should measure, and weigh, and calculate, and say at last 'I understand I have discovered – It is right and true'" (Goldman 1983:87).

This brief recapitulation of some highlights in the theoretical development of classical physics has demonstrated that certain fields of physics can be described by a few relatively simple principles or equations. It has been shown that Kepler, Newton, and Maxwell either were guided by their belief in the inherent simplicity of nature or saw their successes as evidence that such parsimony had an ontological basis. Another issue that has been raised is an aesthetic issue: Physicists seem to have been motivated by at least two different forms of parsimony, which have been called physical parsimony and mathematical parsimony. Let us now turn to that distinction.

Physical parsimony and mathematical parsimony, also termed "mathematical beauty" by Dirac, are two aesthetic approaches, rooted in very different understandings of the relationship between physics and mathematics. Kepler used the metaphor of a "clockwork" for the planetary motions (Caspar 1959:136), and many physicists have regarded the universe as an elaborate mechanism. Thus physicists typically have described phenomena in terms of physical concepts or even mechanical models. Physicists during the past few centuries had increasingly used mathematics as a convenient and precise language for describing mechanical models or physical concepts, but until the twentieth century they usually had viewed mathematics as subordinate to the physical concepts.

Since the advent of quantum mechanics in the 1920s, however, the physical interpretation of modern physics has become more abstruse in some instances, and many physicists have insisted that the mathematical equations have primacy. Another way of expressing this point of view is that the mathematical equations have the essential truth, and though physicists may try to find physical meaning inherent in the equations or perhaps invest the equations with physical import, the existence of such physical models is of secondary importance (Jaki 1966:52–121; Bohm and Hiley 1993:319–320).

According to physical parsimony, a successful theory of physics must agree with the experimental data and be based on an absolute minimal number of principles, each of which should have a clear physical meaning. Newton's laws of motion are among the most outstanding examples of such physical principles, and Newton articulated his position of physical parsimony in his "Rules of Reasoning in Philosophy" in Book III of the *Principia*: "We are to admit no more causes of natural things than such as are both true and sufficient to explain their appearances" (Newton 1964:324). In the twentieth century, Einstein's special relativity may have been the most elegant theory exemplifying physical parsimony. The theory was derived deductively from two physical principles: A fundamental law of physics must have the same form in all inertial frames of reference, and the speed of light has the same value in all inertial frames of reference (Pais 1982:140–153). McAllister (1996:113; 1998) recently enumerated several "forms of simplicity" in scientific theories and used the phrase "logical simplicity" in almost the same sense that "physical parsimony" is used here. McAllister also cited Einstein as a leading proponent of logical simplicity because the great physicist desired that a fundamental physical theory be based on "a small number of independent postulates." Maxwell's theory of electromagnetism is still another example of a theory that originated from a physical model, although it should be recognized that the Maxwell fields were believed to propagate through a mechanical aether, a concept now discarded. Today, the Maxwell equations stand on their own as concise mathematical summaries of physical concepts, based firmly on experimental data and interpreted in the context of classical fields.

It is important to contrast this physical parsimony with mathematical beauty. According to the aesthetic guide of mathematical beauty, fundamental equations must have special symmetries, simplicity, and "beauty," but there is no requirement that an accompanying physical theory be obtained deductively from a few physical principles. Dirac, the discoverer of the fundamental equation of relativistic quantum mechanics for spin-$\frac{1}{2}$ particles, eloquently described that viewpoint about mathematical beauty: "It seems to be one of the fundamental features of nature that fundamental physical laws are described in terms of a mathematical theory of great beauty and power, needing quite a high standard of mathematics for one to understand it.... One could perhaps describe the situation by saying that God is a mathematician of a very high order, and He used very advanced mathematics in constructing the universe" (Dirac 1963:53).

Dirac claimed, more provocatively, that "it is more important to have beauty in one's equations than to have them fit experiment" (Dirac 1963:47). He further explained that the presence of beauty in equations was an excellent indication that progress would result, and the lack of perfect agreement between an equation's predictions and the experimental data could be due to minor difficulties that might be resolved subsequently.

Although the appreciation and applications of mathematical beauty wrought some of the greatest successes in quantum theory in the twentieth century, appeals to mathematical symmetry have had a long history. After all, the insistence by the ancients on the use of circles, instead of ellipses, to describe planetary motion can be attributed to the greater symmetry of the circle. In the twentieth century, uses of mathematical beauty and simplicity had far-reaching effects, such as discovery of the fundamental wave equations of quantum mechanics. Schrödinger, in 1926, was motivated by an analogy between the propagation of light waves and the possible propagation of electrons as waves, and on the basis of symmetry considerations he knew that a plausible wave equation for electrons would have to treat the three spatial coordinates equally. He also assumed that the partial differential equation would have to be linear and have second-order derivatives in the spatial variables. "Only the striving for simplicity leads us to try this to begin with," Schrödinger (1928:27) stated. Those appeals to mathematical simplicity and symmetry led to Schrödinger's discovery of his nonrelativistic wave equation, one of the fundamental expressions of quantum mechanics. The physical understanding of that equation became clearer a few years later. By using mathematical-symmetry properties in an analogous way, Schrödinger earlier had found a relativistic wave equation, now known as the Klein-Gordon equation, and Dirac somewhat later found his celebrated equation.

The ideals of physical parsimony and mathematical beauty have been depicted in sharp contrast in this discussion, and it is interesting that Einstein at times advocated physical parsimony, and at other times in his career he emphasized mathematical symmetry. Einstein's theory of special relativity has been cited as the preeminent use of physical parsimony in the past century. In 1917, when a colleague wrote to Einstein about a certain mathematical symmetry in the Maxwell equations, Einstein responded: "It does seem to me that you highly overrate the value of formal points of view" (Pais 1982:325). Five years later, Einstein further expressed his commitment to physical models and physical parsimony: "I believe that in order to make real progress one must again ferret out some general principle from nature" (Pais 1982:328). By 1933, Einstein apparently was more amenable to appeals to mathematical symmetry, when he acknowledged that "pure mathematical construction" can enable one to find physical concepts and laws of physics (Pais 1982:347).

It must be emphasized that neither the aesthetics of physical parsimony nor those of mathematical beauty should be identified exclusively with a single philosophical position concerning the basis of parsimony, such as parsimony as a methodological axiom or parsimony as an ontological principle. Physicists frequently use simplifying mathematical approximations to further computational progress, and at other times they deliberately neglect physical influences of secondary importance in their physical models. Both of these uses of simplifying assumptions are methodological. On the other hand, Dirac's mathematical beauty, with its accompanying claim that "God is a mathematician of a very high order," is no less a statement of ontology than is Newton's assertion that "Nature is pleased with simplicity."

In Chapter 8, Ockham was cited as a critic of the view that parsimony is an ontological principle. He opposed the argument that nature must be intrinsically simple. Instead, his understanding of parsimony emphasized that *humans* should not choose a more complicated theory unless there was sufficient reason. In the first part of this discussion it was noted that physicists commonly choose simple theories, in conformity with Ockham's razor, but at this point it is time to try to evaluate whether or not parsimony in physics genuinely has an ontological basis.

What is the evidence that all of nature is governed or described by a few relatively simple physical laws or a few "beautiful" equations? The following illustration from Richard Feynman might help to answer the question. Feynman included the Maxwell equations, the Lorentz-force law, Newton's second law of motion (with Einstein's modification of the expression for momentum), and Newton's law of gravitation in a table of equations at the top of a page in his renowned lectures on physics. He then stated that the table has "all that was known of fundamental *classical* physics, that is, the physics that was known by 1905. Here it all is, in one table. With these equations we can understand the complete realm of classical physics" (Feynman et al. 1964:18-5). There are qualifications to Feynman's comments that are noteworthy. He stated that the table includes the "fundamental" equations of classical physics, thereby excluding a number of important phenomenological equations of classical origin, and there are definitions not included in the table. Nonetheless, the most severe critics of an ontological basis for simplicity in physics must concede that an immense quantity of experimental data can be explained in terms of the eight equations of classical physics listed in the table.

Since 1905, the horizons of physics have broadened considerably, and the evidence for simplicity as an ontological principle may not be as convincing today. General relativity, quantum mechanics, quantum electrodynamics, and other quantum field theories have been developed and subjected to experimental tests, and some of those theories have shown agreement with empirical data to unprecedented levels of precision. On the other hand, the assumptions of quantum mechanics are more numerous and less intuitive than those for classical physics. The important equations, assumptions, and parameters of the best-verified quantum field theories might fill a small handbook instead of the brief table for classical physics in Feynman's text. The procedures used in quantum field theories for absorbing infinities into renormalized masses and renormalized charges and for extracting finite answers continue to trouble many physicists philosophically. General relativity and quantum field theory are the two major paradigms in physics today, but a consistent union of these two programs into a quantum theory of gravity that will be free of renormalization troubles is problematic. These developments and difficulties have led some physicists to conclude that the previous apparent simplicity of nature may have been illusory. On the other hand, optimistic proponents of the intrinsic simplicity of nature suggest that such difficulties demonstrate some of the provisional characteristics of today's major theories. According to that point of view, parts of twentieth-century physics may be likened to the planetary model of Ptolemy, and the modern analogues of epicycles and deferents will be cast aside when the next Kepler or Newton appears.

The General Scholium at the end of Newton's *Principia* and some quotations cited in this discussion indicate that in earlier centuries natural philosophers and physicists such as Kepler, Newton, and Maxwell were motivated by their belief in God and his parsimonious design of the universe. Because of their faith they searched for a few principles to describe the physical universe. At the beginning of the twenty-first century, such faith may be less common among physicists and members of society in general. Is there an ontological basis for parsimony in physics? If so, then physicists must address Einstein's observation that "The eternal mystery of the world is its comprehensibility" (Einstein 1936:351).

MOLECULE SHAPE AND DRUG DESIGN

with P. Andrew Karplus

Biochemists and pharmacologists frequently need to know a molecule's shape in order to understand how a natural molecule such as an enzyme functions or to design a new drug (Bugg, Carson, and Montgomery 1993; Gibbs 2001; Nicolaou and Boddy 2001; Ezzell 2002). One major way to gain such structural information is by x-ray crystallography, which allows visualization of the atomic detail of protein molecules containing thousands of atoms. Currently, hundreds of new protein structures are determined annually. Two examples will illustrate the useful information provided by such structural analyses.

The first example is related to understanding disease and concerns the tumor-suppressor protein p53, shown in Figure 9.5. This protein was discovered in 1989, and it has since been shown that about half of all human cancer cases involve cells that have a mutation in the gene for p53 (Cho et al. 1994). In addition, other research has shown that normal p53 can block oncogenic transformation of cells exposed to enzymes that normally cause cancer. The p53 tumor suppressor is thought to work by monitoring the health of the DNA in a cell, and if excessive DNA damage is present, the cell is shut down before cancer can develop. Until the crystal structure was

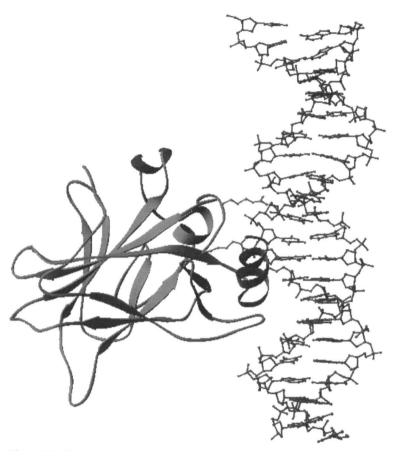

Figure 9.5. The structure of the tumor-suppressor protein p53 in complex with DNA. The protein is drawn as a simplified ribbon that traces the path of the protein chain, and the double-helical DNA is shown on the right as helical ribbons for the phosphate backbone, with horizontal bars representing the base pairs. The six protein side chains shown are hotspots for mutations in p53 that are involved in cancer. The structure reveals that these debilitating mutations are all at the interface involved in DNA binding.

solved, it was a mystery how the many known mutations that were scattered throughout the p53 protein blocked its function. But the crystal structure clearly revealed that all of the most common mutations associated with cancer were clustered in the structure and were involved in recognizing and binding DNA.

The second example is related to drug design and concerns the virus that causes the common cold, shown in Figure 9.6. This virus gains entry into human cells when the viral surface (the capsid) binds to a protein on the

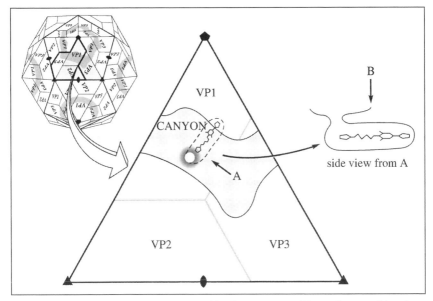

Figure 9.6. The cause of the common cold. The structure of intact human rhinovirus is an icosahedral shell constructed from 20 copies each of three protein chains named VP1, VP2, and VP3 (VP for viral protein). Crystallographic analysis has revealed that the surface topography includes a canyon-like feature on each face of the capsid. This canyon allows the virus to gain entry into human cells by specifically binding to a protein receptor on the surface of the cell. Further analyses have shown that compounds with known antiviral activity bind in a pocket at the bottom of the canyon and disrupt receptor recognition. (Reprinted from Zhao et al., 1996, with kind permission from Michael G. Rossmann.)

surface of a cell and becomes internalized. The concept behind antiviral drugs and vaccines is to block the virus-receptor interaction, either by drug binding or by stimulating the generation of antibodies that block the interaction. The crystal structure of the virus has revealed which part of the capsid binds to the cell surface receptor and has played an important role in guiding the development of effective antiviral agents. In such applications, accuracy is critical. Yet this pursuit of exact structure has subtle pitfalls. Recently, greater emphasis on parsimony and on the use of Bayesian analyses has opened the door to greater accuracy, reliability, and objectivity.

Determination of the shapes of proteins or other large molecules by x-ray crystallography progresses through two stages. First, the basic shape is approximated by manually building the known chemical constituents to get a good fit with the experimental electron-density distribution. Second, this preliminary structure is refined by automated calculations that use several kinds of data. The discussion here focuses on this second stage of refining the model of a molecule's shape, particularly how the model's accuracy is assessed. Refinement is a gradual process, and as errors are found and corrected, it becomes easier to fix remaining errors.

Structure refinement relies on two principal kinds of data. First and foremost are x-ray-diffraction data collected from a crystal of the macromolecule. As the working model of the molecule is revised, a predicted or expected x-ray pattern is calculated. The actual and predicted patterns are then compared by some measure of fit. Proposed tinkerings with the working model are accepted or rejected depending on whether this measure indicates better or worse fit with the x-ray data.

In this comparison of data and model, there are problems, however, with both the data and the model. The data are imperfect, affected by experimental errors and limited in amount. For example, the model of a protein having 2,000 atoms requires defining 8,000 variables (three for the position of each atom and one for its vibrational mobility). Usually the number of x-ray-diffraction observations is not sufficiently in excess of the number of variables, so the x-ray data alone are inadequate to refine a structure satisfactorily. The model is also imperfect, limited by imperfect chemical theory and finite computational power. For example, some water molecules associated with a macromolecule may be ignored because of inability to model them accurately. Yet it is well known that water affects the structure and function of macromolecules in living systems (Gerstein and Levitt 1998).

Accordingly, data of a second kind are also useful: a data bank of typical bond lengths and angles based on previous experiments with smaller molecules that could be measured unusually accurately. These data from other molecules are used to constrain the analysis of the x-ray data for the macromolecule of interest. For example, analysis of the x-ray data is not allowed to yield any C–C bond length that is outrageously short or long. Extensive experience shows that this combination of two kinds of data works much better than either alone. Figure 9.7 depicts the process of structure refinement, starting from an initial model and adding experimental information from known small-molecule structures and from x-ray-diffraction intensities for the molecule of interest.

Given these two kinds of data, however, the question arises how much weight to place on each. Different weights lead to different models of the molecule. And this question must be answered on a case-by-case basis, because each experiment has its own unique limitations and noise level. To determine an objective and ideal weight, scientists need a reliable measure of a model's accuracy. Then the weights that will optimize that goal of accuracy can be chosen. Hence, being able to determine a model's accuracy reliably is important both for guiding choices during model refinement and for knowing what level of detail to take seriously in the final model.

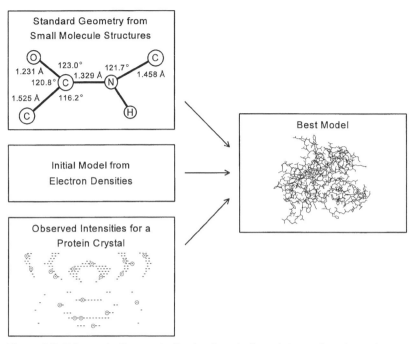

Figure 9.7. Schematic diagram indicating how indirect information about the structures of other molecules and x-ray-diffraction intensities for the molecule of interest are used during model refinement in macromolecular crystallography. The top left panel depicts angles and distances in just one small chain with several carbon, oxygen, nitrogen, and hydrogen atoms, but the actual database provides hundreds of numbers for numerous small molecules. Only average values are shown here, but the complete database also gives standard deviations to describe the ranges in actual values. No values are indicated here for the hydrogen atom because it is too small to diffract x-rays in these studies. The bottom left panel depicts an x-ray-diffraction pattern. Most of the data are allocated to the working set, and about 10% are selected at random for the test set used for cross-validation.

For decades, a measure called the R-factor has been an especially popular measure of the fit between x-ray data and a molecular model. But more recently, crystallographers have discovered that it has serious problems. It is increasingly clear that in some cases there is little correlation between the R-factor and accuracy, and even large errors may go undetected. An extreme example concerns a protein structure that was presumed accurate and was published in the prestigious *Proceedings of the National Academy of Science*, but later it was found that 80% of its atoms had been placed in badly erroneous conformations (Kleywegt and Jones 1995). Also, in flagrant disregard for parsimony, the R-factor can be made arbitrarily "good" merely by adding more parameters to the model. Hence, it has little objective meaning.

In 1990, an influential commentary emphasized that ineffective measurement of accuracy leads to errors and alarming subjectivity (Brändén and Jones 1990). Consequently, crystallographers began to seek more reliable and more objective measures of the accuracy of molecular models.

In 1992, a new concept was introduced that used the statistical method of cross-validation to generate a parsimonious, predictively accurate model (Brünger 1992, 1993, 1997). Briefly, the data are split into two portions: modeling data and validation data, also called the working set and test set. Typically, about 90% of the data are used to construct and refine the model, and the remaining 10%, selected more or less at random, are used to validate or check the predictive accuracy of the model. Often this process is repeated numerous times with different randomizations and the results are averaged. (Recall that this same method of data splitting and cross-validation was cited in Chapter 8 for agricultural yield-trial data.) This new measure of structure accuracy, the R_{free} value, provides reliable assessments of accuracy, as well as an objective criterion for choosing protocols and models during refinement. Furthermore, rather than imposing more work, this new protocol's reliable guidance promotes faster solutions. So a better structure emerges with less work. Parsimony pays!

Using this new R_{free} value, as the weight on the x-ray data is increased (and correspondingly the weight on the data bank of ideal geometries is decreased), an Ockham's hill emerges, so some intermediate weight is best, as depicted in Figure 9.8. "Too few parameters will not fit the data satisfactorily whereas too many parameters might fit noise" (Brünger 1993), just as was demonstrated in several other applications in science in Chapter 8. The most predictively accurate model has intermediate weights that blend the direct x-ray information and the indirect data bank, as in Figure 5 of Brünger (1993), which is analogous to Figure 8.9 in the preceding chapter. By contrast, the old R-factor is misleading, continuously improving as more weight is placed on the x-ray data. Hence, the predictive R_{free} value shows a nice Ockham's hill, but the postdictive R-factor keeps rewarding model complexity, just like the lines for prediction and postdiction in Figure 8.1.

This section's introduction promised commentary on Bayesian analyses, but so far nothing has been said about Bayesian inference. Nevertheless, although implemented informally, the crystallographers' combining of different kinds of data to solve molecular structures has a deeply Bayesian mentality. Recall from Chapter 7 that Bayes's theorem combines new and prior information. In this case, the x-ray-diffraction data on the macromolecule of interest provide the direct or new information, and the data bank of typical bond lengths and angles in other molecules provides the indirect or prior information. Indeed, rearrangement of the usual equations used in refinement

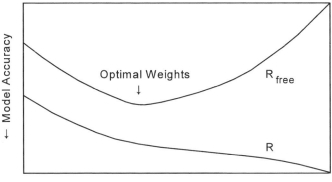

Figure 9.8. Effects of the weights placed on direct and indirect data on two measures of model accuracy. Empirical studies reveal the importance of properly weighting the relative contributions of the direct and indirect information during crystallographic refinement. The R-factor measures the postdictive agreement between the model and the diffraction data, so naturally it continues to improve as more weight is placed on the diffraction data. But R_{free} monitors the predictive accuracy of the model, using the subset of test data that were not used in refinement but rather were saved for cross-validation. The best model is that which minimizes R_{free}. Here model accuracy increases downward, so Ockham's hill appears upside down relative to its usual depiction.

makes more obvious the deeply Bayesian nature of this combining of new and old information (Terwilliger and Berendzen 1996b).

Nevertheless, recent research shows that greater accuracy can be achieved with more explicit and vigorous Bayesian analyses (Bricogne and Carter 1996; Terwilliger and Berendzen 1996a,b; Bourgeois 1999; Cowtan 1999). This makes sense because the task of structure refinement does have extensive and accurate prior information that is well worth exploiting, and furthermore Bayesian answers come in exactly those terms most natural to a crystallographer, namely, the probability that a molecular model is correct within some specified limits. Hoffmann et al. (1996) have provided a wide-ranging overview of parsimony in chemistry, including remarks on Bayesian analysis and remarkably insightful and balanced comments on the ontological and epistemological aspects of parsimony.

But to return to the extensive results already achieved for numerous molecules using R_{free} values to guide an informally Bayesian analysis, it is a clear empirical fact that these parsimonious models are more accurate than overly complex models guided by the old R-factor. Also, the theoretical explanation for this outcome is simple. The data are imperfect and noisy. Parsimonious models account for the real signal in the data while discarding much of the noise. Furthermore, the practical advantages in biological and medical research that result from more accurate molecular structures are manifest and substantial (Adams et al. 1997).

Nevertheless, this success story also has a sad element. The switch to the new paradigm has been slow. A major complaint has been that by setting aside some data not used for model construction, the new protocol wastes data. But that false argument ignores the repeatedly demonstrated fact that R_{free} with 90% of the data outperforms the R-factor with 100% of the data. In other words, it disregards the simple logic that it is better to squeeze all of the juice from 90% of a bag of oranges than to squeeze only half of the juice from the whole bag. Also, after using validation to select the best model, routinely all of the data are recombined for the final calculation, so no data are wasted after all. Anyway, although the new paradigm introduced in 1992 is demonstrably better, it has taken time to catch on. For example, Kleywegt and Brünger (1996) surveyed the May 1996 release of the enormous Protein Data Bank and found that of its 3,657 macromolecular x-ray structures, only 178, or less than 5%, were accompanied by R_{free} values.

But the larger tragedy here is not that implementation has been slow since 1992, but that discovery came only in 1992. Philosophers of science had emphasized parsimony and statisticians had developed cross-validation techniques decades before 1992. Decades were wasted before cross-validation techniques were imported from statistics into this particular application in biochemistry. That unnecessary delay is curious in that surely hundreds of biochemists took courses in statistics. Also, biochemists had already found cross-validation helpful in some closely related applications, such as predicting protein structure from amino-acid sequences. But the old R-factor had been received with such acceptance and complacency for decades that there was no search, no hunger, for a better measure of accuracy.

In a similar vein, recently it has come to light that there is also a problem with the reliability factor that has been used during the past 30 years to judge data quality. That factor's formula is statistically invalid, so that it has systematically given the impression that collecting fewer replicates of the data would be better (Diederichs and Karplus 1997). Consequently, there has been a double compromise of both data quality and refinement procedures. Regrettably, the opportunity cost has been great. Had the relevance of parsimony in determining molecular shapes been appreciated a decade or two before 1992, we might now have better cures for more diseases.

One might hope that were scientists given broader training in science's method and philosophy, more frequent thought would be given to principles such as parsimony and predictive accuracy, complacency about second-rate methods would give way to energetic search for better approaches, valuable ideas would be imported faster from related fields, and effective new paradigms would be accepted quicker. For example, parsimonious models selected by cross-validation have already proved their worth for agricultural yield trials and for molecular shapes (Gauch 1993). But there are a thousand potential applications for this off-the-shelf statistical methodology waiting to be implemented in numerous specialties in science and technology.

ELECTRONICS TESTING

Experiments are expensive, regardless whether performed for university research or for industrial production. This case study will show how a clever experimental design can reduce testing costs and increase profit margins in the manufacture of electronic components.

Over the past several years, engineers at the National Institute of Standards and Technology have developed a new and remarkably efficient method for testing analog-to-digital converters (ADC) and other electronic components (Souders and Stenbakken 1991). A prime example concerns a 13-bit ADC that measures an input voltage and expresses the result as one of 2^{13} or 8,192 different values. Because of uncontrolled variations in the manufacturing process, ADC devices vary in their performance. Accordingly, they must be tested individually and then sorted into performance bins.

The conventional test used automated equipment to check each device at all 8,192 input voltages, rating the device by its maximum measurement error. The problem with that test was that 8,192 measurements were time-consuming and costly, reducing throughput and profit. Often the testing cost was an appreciable fraction of the entire manufacturing cost. They needed a faster, cheaper test that still would maintain good reliability. More specifically, the goal was to find a small, wisely chosen subset of the 8,192 measurements that would allow determination of a device's performance nearly as accurately as would a complete test.

Briefly, the new strategy was to test a small number of units completely, analyze those data to discover a small subset of test voltages that would be maximally revealing of a device's performance, and then in production runs to test an enormous number of devices economically with the quick reduced test. In one study, Souders and Stenbakken (1991) performed an exhaustive test on 50 ADC devices. Those data were analyzed by a statistical model (Ding and Hwang 1999) similar to the AMMI model used for agricultural yield trials cited in Chapter 8. That analysis, together with some knowledge of the device's circuit, showed that just 18 of the 8,192 test voltages were especially informative. Another 46 measurement voltages were added for redundancy, to reduce problems from experimental errors, for a total of 64

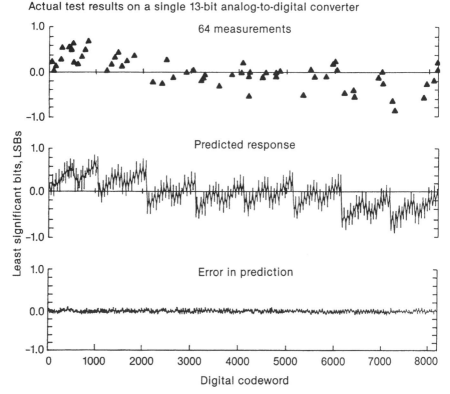

Figure 9.9. Test results from a parsimonious experiment on an analog-to-digital converter. The test requires merely 64 measurements (top) to predict the converter's errors at all 8,192 levels (middle). The differences between these predicted errors and the actual errors are minute (bottom). This quick test reduces testing and manufacturing costs. (Reproduced from Souders and Stenbakken, 1991, with kind permission from the Institute of Electrical and Electronics Engineers. © 1991 IEEE.)

test voltages. So the reduced test had only 64 measurements instead of 8,192, less than 1% of the original task. A computer applied a mathematical model to those 64 measurements to predict all 8,192 responses, and as before, the worst of those (predicted) responses served as the criterion for sorting the device into its appropriate performance bin.

How well did the quick test work? The test was applied to 77 other devices, not part of the original 50, that previously had been sorted by the expensive exhaustive test. The quick test performed virtually as accurately as the conventional test. Yet the testing time was reduced by a factor of 128. Assuming a stable manufacturing process, the reassuring findings from that reasonable statistical sample imply that the new test procedure will be suitable for production runs.

Figure 9.9 shows a typical result for an individual ADC device. Only 64 measurements were taken, from which all 8,192 responses were predicted. Comparisons between the actual voltages and the model's predicted voltages yielded very small errors. So the quick test worked very well.

That particular example of ADC electronic devices is representative of countless manufacturing situations that involve costly testing of each production unit. Clever strategies can replace extensive experiments with parsimonious experiments that can save time and increase profits. Then no more data need be purchased than are really required to get the job done. The long-term profitability and competitiveness of many companies will be strongly affected by whether or not they can learn to use clever tricks to gain efficiency.

STATISTICS IN MEDICINE

Bayesian and frequentist statistical methodologies were explained and compared in Chapter 7, with the conclusion that the Bayesian paradigm often addresses researchers' questions more directly. This section will consider that comparison in the specific arena of medical and pharmacological research. Most medical research has an important statistical component, often requiring a consulting statistician as a member of the research team.

The Bayesian and frequentist paradigms have both philosophical and practical differences. But equally important, they have deep similarities and the same goals in medical applications of improving clinical practice, containing research costs, and meeting ethical requirements. Because of their shared goals, comparisons of Bayesian and frequentist methods are meaningful, with the potential for winners and losers. Furthermore, their philosophical and practical merits are not juxtaposed, as if optimizing the one would necessitate compromising the other. Rather, the legitimate expectation is that good philosophy, good science, and good medicine go together.

Having good statistics always helps scientists, but that is especially urgent in medical research because of the tremendous costs, great suffering, and pressing ethical concerns. Medical researchers and consulting statisticians receive considerable ethical guidance from laws and agencies that regulate experiments with animals and humans (McNeill 1993; Stuart J. Pocock, in Ashby 1993:1459–1475; Kadane 1996). But they still bear great responsibility for determining numerous experimental and statistical details within the guidelines imposed by law. As statistical methods are chosen in medical research, three principal ethical considerations emerge: (1) research efficiency, (2) direct statistical answers to medical researchers' questions, and (3) accommodating the diverse roles and responsibilities of all those involved in medical care.

First and most obviously, medical research needs to be highly efficient, benefiting from optimal statistical designs and analyses. Efficiency can minimize the number of patients in an experimental trial who must suffer through inferior treatments, and it can maximize the number of patients who will receive superior treatments in the future. Also, given the finite resources to be divided among countless disease treatments and research projects, efficiency is crucial for containing costs. It is unethical to waste experimental subjects or to waste research funds because of second-rate statistics.

An effective way to demonstrate the urgency of efficient medical research is to scan reports on medical experiments and realize the suffering involved. For instance, one can read of 28 dogs experimentally subjected to cardiac arrest in order to compare two resuscitation protocols (Niemann et al. 1992). The purpose of such research is to gain knowledge that can save human lives. Nevertheless, all 28 dogs suffered, and 16 of them died. Although some people feel that making dogs suffer for humans' benefit is wrong, most people believe that animal research is justified, and the law reflects that majority view. Anyway, that ethical debate over animal rights is not engaged here. What can be said here, and on which everyone should agree, is that any such research must be efficient, using as few dogs as are absolutely necessary to obtain adequate information. And the importance of efficiency only increases as research moves from animals to human subjects. Anyone who has visited hospitalized cancer patients in a clinical trial will have grasped something of their immense suffering. Medical research is serious business. Truly, in medical research, efficiency counts.

Second and less obviously, but not less importantly, the questions asked in medical research must be expressed and understood precisely. Medical researchers must communicate their research questions clearly to consulting statisticians, and in turn statisticians must respond with direct answers. Statisticians must not resort to indirect and convoluted answers that can be easily misinterpreted by medical researchers merely because the statisticians want to work within some familiar paradigm or have easy calculations. Statisticians must communicate not only their numerical results but also the exact meaning of those numbers. One of the most detrimental outcomes from failure to communicate clearly occurs all too frequently: Medical researchers will receive a complex but familiar-sounding statistical report and comfortably interpret it as a straightforward answer to their question, whereas a correct interpretation of the actual meaning of the statistical analysis would lead to a different assessment of the experimental evidence and even to different recommendations for patient care. It is unethical for statisticians to take a casual approach and fail to fully understand researchers' questions or to offer some "answer" that answers the wrong question or invites misinterpretation.

Furthermore, obtuse and misleading statistical replies are unforgivable because there is no mystery about the common questions asked in medical research and practice. What would constitute an efficient design and analysis for a given medical experiment? Is some new treatment better than the conventional treatment? Which is the best treatment to administer to a particular new patient? Clearly, given such questions as these, any statistical paradigm that asks deduction to do induction's job or that asks postdiction to do prediction's job is in danger of offering confusing answers to medical researchers' questions. Also, these first and second ethical concerns are related, because imprecise questions and unclear answers can result in inefficient research.

Third and finally, statistical paradigms must accommodate the diverse roles and responsibilities of all those who have a stake in medical care, including researchers, physicians, regulators, and patients. The motive for a clinical trial is to improve clinical practice. However, to improve patient care, researchers not only must satisfy themselves that a new treatment is better but also must produce sufficient evidence to convince the physicians who treat patients. Naturally, physicians are individuals, and they come to such new evidence with a range of predispositions and opinions about new treatments, with some physicians being relatively optimistic and hence easy to convince, but others being relatively skeptical and hence demanding strong evidence. Consequently, to influence and improve clinical practice substantially, rather strong evidence may be needed to convince most physicians. Likewise, before the developers of a new drug can market their product (or even conduct a clinical trial with human subjects), they must build a strong case that will convince the government regulators of the pharmaceutical industry. Patients also have a right to know the benefits and risks of various treatment options, insofar as that can be quantified by the best available research. For these reasons, statistical procedures for medical research must recognize that medicine has many stakeholders, with a diversity of concerns and a spectrum of beliefs about different kinds of treatments.

So, good efficiency, direct answers to precise questions, and accommodation of various stakeholders are paramount ethical requirements for statistical applications in medical research. With these ethical considerations in mind, this section continues with philosophical and then practical comparisons of the Bayesian and frequentist paradigms.

For decades, the frequentist paradigm has dominated in medical research. Why? There seem to be two principal reasons, one well recognized and the other not. The well-recognized reason is that Bayesian methods often have been perceived to be subjective, but frequentist methods have been considered admirably objective. Indeed, the Bayesian approach to probability has often been termed "subjective" probability, which has strongly reinforced that perception and implied that the frequentist alternative must be objective probability.

For instance, on reading a Bayesian paper that analyzed a medical data set with four different priors, thus giving four somewhat different answers, a researcher of frequentist orientation was especially concerned about cases where one prior might indicate significance at the 5% level, but another not: "In this case, what would be the regulatory agency's conclusion...? My own view is that where one party (manufacturer) is in the position of having to prove a point to the satisfaction of a second party (regulatory agency), arbitrary assumptions, on which the parties could disagree, should be kept to a minimum. The Bayesian approach introduces an arbitrary assumption [through its prior] that the other [frequentist] methods...do not require" (Wilfred J. Westlake, in Peace 1988:347). To many readers, that could seem like a mighty convincing argument against Bayesian and for frequentist statistics. What do you think?

Contrary to the foregoing quotation, Berry (1987) cited an interesting example in which frequentist analyses of a given data set with two different stopping rules produced one test that was significant at the 5% level and one that was not. So the frequentist had the same problem! Such influence of imaginary data induced by arbitrary stopping rules, as explained in Chapter 7, is a rather objectionable kind of subjectivity.

So what is the truth about statistics and subjectivity? What is reality, and what is illusion? Briefly, there seems to be what one might call a Law of Conservation of Subjectivity. The difference between the Bayesian and frequentist paradigms is not that one is subjective and the other objective, but rather that subjective inputs are explicit and direct in the Bayesian paradigm, whereas they are implicit and obtuse in the frequentist paradigm. Unfortunate indeed is any rhetoric that penalizes the Bayesian paradigm for being direct and rewards the frequentist paradigm for obscuring and seemingly disowning its equally subjective inputs!

Anyway, the ultimate source of this Law of Conservation of Subjectivity is nothing less noble or escapable than science's Law of Total Evidence, saying that all relevant and important information should be considered, which includes the current experiment and other prior experiments and knowledge. Also, the ultimate source of objectivity is the strength of the data, not fancy statistics. Frequently, the only way to demonstrate objective and compelling conclusions is to gather enough data to make the truth evident to everyone concerned.

To reach a rational preference between Bayesian and frequentist statistics, it is imperative to understand the issue of subjectivity correctly. If frequentist statistics really were more objective, but Bayesian statistics really were more efficient, then researchers would face a tough choice, because they would want both objectivity and efficiency. Inevitably, different researchers would place different weights on those incommensurable goods, leading to different preferences, with no possibility of resolution, because the discrepancy would arise at the very deepest level of assigning weights to diverse goods such as objectivity and efficiency. Consequently, a conflict between objectivity and efficiency, if it were real, would induce what might be termed "deep subjectivity," which is quite incurable. But once the illusion that frequentist statistics are more objective is dispelled, unfettered thought can be given to efficiency. Indeed, a good argument can be made that Bayesian subjectivity is more natural and more desirable than frequentist subjectivity. Furthermore, in a profound sense, objectivity and efficiency are not two separate and incommensurable criteria, but rather they become unified for a single purpose, namely, to extract sufficient information from limited data such that virtually everyone can reach the same conclusions.

The poorly recognized reason why frequentist statistics dominate has to do with a philosophical misunderstanding: Few scientists clearly understand the essential nature of experiments – that they provide likelihood information rather than probability information. That blunder allows scientists to continue to believe that the Bayesian's prior is dispensable and that frequentist statistics can deliver pristine objectivity.

Nevertheless, remarkably little clear thinking will suffice to dispel those illusions. Recall the salient feature of the marble experiment analyzed in Table 7.1, that drawing a blue marble makes H_B three times more likely to be true than H_W, and contrariwise that a white draw makes H_W three times more likely than H_B. Now focus on a very simple experiment, one draw, and suppose that this draw is a blue marble. What, precisely, is the import of this experiment?

Well, H_B is now three times more likely than H_W. But more likely than what? Clearly, it is more likely than what it was before this experiment was performed! So a complete and precise statement of the import of the experimental outcome of one blue draw is that H_B is now three times more likely than H_W was before this experimental outcome arrived. Consequently, to obtain the probability that H_B is true, its prior probability is also required, as is manifest in Table 7.1, in which the probability of H_B after any outcome of a blue draw also depends on the available prior information. Thus the prior is indispensable and ineliminable.

A further proof of the Law of Conservation of Subjectivity or of the indispensability of the prior information is that inevitably any statistical paradigm that feigns avoidance of a prior winds up by simply shifting subjective inputs elsewhere, such as placing weird emphasis on subjective stopping rules and imaginary data. Subjectivity can be moved around from one part of a statistical analysis to another, and it can be disguised, but it cannot be eliminated. Experiments render hypotheses more likely or less likely *than they were beforehand*, which is an intrinsic feature of experiments that cannot be changed by statistical paradigms. And that is precisely the understanding of experimental evidence that is implemented in the Bayesian paradigm.

Wanting an experimental conclusion with no prior in it is in the same category as wanting water with no hydrogen in it – it is absolutely impossible in principle. But the situation is not alarming if one realizes that subjectivity is not an all-or-nothing affair, but rather a matter of degree. Data that are extensive and strong will support objective conclusions that will be clear to everyone concerned, provided only that researchers are willing to ask new questions and that their statistical methods are tolerably sensible.

The foregoing must suffice as a philosophical comparison of the Bayesian and frequentist paradigms. Let us progress next to a brief practical comparison, applying both paradigms to several medical data sets to see which performs best.

Medical research involves an almost unimaginable diversity of diseases and experiments, but for the sake of focus, this discussion will address a particularly common and influential kind of experiment, the randomized clinical trial used to compare treatment options (Zivin 2000). Most clinical trials require periodic monitoring of the accumulating data so as to minimize the number of experimental patients who will continue with the inferior treatments and to expedite the delivery of superior new treatments to future patients.

For a first example, consider a clinical trial that concerned a very serious human disease, acute leukemia (Freireich et al. 1963). There were two treatments: 6-mercaptopurine (6-MP) and placebo. As patients were recruited into the trial, they were randomized in pairs to receive either 6-MP or placebo. Investigators recorded which patient of each pair remained in remission longer. If it was the patient receiving 6-MP, that was termed a success (S); otherwise, a failure (F). There were 21 patient pairs, and the data are given in the second column of Table 9.1.

Ethical considerations for such a serious human disease motivated researchers to adopt a fully sequential analysis, meaning that the data were analyzed anew each time another outcome became available. If 6-MP were proving to be effective, it would be unethical to continue giving the placebo to half of the experimental patients. Those data will be analyzed by frequentist methods and then Bayesian methods.

The fully sequential frequentist analysis used in that clinical trial had a stopping rule that would be activated when the excess of successes over failures (or the reverse) became sufficiently large to reach the 5% significance level, or else when sufficient data indicated no appreciable treatment difference. Depending on the observed data, that rule was bound to stop the experiment after results had been reported for somewhere between 9 and 67 patient pairs. As it happened, superiority of 6-MP over placebo was detected

Table 9.1. Frequentist and Bayesian sequential analyses of the same data, from a clinical trial involving two treatments for human acute leukemia

Patient pair	Outcome	Frequentist result	Bayesian result
1	S	—	0.25
2	F	—	0.50
3	S	—	0.31
4	S	—	0.19
5	S	—	0.11
6	F	—	0.23
7	S	—	0.14
8	S	—	0.090
9	S	—	0.055
10	S	—	0.033 *
11	S	—	0.019
12	S	—	0.011
13	S	—	0.0065
14	F	—	0.018
15	S	—	0.011
16	S	—	0.0064
17	S	—	0.0038
18	S	0.05*	0.0022
19	S	—	0.0013
20	S	—	0.0008
21	S	—	0.0005

Note: Asterisks show the frequentist analysis reaching a significant result at the 0.05 level after 18 patient pairs, but the Bayesian analysis reaching that significance level after only 10 patient pairs. Hence, the frequentist analysis needs almost twice as much data – and almost twice as many humans in a clinical trial – to reach a conclusion with a comparable level of confidence. The Bayesian analysis is more efficient.
Source: Adapted from Berry (1987). These data are reproduced with kind permission from *The American Statistician.* © 1987 American Statistical Association, all rights reserved.

at the 5% significance level after testing 18 patient pairs, as shown in the third column of Table 9.1. At that time, recruitment of new patients into the trial was stopped, but 3 more pairs had already begun treatment, and they were followed to observe the outcomes, so the table shows results for 21 patient pairs all told.

The data were reanalyzed using a fully sequential Bayesian analysis with a noninformative uniform prior, as shown in the last column of Table 9.1 (Berry 1987; also see Donald A. Berry, in Ashby 1993:1377–1393). The

Bayesian analysis reached the 5% level (meaning that there was only a 5% probability that the placebo was better than 6-MP) after only 10 treatment pairs – about half as many patients as in the frequentist analysis. The main reason for that sizable difference is that frequentist methods penalize final conclusions for interim looks at the data, whereas Bayesian methods do not.

So Bayesian analysis reached a significant conclusion almost twice as fast as frequentist analysis in that case. Interestingly, Berry (1987) showed the results of those two analyses (plus another two analyses not discussed here) to "hundreds of statisticians and physicians" and reported that "Most wanted to stop the trial between the 8th and 12th pairs; only a few wanted to keep it going to the 18th pair," even though most of those statisticians were avowed frequentists. He then commented that "I take this as evidence that one's intuition adheres to the conditional [Bayesian] view even though one's training takes one elsewhere," and that from extensive experience, one learns "to take data at face value." Anyway, in that case of a clinical trial for leukemia that gave a good treatment to half of the patients and a bad treatment to the other half, there was a high price to pay for running the experiment too long because of inefficient statistics.

For a second example of Bayesian and frequentist analyses of the same data, consider a randomized trial of a *Hemophilus influenzae* type B (HIB) vaccine and a placebo vaccine, conducted among the Navajo people (Donald A. Berry et al., in Bernardo et al. 1992:79–96). The HIB bacterium can cause septicemia, meningitis, arthritis, and other problems, including death, primarily in young children. Certain tribes of native Americans face a tenfold or greater risk of such infection as compared with other American populations. Because the incidences of some important diseases, such as those caused by HIB bacteria, are low, most subjects given placebo or ineffective vaccines will remain disease-free, so demonstrating that a vaccine is effective can require rather large trials.

The frequentist design for that HIB vaccine trial called for a fixed total of 5,000 subjects to complete the trial (which meant about 5,600 subjects to begin the trial, to cover some attrition). A Bayesian analysis compared the predicted numbers of future HIB cases if the trial were stopped and if it were continued, and it was determined that the fewest HIB cases would result from stopping the trial slightly earlier, after about 4,600 subjects had completed the trial. At that point, infants given the new vaccine had developed only 1 case of HIB, whereas those given the placebo had incurred 18 cases. In that study, the Bayesian and frequentist analyses required roughly the same sample size, and both paradigms clearly detected the efficacy of the new vaccine. But the Bayesian analysis was remarkable for offering useful predictions about the future consequences of various choices.

For a third and final example, consider a randomized trial comparing two resuscitation treatments for dogs with experimentally induced cardiac arrest (Niemann et al. 1992; Lewis and Berry 1994). One treatment was the standard protocol for humans suffering cardiac arrest due to disorganized electrical activity, which was to deliver an electric shock to the heart (defibrillation) immediately. Unfortunately, with cardiac arrest the recent lack of blood flow in the heart muscle will have created a nonresponsive situation, with a low survival rate. The experimental new treatment, being considered for use in humans, was to administer epinephrine and perform cardiopulmonary resuscitation prior to defibrillation in the hope of first eliciting some cardiac blood flow and thereby making the heart more responsive to electric-shock treatment. Pairs of dogs were randomized, with one assigned to each treatment. Ethical considerations motivated a fully sequential statistical analysis to minimize the number of experimental dogs and to expedite movement of promising results to human medicine.

As in the preceding two examples, Bayesian and frequentist analyses were applied to the same data for comparison. An intriguing aspect of this example, however, is that the performance of the Bayesian method was also evaluated according to frequentist criteria to provide a direct head-to-head comparison. Such comparison is possible because a trial terminated by a Bayesian stopping rule can be analyzed from either a Bayesian or frequentist viewpoint. Bayesians think about sample sizes in terms of probabilities of getting at the truth, whereas frequentists think about sample sizes in terms of the long-run error rates that would obtain were the experiment repeated numerous times. Of course, both paradigms will perform better with more data, but the issue is just how much data each statistical paradigm needs to achieve adequate results.

Average sample sizes were determined by Lewis and Berry (1994), for both Bayesian and frequentist stopping rules, under a wide range of scenarios about the efficacies of the standard treatment and the new treatment. The Bayesian trial design always outperformed the frequentist design, using fewer animals to reach the same level of statistical significance. The advantage of the Bayesian design was slight in some scenarios, but a twofold or greater reduction in numbers of experimental animals was achieved for most scenarios, and even a tenfold reduction in some extreme cases. Hence, even by the frequentists' own criteria, the Bayesian approach always performed better, requiring fewer dogs to reach adequate conclusions.

The clear conclusion from these philosophical and practical comparisons of the Bayesian and frequentist paradigms is that Bayesian methods frequently outperform frequentist methods. Yet despite the ethical significance of that finding, frequentist analyses continue to dominate in medical research.

What can be done about that? Medical experiments generally involve a team of researchers, statisticians, and ethicists. Accordingly, a word may be said to each of those groups.

First, the word to medical researchers is to make their questions crystal clear to consulting statisticians and to verify that the answers offered are meaningful and direct. Medical researchers should ponder, again and again and again, the formula that Inference $\neq$ Prediction $\neq$ Decision. They must not use an inference procedure for a prediction problem or a decision problem, nor, worse yet, use a second-rate inference procedure. That would be like using a screwdriver to do the job of a hammer or saw, and a poor screwdriver at that.

Second, the word to statisticians is to cease their superficial and misleading rhetoric about Bayesian methods being subjective and frequentist methods being objective. Rather, address the real issue about which paradigm's kind of subjectivity better suits medical research. The Bayesian's pursuit of prior or pre-experimental information is subjective in that different individuals may possess different information, apart from that in the one experiment at hand. Nevertheless, the opportunities to gain efficiency and reliability from using all available data are frequent and sizable, and in any case, consideration of other data from outside the current experiment invites something like a sub-stantive scientific discussion. But the frequentist's consideration of imaginary data invites a rather obscure subjectivity. There is no doubt that imaginary outcomes are indispensable for computing p-values, but often there is doubt that this statistic is answering the actual question that medical researchers are asking.

Third and finally, the word to ethicists is that the choice of a statistical paradigm will affect research efficiency, costs, progress, and particularly the number of animals or humans studied before a clinical trial can deliver adequate results. Although ethicists may not follow every detail of the re-searchers' procedures or the statisticians' equations, it is incumbent on them to verify that other members of the research team have considered several options for the project's statistics and have good reasons for selecting a par-ticular one. Ethicists should ask questions such as these: Have other statistical methods, besides the recommended one, been considered? Have published papers about similar experiments used other statistical methods that have not yet been mentioned and weighed? What evidence and arguments indicate that the recommended statistical method will best serve the experiment's ethical requirements of improving clinical practice, not wasting experimen-tal subjects, and not wasting money?

Lastly, the Bayesian–frequentist debate has absorbed this section's atten-tion because frequentist methods currently dominate medical research and yet Bayesian methods often are demonstrably superior (Berry 1990). Many medical researchers have received virtually no training in Bayesian methods and have little understanding of the conceptual or practical differences between Bayesian and frequentist methods. But that is not the only count on which statistical practice could be improved in medical research. Perhaps the greatest issue is to distinguish among inference, prediction, and decision problems and to use an appropriate tool for each task, rather than pressing second-rate inference procedures into service for everything. A third problem is that medical researchers rarely understand the hindrances imposed by noisy data and the opportunities for parsimonious models to reduce noise and gain efficiency. Another serious problem is that statistical training emphasizes univariate and bivariate methods to the relative neglect of multivariate methods, and yet much, if not most, medical data are multivariate. Multivariate data require multivariate analyses to really show what is going on. On these four counts, there is tremendous need for consulting statisticians to bring to medical researchers the very best that their craft has to offer.

DISCUSSION

The first of the foregoing case studies was aimed at enhancing science's credibility. It concerned the motions of objects in familiar circumstances, showing that anyone can easily perform relevant little experiments with physical reality. Some experience and thought about that elementary case study will reveal all that really can be offered by way of reply to skeptical, relativistic, or postmodern attacks on science's credibility. If that little exemplar of science makes sense, the whole scientific enterprise makes sense; otherwise, not. It would seem to be terribly difficult to bring to bear any criticism of that rudimentary learning about reality that would be both serious and sincere.

The remainder of these case studies involved substantial scientific problems in which an exemplary grasp of scientific method was shown to improve productivity. Looking over these case studies, two lessons should be learned. The immediate lesson is that all too often, overspecialization renders scientists poor importers of useful ideas and procedures. For example, philosophers and statisticians have valued parsimony, yet the case study on molecule shapes revealed very slow incorporation of that criterion. Likewise, cross-validation and Bayesian methods are off-the-shelf statistical tools, and yet their applications in molecular and medical research have developed slowly. Many great advances in science and technology have resulted from long-overdue combinings of off-the-shelf tools that had long been sitting around in several different disciplines. One valuable remedy would seem to be to complement extensive specialized training with modest education in the basic, pervasive principles of scientific method.

But the larger lesson is that all too often, even remarkable advances catch on slowly. Why? One factor is simple complacency, the assumption that things are already good enough. Yet the deeper cause probably is that most scientists have an inadequate grasp of the fundamental principles of scientific method. For instance, even when someone bothers to incorporate superior inductive procedures or to employ parsimony, there is no intuitive sense that better induction or parsimony resonates with the fundamental principles that make science work. There is no quick and lively sense that such advances point to the way ahead, because there is no adequate grasp of science's narrow road and exacting method.

If scientists had a better grasp of fundamental methodological principles, they would more instinctively ask themselves some penetrating questions that would disturb complacency and increase productivity; and even if that failed, at least they would recognize when someone else had proposed a significant refinement or a better paradigm. It is hoped that reflection on this chapter's case studies may prompt researchers in various specialties to see the importance of the general principles of scientific method in their own work, both to correct entrenched misunderstandings and to stimulate new advances.

CHAPTER TEN

SCIENCE'S POWERS AND LIMITS

What are science's powers and limits? That is, where is the boundary between what science is and is not able to discover? The American Association for the Advancement of Science has identified that issue as a critical component of science literacy: "Being liberally educated requires an awareness not only of the power of scientific knowledge but also of its limitations," and learning science's limits "should be a goal in all science courses" (AAAS 1990:20–21; also see p. xv). As Caldin (1949:vii) warned, "The prestige of science to-day is very great; scientists have therefore a certain responsibility to consider carefully the scope and limitations of their professional activities."

This question about science's reach has elicited great interest. For instance, Horgan (1996) interviewed dozens of leading scientists regarding the limits of scientific knowledge, and his book became a runaway best-seller and made the front page of the *New York Times* book-review section (30 June 1996). Likewise, the end-of-the-millennium special issue of *Scientific American* offered fascinating reading on "What Science Will Know in 2050" (December 1999).

People's motivations for exploring the limits of science can easily be misconstrued, so they should be made abundantly clear from the outset. Unfortunately, for some authors writing about science's limits, the motivation has been to exaggerate the limitations in order to cut science down, support anti-scientific sentiments, or make more room for philosophy or religion. For others, the motivation has been to downplay science's limitations in order to enthrone science as the one and only source of real knowledge and truth. Neither of those excesses represents my intentions. I do not intend to fabricate specious problems to shrink science's domain, nor do I intend to ignore actual limitations to aggrandize science's claims. Rather, the motivation here is simply to find the actual boundary between what science can do and cannot do.

Because this book is about scientific method, this chapter will explore the implications of method for what science can and cannot investigate, as contrasted with the implications of ethics for what science should and should not investigate, or as contrasted with the implications of budgets for what science can and cannot afford to investigate. Ethics and budgets are enormously important and have been capably considered elsewhere; it is beyond the scope of this brief chapter to explore those complex matters. Here it will be challenge enough to tackle just one focused question. Given science's method, what powers and limits result regarding what science is able or unable to investigate?

OBVIOUS LIMITATIONS

The most obvious limitation is that scientists will never observe, know, and explain everything about the physical world. The fancy version of this critique points to Heisenberg's uncertainty principle, Gödel's theorem, and chaos theory as fundamental limits – not to mention numerous practical restrictions due to finite time and money. The plain version of this story points to one little pebble and concludes that were all of humanity devoted to the study of this pebble until our sun grows cold, we still would not know everything about its atomic and subatomic makeup, geological history, and so on. The inherent complexity of the world is immense relative to the reach of human understanding. Furthermore, each answer tends to raise ten new questions, so science's questions increase faster than its answers. Indeed, an increase in precise and productive questions might be the greatest evidence of scientific progress.

Another familiar limitation is that science cannot completely supply its own ethical requirements (AAAS 1989:25; 1990:25–26). The scientific enterprise needs ethical guidelines, especially for sensitive issues like animal and human experimentation and weapons research. More generally, scientific priorities must reflect the goals of a society that pays for research through taxes and other means, and ultimately these priorities reflect an ethical vision about what is good. Admittedly, scientific knowledge frequently informs ethical decisions, such as prioritizing medical research by quantifying the numbers of persons suffering from various serious diseases. But clearly, a full-blown system of ethics cannot emerge from scientists' test tubes, telescopes, and data. Beneficial science that promotes the good must look beyond science itself to obtain needed vision and ethics, as Caldin (1949) and others have emphasized.

Several books have discussed the limits of science (Caldin 1949; Medawar 1984; Trigg 1993; Horgan 1996; Rescher 1999). Most of the limitations those books discuss are rather familiar. Accordingly, the remainder of this chapter will explore power and limits that are not so obvious or frequently discussed.

SCIENCE AND ITS PRECONDITIONS

The greatest difference between a naive view and a considered view of science is ignorance of versus recognition of science's presuppositions. Clearly, science's presuppositions must be true in order for its goals to be meaningful and its methods to be rational. For instance, if the physical world did not exist, or if it were incomprehensible to humans, then science would be nullified. Yet science is forever completely incapable of proving its presuppositions. Consequently, a striking limitation of science is that it cannot provide its own foundation. Science is based on non-science.

The roster of science's presuppositions, as explored in Chapter 4, need not be repeated here. Instead, it suffices to perform a simple experiment to emphasize how pervasive those presuppositions are. As an exceedingly simple and yet relevant example of scientific experimentation, the reader is asked to determine the height of this book. If a ruler is at hand, measure this book's height; if not, imagine performing the measurement and make an estimate.

What must be true of the world for this measurement to be meaningful and correct? The book and ruler must exist. Human sense perceptions must be reliable enough to permit a valid observation of this book's height. Nature must be uniform, and more specifically the book must be a rather rigid and stable object with a single and definite height and the ruler must not change its shape or length substantially as it is moved through space and across time from its original location to the book.

Now the alternative hypotheses, that "This book's height is several centimeters" or else that "This book's height is several kilometers," can be put to test by easy scientific experiments, which will render the first hypothesis a definite conclusion. But the underlying presupposition that "The physical world, including this book and ruler, exists" is held in common by both hypotheses and cannot be put to any scientific test. This same limitation applies to any and all of science's presuppositions. For instance, Caldin (1949:61) said that "This belief that there is order in nature is not a conclusion, but a presupposition of science."

So why would a scientist believe all these extra-scientific beliefs that science cannot possibly prove? Why believe that the physical world exists? Why believe that sense perceptions are generally reliable? Why believe that nature is orderly and uniform, including that a ruler is reasonably rigid and stable? In a word, why believe that science makes any sense?

If those questions are asked but once, the answer must be an appeal to common sense, saying that science presupposes nothing more nor other than what we presuppose in claiming that "Moving cars are hazardous to pedestrians" is a rational, true, objective, realistic, and certain belief. If one rudimentary scrap of common sense is acceptable, then science flourishes; but if not, science languishes.

But if those questions are asked twice, the answer must be an appeal to some religion, philosophy, or worldview – some general and basic story about the way things are. Even if one does believe *that* the physical world exists, one may still ask *why* or *how* it exists. Even if one does believe *that* sense perceptions are generally reliable, one may still wonder *how* we and our world came to be this way. And even if one does believe *that* nature is uniform and comprehensible, one may still ask *why* this is so. Perhaps the biggest question of all is simply why there is anything rather than nothing.

The traditional understanding is that the search for answers to life's big questions involves not only the sciences but also the humanities: "The ... method of science is an imperfect means of investigating matters that lie outside its scope, and leads to false conclusions if the attempt is made to apply it to them; so that its limitations ought to be explained as carefully as its merits. ... From natural science we cannot learn what material nature is for, how and why it exists at all, and why it has any laws. In so far as we can answer these questions we do it in terms of a wider survey, made explicit in philosophy and theology, expressed concretely in poetry, felt vividly in the nature-ecstasies of a few" (Caldin 1949:172, 130).

SCIENCE AND WORLDVIEWS

By all but the most skeptical and peculiar accounts, science's powers include the capacity to answer routine questions, such as which brand of light bulbs will last longest. But can science also tackle big questions with worldview or religious import, such as whether or not God exists? This is the most controversial – and yet perhaps the most significant – aspect of the boundary between science's powers and limits. This section will explore this huge issue from the focused perspective of this book's topic, scientific method. It will be argued that an answer is *not* forthcoming from considerations of scientific method.

The opposite view has been expressed by the AAAS (1989:26), that scientific method definitely is not applicable to worldview questions: "There are many matters that cannot usefully be examined in a scientific way. There are, for instance, beliefs that – by their very nature – cannot be proved or disproved (such as the existence of supernatural powers and beings, or the true purposes of life)."

It is most perplexing, however, that another AAAS position paper (1990:xiii; also see p. 24) claims that science does have some worldview answers. "There can be no understanding of science without understanding change and the fact that we live in a directional, though not teleological, universe." Now "teleological" just means "purposeful," so here the AAAS is boldly declaring as fact that we live in a purposeless universe. Although a scientific assertion could hardly be worded more strongly, regrettably not a shred of argumentation or evidence is offered. Anyway, because all scientific facts result from applications of scientific method, the implication necessarily follows that scientific method is applicable to worldview questions and has produced this fact of a purposeless universe.

So one AAAS document says that the purposes of life are outside science's purview, whereas another says that a purposeless universe is science's conclusion. Clearly, those two statements are contradictory. Consequently, this is one of those rare instances in which AAAS statements have not provided reliable guidance.

But fortunately, the AAAS (1989:139) has provided the remedy for those confused and contradictory opinions by their advice to check an argument's presuppositions. They list as one of the "symptoms of doubtful assertions and arguments" the problem that "The premises of the argument are not made explicit." Accordingly, a key question must be raised here: What must be presupposed in order for the first AAAS claim to be true, that scientific method cannot prove or disprove the existence of supernatural beings? The required premise is that there are no observable interactions between natural and supernatural entities. (Incidentally, that might be simply because no supernatural beings exist, or because such beings do exist but they do not interact with physical things in ways that we can observe.)

To see why this premise is necessary, an analogy may be instructive. The most common type of agricultural yield-trial experiment produces yield measurements for a number of genotypes (G) grown in a number of environments (E, which can be different locations or years or both). Although such an experiment has *two* factors, genotypes and environments, it has *three* sources of variation. The obvious sources of variation in yield are the genotypes and the environments, but the additional one is the genotype-by-environment interaction (G × E). The key feature of this analogy for present concerns is that various agricultural specialists have different stakes. Agronomists work on improving environments, so their stake is E, but also G × E, because it also involves environments. And plant breeders work on improving genotypes, so their stake is G, but also G × E, because it also involves genotypes. Interestingly, the variability in yield associated with G × E is usually larger than G, so plant breeders can make even larger gains by handling G × E than by handling G. In getting the whole story, interactions are important!

Returning now to the earlier context, an analogous situation holds for the supernatural (S), of concern to religion, and the natural (N), of concern to science, that an additional factor is the supernatural-by-natural interaction

(S × N). Scientists' stake is N, but also S × N if it exists, because it also involves the natural world. Consequently, only if there are no interactions between supernatural and natural entities can it be presumed that no observations by scientific methods of natural entities could occur that would have supernatural explanations and worldview import.

What could possibly legitimate the implicit AAAS presupposition of no such interactions? Recall that in Chapter 4 science's presuppositions were installed by philosophical analysis of a mere scrap of common sense, the reality check, that was adopted by faith. Certainly that installation provides no verdict whatsoever on whether or not supernatural beings exist and, if so, whether or not they interact with natural entities. Furthermore, to preserve science's status as a public institution, its presuppositions must suit a worldview forum that includes all worldviews (except radical skepticism). In science's worldview forum, beliefs about supernatural beings are simply controversial (Easterbrook 1997; Larson and Witham 1999). So the implicit AAAS presupposition is problematic, constituting a nasty blow to science's public status. That is exactly the problem that the recommended installation of presuppositions avoids by relying instead on a shared, worldview-independent scrap of common sense. Scientific method, as grounded in presuppositions that are worldview-independent and that preserve science's public status, implicates no verdict whatsoever regarding whether or not scientific observations might reveal supernatural beings. Maybe, and maybe not.

If the world is as some worldviews would have it, then no traces of supernatural activity will be found in scientific observations. But if the world is as some other worldviews would have it, then some scientific or physical observations that require supernatural explanations are to be expected. And if the world is as still other worldviews would have it, there may be no clear expectation either for or against observable S × N interactions. In any case, the important point from a methodological perspective is that precisely because hypotheses about physical reality interacting or not interacting with nonphysical reality are not among science's (legitimate) presuppositions, such hypotheses retain eligibility to be considered in light of the data if admissible and relevant data can be identified and collected.

More generally, such openness to reality is a requirement not only for scientific inquiry, but for any kind of rational inquiry (Earman 2000:4). Historically, various thinkers (including David Hume) have attempted to reach a verdict on the existence or credibility of S × N interactions from theoretical methodological considerations, rather than from arduous empirical investigations, but Earman (2000) has shown that such misdirection is bound to end in abject failure.

Then if the action is not found in science's method, what about its data? Certainly, several possibilities for relevant data – empirical, public, and equally able to count either for or against S × N interactions, depending on the facts – have interested some scientists and philosophers. They have included prayer (Medawar 1969:3–7; Koenig, McCullough, and Larson 2001) and prophecy (J. L. Mackie, in Swinburne 1989:90; David Hume, in Earman 2000:153).

But exploring such data and their interpretation is outside the domain of a book on scientific method. Instead, what is relevant and important for the topic of scientific method is just to make it clear that any resolution of worldview-level questions pursued by science lies not in the methods and not in the presuppositions, but rather in the data. At issue is a contingent feature of the world that must be investigated by looking at the world to see what happens. The value of methodological clarity, which is all that this brief section can pursue, is precisely that proper methods allow empirical data to have their rightful influence in honest inquiries.

That the scientific community is also interested in larger questions than which brand of light bulbs will last longest is testimony to its vitality and courage. Doubtless, such ambitious inquiries involve some challenges, controversies, risks, and mistakes, but they also have had some successes and benefits. The reflections of any individual scientist have considerable limitations, but the labors of the entire scientific community are reasonably promising. For instance, the AAAS has exhibited fine leadership by offering in its position papers numerous, mostly sensible perspectives on religion, God, the Bible, clergy, prayer, and miracles (AAAS 1989, 1990). Although the sciences on their own can offer some insights into life's big questions, the sciences in concert with the humanities can offer even more. In that regard, one can look forward to the ongoing events and publications from the Dialogue on Science, Ethics, and Religion (DoSER) program of the AAAS (visit www.aaas.org/spp/dser).

PERSONAL REWARDS FROM SCIENCE

The intellectual, technological, and economic benefits from science have already been touted herein and are acknowledged by society in general. Likewise, the importance of science education for good citizenship in a scientific and technological age is widely appreciated. But an important topic receives far too little attention: the personal rewards of science, that is, the potentially beneficial effects of science on scientists' personal character and experiences of life. As this chapter on science's powers and limits draws toward a close, these powers merit attention. Caldin (1949) explored this topic with rare wisdom and charm, so the following remarks draw much from him. Additional reflections can be found in the work of Weaver (1948), Midgley (1992), and Trigg (1993).

To see science's value for exercising and refining character is to see how appealing science really is. Unfortunately, "the place of science in society is too often considered in the narrow setting of economic welfare alone, so that the potential contribution of science to the growth of the mind and will is under-estimated" (Caldin 1949:155).

Science by itself cannot initiate or guarantee the good life. But presuming that life experiences have already shown the dignity of persons and the value of reason, science can stimulate and strengthen those factors contributing toward the good life:

> We have argued that man has a certain definite nature, with definite potentialities; he is rational, capable of knowing truths and desiring goods, and of acting accordingly; he is also an animal, equipped with animal instincts; and he is a social being, not an isolated individual.... Truthfulness and wisdom are ends befitting a being equipped with intellect; love and self-discipline become a being equipped with a rational appetite for the good; justice, honesty, faithfulness, altruism and mutual love are fitting in human beings because they are also social by nature.... The good life is concerned with pursuing ends consonant with our nature; with realising, fulfilling, what we are made for. (Caldin 1949:138)

One reward from science is stimulation of rationality and wisdom:

> Now a knowledge of nature is part of wisdom, and we need it to live properly.... Science is, therefore, good "in itself," if by that we mean that it can contribute directly to personal virtue and wisdom; it is not just a means to welfare, but part of welfare itself.... Scientific life is a version of life lived according to right reason.... Consequently, the practice of science requires both personal integrity and respect for one's colleagues; tolerance for others' opinions and determination to improve one's own; and care not to overstate one's case nor to underrate that of others.... By studying science and becoming familiar with that form of rational activity, one is helped to understand rational procedure in general; it becomes easier to grasp the principles of all rational life through practice of one form of it, and so to adapt those principles to other studies and to life in general. Scientific work, in short, should be a school of rational life. (Caldin 1949:133–135)

Still another personal reward from science is cultivation of discipline, character, realism, and humility:

> It is not only the intellect that can be developed by scientific life, but the will as well. Science imposes a discipline that can leave a strong mark on the character as can its stimulation of the intellect. All who have been engaged in scientific research know the need for patience and buoyancy and good humour; science, like all intellectual work, demands (to quote von Hügel)

"courage, patience, perseverance, candour, simplicity, self-oblivion, continuous generosity towards others, willing correction of even one's most
cherished views." Again, like all learning, science demands a twofold attention, to hard facts and to the synthetic interpretation of them; and so
it forbids a man to sink into himself and his selfish claims, and shifts the
centre of interest from within himself to outside. But for scientists there
is a special and peculiar discipline. Matter is perverse and it is difficult to
make it behave as one wants; the technique of experimental investigation
is a hard and chastening battle. Experimental findings, too, are often unex-
pected and compel radical revision of theories hitherto respectable. It is in
this contact with "brute fact and iron law" that von Hügel found the basis
of a modern and scientific asceticism, and in submission to this discipline
that he found the detaching, de-subjectifying force that he believed so necessary to the good life. The constant friction and effort, the submission to
the brute facts and iron laws of nature, can give rise to that humility and
selflessness and detachment which ought to mark out the scientist. (Caldin
1949:135–136)

Though science comes laden with rich rewards, certain excessive views
can spoil its benefits. As Augustine warned, every excess becomes its own
punishment. The potentially most dangerous excess is uncritical faith that
science offers the only route and a complete method for finding truth and
meaning, which can poison the very wellsprings that otherwise could nourish
science.

Many people have been led to think that the procedure of natural science
is the royal road to truth in every field, and that what cannot be proved
by science cannot be true.... The new element that characterises so much
modern thinking is not so much science, as an attitude towards science;
not scientific method, but faith in the unlimited applicability of scientific
method; not the conclusions of science, but faith in the relevance of science to every kind of problem. Such a faith calls for examination.... For
science will not give us the conceptions that we need in dealing with the
fundamentals of life – the great conceptions of personality, justice, love,
for instance, must be drawn from elsewhere; moreover, science gives us
only part of the mental training for using them. Again, although science,
given its appropriate setting, plays its part in developing both intellectual
and moral virtues, it does not originate those virtues; it only supports them
where they exist already. It favors an intellectual climate where persons are
respected; but it does not create these values, it presupposes them.... In
relation to the great social purposes – such as peace, justice, liberty – science
is instrumental; it is not normative, does not lay down what ought to be.
If scientific method is applied outside its own field, if the attempt is made
to use one rational method for work appropriate to another, the result is
always confusion and, at worst, disaster. To make its full contribution in

society, science must respect the fields of other studies; conversely, it relies upon those studies being in a healthy condition. In concrete terms, this means that a scientist who is ignorant of philosophy, history, art and literature, will not be able to speak for reason with the power that he should. (Caldin 1949:6, 169, 159)

In conclusion, science suits the good life: "Science and its technical application can both be integrated into the pattern of the good life; science forms an integral part of the good life for a scientist, and the applicability of science can be turned to good account. Potentially, then (whatever we may think of it in practice), natural science is an ally of wisdom and good living.... The battles against disease, destitution and ignorance are always with us, and science is a powerful ally; perhaps the most necessary ally after love and understanding and common sense" (Caldin 1949:175, 161).

SUMMARY

Understanding the boundary between science's powers and limits is a core component of scientific literacy. Some limitations are rather obvious, such as that science cannot explain everything about the world and that it cannot provide its own needed ethics. But other limitations are more subtle and merit more careful thought. Another limitation of science is that it cannot prove its own needed presuppositions.

Many scientists and philosophers have debated whether helping to inform larger worldview issues, such as the purpose of life, is among science's powers or beyond its limits. The topic in this book is scientific method, and here it is argued that the answer to that question is not forthcoming from science's method or presuppositions. Instead, empirical investigations are needed to test science's reach.

Among science's powers is a considerable ability to be of benefit to scientists' personal character and experiences of life. The essence of this benefit is the selflessness, detachment, and humility that result from deliberate and outward-looking attention given to the physical world.

CHAPTER ELEVEN

SCIENCE EDUCATION

Science education is absolutely crucial for the scientific enterprise. Without education and culture, we might today be celebrating the discovery of fire and invention of the wheel, rather than exploring space and interpreting genomes. Also, most scientists employed by universities are called upon to be educators as well as researchers. Accordingly, this chapter will consider the interactions between scientists and educators, especially as regards a fundamental element in the science curriculum, namely, scientific method, or, somewhat more broadly, the nature of science.

Science educators typically place the topic of scientific method within a broader context: "For science educators the phrase 'nature of science,' is used to describe the intersection of issues addressed by the philosophy, history, sociology, and psychology of science as they apply to and potentially impact science teaching and learning" (William F. McComas et al., in McComas 1998:5). Figure 11.1 depicts these four disciplines that contribute to our description and understanding of science. The various sizes of these circles represent the relative importance of these sources.

Educators and scientists alike perceive the centrality of the nature of science (NOS) in the science curriculum. Indeed, the NOS constitutes the very first chapter or topic in most documents specifying national standards for science education, because the NOS is seen as "a vital foundation for all future science learning" (William F. McComas and Joanne K. Olson, in McComas 1998:51). The view taken here is that the NOS is the center of science, and scientific method is the center of NOS (Gruender 2001).

The main purpose of this chapter is to document the benefits from teaching and learning the NOS. Science educators have conducted hundreds of empirical studies, so the education community is well positioned to inform the scientific community about these benefits. This chapter will also take a quick look at the prevalent educational philosophy, constructivism. It will

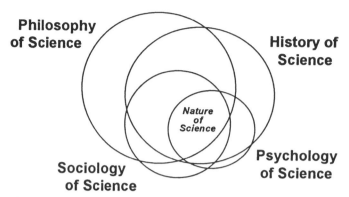

Figure 11.1. Four disciplines that contribute content to the nature of science. (Reproduced from William F. McComas and Joanne K. Olson, in McComas, 1998:50, with kind permission from William F. McComas and Kluwer Academic Publishers.)

conclude by mentioning several exciting frontiers in science education that are likely to develop in the near future.

SIX BENEFITS

The introductory chapter mentioned six benefits that result from study of the NOS in general and scientific method in particular. In this chapter, these benefits are documented and discussed.

At the outset, however, it must be emphasized that these proposed benefits apply to NOS material that is properly focused for a given audience and purpose; otherwise, all bets are off. For instance, William F. McComas et al. (in McComas 1998:3–39), Robinson (1998), and Gruender (2001) have argued persuasively that although the history and philosophy of science inform the NOS (as Figure 11.1 indicates), standard courses in the history and philosophy of science are not the right model for NOS instruction for science educators. The NOS can be studied to make philosophers better philosophers, or historians better historians, or sociologists better sociologists, or educators better educators. Here the emphasis is on making scientists better scientists.

Likewise, these proposed benefits require explicit instruction in the NOS, for "the nature of science, like any other cognitive outcome, should be addressed explicitly" (Bell, Lederman, and Abd-El-Khalick 1998; also see Abd-El-Khalick, Bell, and Lederman 1998; McComas et al., in McComas 1998:27; Lavoie 1999; Bevilacqua, Giannetto, and Matthews 2001; Hogan and Maglienti 2001). This instruction may occur in separate modules and courses or may be integrated with science content, or both. But it needs to be explicit. Otherwise, the likely outcome from implicit impressions will be a mixture of incoherent ideas. Koulaidis and Ogborn (1995) reviewed 26 positions on the NOS and found that 21 were based on implicit and vague ideas, which invariably resulted in "idiosyncratic collages from various philosophical systems."

Despite many intriguing or sometimes tiresome debates, sufficient consensus about the NOS exists to include a mainstream version in the science curriculum. Alters (1997) has argued otherwise. But the response from Smith et al. (1997) seems much more compelling, that "We agree with Matthews (1994[:8]) that although there is not unanimity, 'there is a reasonable consensus on many lower-level points' about the NOS," and "Too much is being made of disagreements concerning the NOS that involve tenets that are esoteric, inaccessible, and probably inappropriate for most K–12 instruction." Furthermore, "the U.S. National Academy of Science has been able to produce a series of consensus statements about the NOS for inclusion in the National Science Education Standards (NRC, 1996) based on input from thousands of scientists, teachers, science educators, philosophers and sociologists of science, and others." Accordingly, their advice is to go with the mainstream consensus in documents from the world's major scientific organizations. William F. McComas et al. (in McComas 1998:3–39) and William F. McComas and Joanne K. Olson (in McComas 1998:41–52) found considerable agreement regarding the NOS among international documents.

The evidence for this section's six benefits from NOS instruction is extremely strong. Of course, occasional studies of NOS instruction, such as those by Irwin (2000) and Moss, Adams, and Robb (2001), have detected little if any benefit. That is not surprising, because educational studies vary in their implementation of NOS instruction, the kinds of putative benefits measured, sample sizes, and other factors. Similarly, an occasional study may fail to detect detrimental effects of smoking or asbestos. However, the overwhelming majority of studies find significant benefits from instruction in the NOS, and the cumulative case is compelling.

(1) Better Comprehension. Students must master scientific method and the NOS in order to make scientific knowledge their own: "As Plato insisted so long ago, education is not just the having of correct beliefs, it is the having of adequate reasons for these beliefs" (Matthews 1998b). To describe the opposite situation for a student learning only science's facts from other persons who also know science's process, he quotes memorable words from John Locke: "Such borrowed wealth, like fairy money, though it be gold in the hand from which he received it, will be but leaves and dust when it comes to use."

One item among the "modest goals" for NOS instruction advocated by Matthews (1998b) is logic: "Studies show that all the standard logical fallacies known since Aristotle's day are routinely committed by science students.

Given that being able to reason clearly is an obvious component of being scientific, then one of the low-level nature of science objectives might be to teach some elementary formal and informal logic." Basic instruction in deductive and inductive logic can reduce common fallacies and misconceptions and can enhance reasoning, particularly the kind of reasoning needed to evaluate competing hypotheses (Lawson 1990; Piburn 1990; Driver, Newton, and Osborne 2000; Hogan and Maglienti 2001).

Of course, in general, both reasoning ability and prior knowledge of a subject help students comprehend more advanced material, and yet surprisingly the former is often more important. For instance, what factors influence students' ability to comprehend college biology? Johnson and Lawson (1998) pretested students to determine their reasoning ability and prior knowledge. Then some classes received "expository instruction" focused on facts and concepts, whereas others received "inquiry instruction" giving more attention to science's logic, method, and process. The results were that "Reasoning ability but not prior knowledge or number of previous biology courses accounted for a significant amount of variance in final examination score in both instructional methods." Furthermore, the inquiry classes showed better improvement in both reasoning skills and biology comprehension.

Accordingly, those authors concluded that

> Given that reasoning ability appears to be a significant predictor of achievement in introductory level college biology, more so than the number of previous biology courses completed or a pretest measure of domain-specific knowledge, high school biology instructors would be well advised to be more concerned with the development of their students' reasoning abilities than with making certain that they cover a wide range of specific biology concepts. The same advice would also seem to be appropriate for college level instructors who may be concerned about their students' performance in more advanced college courses.... The inquiry students in the present study not only showed greater improvement in reasoning ability during the semester than the expository students, they also did better on the measures of biology achievement. In other words, nothing of importance seems to be lost by switching to inquiry instruction, and much seems to be gained. (Johnson and Lawson 1998)

A similar win-win situation for NOS and science by integrating these subjects was found by Rief (1995) and Wyckoff (2001) in teaching physics. And Cartier and Stewart (2000) found the same in teaching genetics.

Those empirical findings prompt some commentary on the *National Science Education Standards* (NRC 1996). The *Standards* include prominent curriculum requirements for the NOS, or in their terminology, for "science as inquiry." That is commendable, given the findings of science educators. However, the table on changing emphases is organized under two headings:

"Less emphasis on" and "More emphasis on" (NRC 1996:113). The very first item is less emphasis on "Knowing scientific facts and information" and more emphasis on "Understanding scientific concepts and developing abilities of inquiry." This might be called the trade-off model, or the win-lose model, in which inquiry wins and information loses. Admittedly, given a fixed schedule, an hour more for the NOS does mean an hour less for science content. But that does not necessarily imply less content, because NOS instruction can pave the way for more rapid coverage of more deeply comprehended material. By analogy, first sharpening a dull ax means less time spent chopping wood, but that does not necessarily mean less firewood, and indeed the more likely outcome is more firewood.

The studies cited earlier suggest that when NOS instruction is properly integrated within the science curriculum, a win-win situation for NOS and science can emerge. Cartier and Stewart (2000) are quite explicit about an extremely positive educational outcome: "In contrast to what is popularly believed, focusing on epistemological issues in the classroom does not have to mean teaching less subject matter in exchange. In fact, helping students to appreciate the epistemological structure of knowledge in a discipline can provide them with the means to more fully understand particular knowledge claims and their interrelationships." So perhaps a better model, and a realistic expectation as well, would be to put both "Knowing scientific facts and information" and "Understanding scientific concepts and developing abilities of inquiry" under the column "More emphasis on."

What sort of NOS instruction can create such a win-win situation? As mentioned near the outset of this section, effective NOS instruction must be properly focused for a given audience and purpose and must be explicit. Beyond that, educators can learn principles and practices from programs already in place that have a track record of exemplary success.

(2) Greater Adaptability. Science and technology are experiencing rapid and pervasive changes, requiring scientists to be increasingly adaptable and flexible. Young scientists can expect to change jobs and assignments more frequently than their predecessors. They can also expect much of what they learn to have a short half-life of only a few years. This has profound implications for science education.

Accordingly, the NSF (1996) and NAS (1995) position papers for reshaping undergraduate and graduate science education, respectively, promote adaptability as a prominent goal. Indeed, the first of three general recommendations from the NAS (1995:76) is that "graduate programs should add emphasis on versatility; we need to make our students more adaptable to changing conditions." Versatility has become so essential that the NAS recommends that evaluation criteria for funding education/training grants "include a proposer's plan to improve the versatility of students" (NAS 1995:80).

Although adaptability clearly is desirable, that goal faces three powerful hindrances that must be overcome. First, overspecialization is the direct opposite of adaptability. Yet there is tremendous temptation to overspecialize because it can seem so advantageous in the short run (NAS 1995:77–78). Increasingly, scientists are finding employment in nonacademic settings, and "such employers complain that new PhDs are often too specialized for the range of tasks that they will confront" (NAS 1995:3). Second, methodology is being neglected: "The textbooks...are large collections of facts. What I see really missing from these textbooks is the *process* of science" (NSF 1996:4). Third, bodies of knowledge remain isolated. Knowledge transfer is poor, thereby decreasing adaptability. Knowledge learned within a given specialty tends to stay locked up within those narrow confines, even if it could be useful elsewhere: "We know that students rarely realize the applicability of knowledge from one context to another" (NSF 1996:3).

Instruction in scientific method in particular, and NOS more generally, addresses these three challenges. Whereas much scientific knowledge will be having shorter and shorter half-lives, the general principles of scientific method are refreshingly enduring. Whereas most scientific knowledge is highly specialized in subfields, such as radio astronomy or microbial ecology, the general principles of scientific method are refreshingly cosmopolitan, relevant throughout the scientific enterprise. As Machamer (1998) remarked, "Reflecting about the goals and procedures of problem solving helps one solve problems better. It also enables one to adapt and restructure old goals and procedures to new environments and problems."

Indeed, perception of instances of general principles at work is precisely what allows students to unlock knowledge and transfer learning from one context to another, enabling a scientist to adapt more quickly to a new task. Zoller et al. (2000) take the interesting perspective that "One of the ultimate goals of higher-order cognitive skills...and...critical thinking...in science education is the transfer of capability across disciplines and domains," which they term "teaching for transfer." Likewise, Lawson et al. (2000) found that general reasoning skills, but not subject-matter knowledge, accounted for a significant amount of the variance in the ability to solve a new "transfer" problem.

(3) Greater Interest. Science educators find that science in its historical and philosophical context just makes more sense and has more appeal than does science abstracted from its context (Carson 1997). "Incorporating the nature of science while teaching science content humanizes the sciences and conveys a great adventure rather than memorizing trivial outcomes of the process. The purpose is not to teach students philosophy of science as a pure discipline but to help them be aware of the processes in the development of scientific knowledge" (William F. McComas, in McComas 1998:13).

Including the NOS is an effective measure to increase interest and retention among science students, especially for groups that otherwise have tended to be underrepresented. Indeed, "a number of potential university science students...lament that science classes ignore the historical, philosophical, and sociological foundations of science" (William F. McComas, in McComas 1998:13; also see Tobias 1990).

Through major funding from the NSF, Becker (2000) produced a sequence of "history-based" science lessons particularly to reach "those students identified as traditionally alienated from the world of science and technology." She reported that "Students who took the course found it satisfying, diverse, historical, philosophical, humanitarian, and social. They...found the historical approach to be interesting and the text enjoyable to read." The material engaged "learners of varying abilities" from a "socio-economically, ethnically, and linguistically diverse student population."

Another example is the Harvard Project Physics course, which incorporates some history and philosophy of science. Over 60 studies of the effectiveness of that course have been published, all positive: "Measures such as retention [of students] in science, participation by women, improvement on critical thinking tests and understanding of subject matter all showed improvement where the Project Physics curriculum was adopted" (Matthews 1994:6).

Mauldin and Lonney (1999) studied the NOS content in a course that fulfilled a general education requirement for students who were not science majors. It explained the basic components of scientific method so that nonscientists could achieve personally relevant goals that genuinely interested them, such as accurately analyzing "scientific reports as presented in popular magazines and newspapers." Also see the study by Deeds et al. (2000).

A particularly important part of the NOS for interesting a diversity of students in science is its simplest, common-sense elements. Warren et al. (2001) contrasted perspectives on the relationship between science and common sense as being either "fundamentally discontinuous" or "fundamentally continuous" and found the latter much more conducive to learning science: "Our work ... shows that children, regardless of their national language or dialect, use their everyday language routinely and creatively to negotiate the complex dilemmas of their lives and the larger world. Likewise, in the science classroom children's questions and their familiar ways of discussing them do not lack complexity ... or precision; rather, they constitute invaluable intellectual resources which can support children as they think about and learn to explain the world around them scientifically." Culturally sensitive "everyday sensemaking" is a powerful bridge to scientific interests and patterns of thought.

Many other reforms have failed to reduce the achievement gap (Lynch 2001). Therefore, it is significant to note that including the NOS in science instruction *is* an effective measure for stimulating interest and retention in science.

Finally, although the skills involved in understanding the scientific method and the NOS provide the gateway to understanding science, at a deeper level one's interest in cultivating those skills must be promoted by a desire to seek the truth. An important study by Ben-Chaim, Ron, and Zoller (2000) explored numerous skills and dispositions related to scientific thinking. They reported that "truth-seeking is courageous intellectual honesty, a major attribute of CT [critical thinking], which, in turn, is an important component of HOCS [higher-order cognitive skills]." Likewise, Machamer (1998) emphasized the role of curiosity in motivating the study of both science itself and the NOS: "First and foremost, I believe, scientists, especially the good ones, engage in science because they are curious and have fun assuaging that curiosity through doing science. . . . Now part of the fun of science, as in most interesting human activities, lies in thinking about how and why it is done, and how it might be done better. In this way, philosophy is the discipline that studies the history and structure of inquiry, for asking critical questions that any curious and self-conscious practitioner would be asking."

(4) More Realism. Quite sensibly, awareness of science's powers and limits has been widely recognized as a key component of scientific literacy (AAAS 1989:26; 1990:20–21; NRC 1996:21). For science and technology undergraduates, the NRC (1999:34) posed six specific questions that they should be able to answer: "How are the approaches that scientists employ to view and understand the universe similar to, and different from, the approaches taken by scholars in other disciplines outside of the natural sciences? What kinds of questions can be answered by the scientific and engineering methods, and what kinds of questions lie outside of these realms of knowledge? How does one distinguish between science and pseudoscience? Why are scientists often unable to provide definitive answers to questions they investigate? What are risk and probability, and what roles do they play when one is trying to provide scientific answers to questions? What is the difference between correlation and causation?"

What science educators find is that an understanding of science's method powerfully promotes the educational goal of developing realistic ideas about science's powers and limits. This makes sense given that powers and limits are largely consequences of methods: "Understanding how science operates is imperative for evaluating the strengths and limitations of science" (William F. McComas, in McComas 1998:12; also see Gruender 2001). Also, "The ability to distinguish good science from parodies and pseudoscience depends on a grasp of the nature of science" (Matthews 2000:326; also see Keeports and Morier 1994 and Machamer 1998).

Clear methodology also promotes confidence. When considering alternative hypotheses, students who use reflective reasoning reach conclusions with firmer conviction than do those with a more intuitive approach (Lawson and Weser 1990; also see Lefevre, Escaut, and Bouldoires 1997).

(5) Better Researchers. Research competence for undergraduate and graduate science majors obviously is a major educational objective. Again, my thesis is that the winning combination for scientists is strength in both the research technicalities of a given specialty and the general principles of scientific method. Certainly, what has already been documented here regarding better comprehension, adaptability, and interest makes for better researchers.

Ryder and Leach (1999) conducted an empirical investigation of research competence among science undergraduates as a function of their understanding of the general principles of scientific method. They found that such understanding "can have a major impact on students' activities during investigative [research] projects," specifically on "students' decisions about whether to repeat an experiment, whether to question an earlier interpretation of their data as a result of new evidence, where to look for other scientific work related to their project and what counts as a conclusion to their investigation." On the other hand, inadequate or naive conceptions of scientific method "constrain student learning" and made it impossible for students to achieve some of their research objectives. More exactly, they found that mere diligence sufficed for students to function as competent technicians, collecting accurate data by means of a prescribed protocol, but real understanding of scientific method was needed for students to function as creative scientists, understanding how data and theory interact to support knowledge claims.

Interestingly, they investigated not only the students but also the students' research supervisors. For one student with a particularly naive and limiting view of scientific method, they found no evidence that the student's supervisor recognized the problem or implemented any strategy to remedy the deficiency. Professors and supervisors must realize that teaching more and more about research technicalities is not the required remedy when the real problem is lack of understanding of some basic principle of scientific inquiry. And clearly, problems of the latter sort are quite common (Lederman 1992; Meyling 1997; Cartier and Stewart 2000; Driver et al. 2000).

(6) Better Teachers. Regarding science and technology faculty, the NSF (1996:ii) advises that "It is important to assist them to learn [and teach] not only science facts, but, just as important, the methods and processes of research." Likewise, the NRC (1996:21) requires that science teachers understand the NOS, including the "modes of scientific inquiry, rules of evidence, ways of formulating questions, and ways of proposing explanations," and the NRC (1999:41) reaffirms those requirements. Summarizing the positions of the AAAS, NRC, and NSF, Lawson (1999) said that "Teaching in ways that help students understand the nature of science and how to use scientific reasoning patterns [has] long been [a central goal] of science education."

Matthews (1994:199–213; 2000:321–351) has documented that knowledge of the history, philosophy, and method of science strengthens teachers. Such knowledge "can improve teacher education by assisting teachers to develop a richer and more authentic understanding of science, . . . can assist teachers [to] appreciate the learning difficulties of students, because it alerts them to the historic difficulties of scientific development and conceptual change, . . . [and] can contribute to the clearer appraisal of many contemporary educational debates that engage science teachers and curriculum planners" (Matthews 1994:7). But unfortunately, science teachers often lack a sophisticated and consistent understanding of scientific reasoning (Lederman 1992; Abd-El-Khalick et al. 1998; Nott and Wellington 1998).

An especially important and difficult element in teacher preparation is to ensure that the professors are properly equipped to detect and correct their students' misconceptions. Students' prior beliefs form an interpretive framework for receiving new material, and mistaken notions can be quite resistant to correction. One helpful means is to acquaint students with the history of science, for when they see that incorrect or incomplete ideas have had to be revised in light of subsequent experiments, they see better that their own ideas may require revision, and they also see exactly how data prompt conceptual changes. Becker (2000) reported that "basing science instruction on historic episodes can open up opportunities for students to identify their own untutored beliefs about the workings of the natural world, to examine them critically in the light of considered historical debate, and to confront these beliefs in a way that results in positive, long-lasting conceptual change." Another helpful resource, as mentioned earlier, is the study of logic. But teaching science's history and logic to students must begin with teaching them to teachers and professors.

Some practical dimensions of NOS instruction require greater attention: "Teachers are clearly asking for classroom resources which they can use virtually 'off the shelf' and slot into relevant places in present science courses. Of prime concern is the reading age and accessibility of the materials for the pupils" (Lakin and Wellington 1994). Another challenge is that typically the documents calling for NOS instruction are worded at a rather general level, respectfully leaving the specifics to others (Shiland 1998). That means that individual instructors must really master the NOS, or at least have access to excellent teaching materials.

A longer-term benefit from clear understanding of the NOS is better sorting of educational theories and debates. Indeed, "educational researchers attempting to use and evaluate innovations would benefit from deeper knowledge of their philosophical and psychological origins," because "those best prepared with an understanding of the origin and basis of philosophical positions on the nature of science are better able to fend off impostors, quick fixes, or poorly grounded innovations" (Loving 1997). Trendy educational fads can lead to disaster, especially for disadvantaged students and minorities (Matthews 2000:340).

Increasingly, many nations and states are mandating that science educators take courses in the NOS (Matthews 2000:321–351). There is also a trend toward mandating that assessments of scientific literacy include questions about the NOS (Norman G. Lederman et al., in McComas 1998:331–350). But such mandates must be implemented carefully: "I have become more convinced . . . that improving evaluation of science instruction is probably the most important part of the equation for improving science education. But, if evaluative processes and materials are not built upon the best knowledge about the nature of science and applied appropriately for the developmental level of students, through school and university, they will fall short of what is needed" (Robinson 1998). In other words, proper knowledge of the NOS is crucial for the government and university administrators and accreditation officials who set the evaluation criteria for science education and research.

Finally, demonstration of benefit will motivate teachers. As mentioned earlier, national standards for science education, mandated assessments of scientific literacy, and other factors are sure to increase the attention given to the NOS. But for science educators, true motivation, as contrasted with perfunctory compliance, arises from known benefits. Indeed, "reports indicating positive changes in students' views and actions regarding the nature of science are needed to bolster teachers' confidence that attention to these issues will reap the desired effects" (William F. McComas et al., in McComas 1998:29). Therefore, to paraphrase that last point's first sentence, it is important to assist science faculty to learn not only the methods and processes of science but also the benefits that result from that learning.

THE GOOD, THE BAD, AND THE UGLY

The science education community is distinguished by its open-mindedness to diverse ideas and vigorous self-criticism. For instance, one can but admire the candor of educators as represented by the article "The Good, the Bad, and the Ugly: The Many Faces of Constructivism" by Phillips (1995). Drawing on that article by Phillips and the work of other educators, as well as reactions from scientists and parents, this section will examine some good aspects of constructivism, then bad aspects, and finally ugly aspects.

Constructivism is the dominant paradigm in science education today. It has also affected scientists themselves. For instance, the *National Science Education Standards* (NRC 1996) document is tremendously influential. The 1992 draft of *Standards* endorsed "A more contemporary approach, often called postmodernism [which] questions the objectivity of observation and the truth of scientific knowledge" and depicted science as "a mental representation constructed by the individual" (Matthews 2000:327). However, that endorsement of postmodern and constuctivist ideas drew much criticism (Gerald Holton, in Gross et al. 1996:551–560; Good and Shymansky 2001), so the appendix containing that material was removed from the 1994 draft and the following 1996 document. But Matthews (2000:327) has shown that "its constructivist content was not rejected, merely relocated."

As a concise introduction to constructivism, the main variants of that paradigm have three key principles (David F. Treagust et al., in Treagust, Duit, and Fraser 1996:1–14). The first principle, from which constructivism draws its name, states that learners individually construct knowledge and meaning from their own experiences. An important implication of that principle is that what the learner already knows or believes will greatly influence how new information will be received and interpreted (or misinterpreted). Also, that principle challenges much ineffective traditional or "transmission" education and rote learning. The second principle states that the function of cognition is adaptation to life's circumstances, constructing viable explanations for our experiences. An extension of that principle, found in some strains of constructivism, says that that is the only function of cognition, hence denying that objective truth about an external physical reality is a legitimate goal or claim of science. The third and final principle states that knowledge is not only personally constructed but also socially mediated, especially in the teacher-student relationship. Hence, constructivism recognizes both individual and social aspects of knowledge construction.

But having cited three core elements of constructivism, it must also be said that constructivism is not a unified movement, but rather an extremely heterogeneous school with diverse historical influences and educational manifestations. Phillips (1995) described three main areas of divergence among constructivists: individual versus social aspects of learning, large versus small roles for nature in constraining human knowledge, and philosophical versus sociological explanations of rationality. Antonio Bettencourt (in Tobin 1993:39–50), Matthews (1994:137–161), and Nola (1997) have described numerous variants. Diverse views on constructivism can be found in the volumes edited by Tobin (1993), Larochelle, Bednarz, and Garrison (1998), Matthews (1998a), and Mintzes, Wandersee, and Novak (1998) and in the November 2000 special issue of *Science & Education*. Also see the February 1998 special issue of the *Journal of Research in Science Teaching* on the epistemological and ontological underpinnings of science education, the

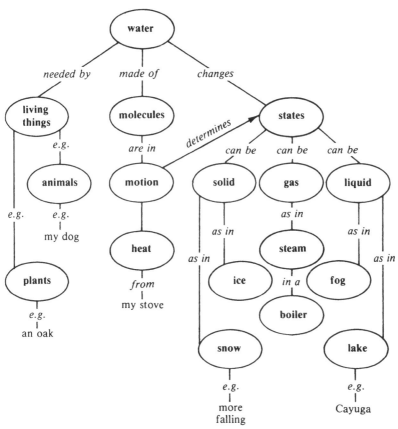

Figure 11.2. Concept map for water. A concept map shows related concepts and propositions, as well as specific examples of events and objects. (Reproduced from Novak and Gowin, 1984:16, with kind permissions from Joseph D. Novak and Cambridge University Press.)

July 1997 and November 1998 special issues of *Science & Education* on the NOS and science education, and the concise overview for educators of the philosophy of science by Machamer (1998).

Turning first to the good in constructivism, scientists can welcome great strides in pedagogy. The traditional education that preceded constructivism had some serious deficiencies, and constructivist educators have developed many exciting and effective tools for teachers and learners. One example is the concept map, which helps students articulate complex ideas (Novak and Gowin 1984; Joseph D. Novak, in Treagust et al. 1996:32–43). Figure 11.2 shows a concept map for water.

Much scientific knowledge involves relationships among things or events, rather than entities in isolation. Concept maps are ideal for displaying such related concepts. Not only can they convey correct information, but also concept maps drawn by students can reveal their misconceptions, which is the essential prelude to correcting misconceptions. Concept maps can also be helpful research tools, sometimes prompting recognition of new relationships or identification of knowledge gaps.

On balance, however, it must be said that much of the fine pedagogy commonly attributed to constructivists is far older and even commonplace: "A great deal of constructivism is laudable – the emphasis on teaching for understanding, on student engagement in learning, on cooperative learning, on being cognizant of the ideas and prior understanding of students, on developing epistemological sophistication, and so forth. However all of these laudable aspects are really ancient educational verities – most can be found in Socrates' dialogues" (Matthews 2000:343).

A welcome direction in the constructivist education literature, especially from the perspective of this book on scientific method, is the emphasis on the NOS and scientific method. That theme often goes under the slogan "Less is more." For instance, in a document published by the U.S. Department of Education, Anderson et al. (1994:2–3, 32) observed that "Some information is more important than other information in developing . . . sophisticated understanding of science and mathematics. It is not just a matter of learning more, it is a matter of learning that which will help build the desired overall conceptual picture. . . . This greater selective attention to the most important conceptual understandings is the foundation of effective learning, thus the idea that 'Less is more.' . . . In constructivist teaching and learning, more emphasis is placed on learning-how-to-learn than on an accumulation of facts. . . . It is most important to give students an understanding of how science works." Certainly the U.S. national standards for science education also emphasize "An appreciation of 'how we know' what we know in science" (NRC 1996:105).

Turning next to the bad in constructivism, from time to time and in various places the constructivist ideas and practices that emerge tend to undermine science or pedagogy or both. Indeed, rather than emphasizing the NOS as just mentioned, some strains of constructivism have the opposite tendency of marginalizing the NOS. Constructivism has had a "mixed impact" on NOS instruction, partly positive, "But more commonly, where constructivism is embraced, teacher education programs place more emphasis on the 'child as sense-maker,' the classroom as a 'knowledge construction zone,' and 'empowerment' of the teacher so [as to] become a 'facilitator' of children's learning. The subject matter being taught, its conceptual structure, and its history and philosophy, are marginalised" (Matthews 2000:337–338). Those problems mainly concern only one of the many faces of constructivism, radical constructivism.

Ernst von Glasersfeld is a particularly outspoken proponent of radical constructivism. He has expressed his views concisely in the volume edited by

Larochelle et al. (1998:23–28). He despises what he calls "trivial" construc-tivism, which goes only so far as to agree with tame theses like "It's obvious, after all, the children don't simply swallow all adult knowledge whole, they have to construct it!" His primordial ontological claim, from which every-thing else flows, is that humans are so constituted as to have no access to "an ontological reality that lies beyond all knowing." Consequently, we should back away from the traditional pursuit of truth about an objective and exter-nal reality and instead settle for viable ideas about "the lived, tangible reality of our experience."

In that same volume, another radical constructivist claims similarly that "the world ... has no preestablished form and hence does not admit of direct perception or knowledge.... The visible world does not exist as such but assumes a form when it is constructed by the eye. Or again, the audible world does not exist as such but is the form the world takes on when it is known by the ear" (Yvon Pépin, in Larochelle et al. 1998:175).

In a review of that volume in *Science & Education* (2000:641–642), how-ever, another educator lost patience with those sentiments: "Well this is heady idealist stuff, and takes us right back to Bishop Berkeley, but is it sensible stuff? Do we, and especially science teachers, wish to say that continental structures, atomic structure, DNA structure, evolutionary processes, and so on, came into existence in virtue of human observation?"

Lest there be any doubt about the extent of constructivist skepticism, Ernst von Glasersfeld (in Tobin 1993:26) said that "To conclude that, because we have a perceptual experience which we call 'chair,' there must be a chair in the 'real' world is to commit the realist fallacy. We have no way of knowing what is or could be beyond our experiential interface. If we can reliably repeat the chair experience, we can only conclude that, under the circumstance, it is a viable construct." But a philosopher, Nola (1997), was impatient with such skepticism, replying that "In claiming that ordinary objects are merely our viable constructions, von Glasersfeld becomes as radical a philosophical idealist as Bishop Berkeley." What von Glasersfeld calls the "realist fallacy" others have called the "correspondence theory of truth." Anyway, an eighth-grade student quoted by Hogan (1999) seems much more sensible, saying that "in science we're studying things that are there – we didn't make a frog just by looking at it."

As Matthews (1998b) has argued, "constructivism is at its core ... an epis-temological doctrine; and it is standardly coupled with commitments to certain postpositivist, postmodernist, antirealist, and instrumentalist views about the nature of science." For instance, a constructivist educator, Meyling (1997), mentioned a school student who began with a common-sensical, re-alist view of science, but in the end she accepted the teacher's "fallibilistic-pluralistic model of epistemology." He quoted her saying that "Truth is relative, we have to get used to that, there are only things that are more correct than others, but there is nothing that is absolutely correct. . . . When you think you know the truth, you force others to think and live that way. . . . This is a claim on absoluteness that cannot be justified – by no one and by no theory." He commented that "I believe that this recognition is far more important than the knowledge about a whole set of scientific 'facts'," and he was particularly pleased that the student extended her new skeptical epistemology to the "ethical level." He then added that "Sir Karl R. Popper was very pleased with this quote" in a letter written to the author (Meyling 1997). Well, Sir Karl's praise notwithstanding, comparisons and criticisms of both one's own and others' philosophical orientations may advance the education of college and university students, but mainstream views are more appropriate for school students.

Certainly, to found an educational scheme on philosophical views that have won a small minority of philosophers from antiquity to the present seems idiosyncratic. And certainly such views are outside the scientific mainstream, as represented by the directly opposing view from the AAAS (1989:25; 1990:16) that humans do find nature observable and comprehensible, as quoted earlier in Chapter 4. So all this radical constructivism with its radical skepticism is not a successful move for anyone wishing to exhibit philosophical respectability and sophistication. Rather, it is an unfortunate move that attempts to dislocate education outside the mainstream in philosophy and outside the mainstream in science.

The lengths to which constructivist ideology can been taken are disturbing. For example, Cromer (1997:11) reported that "New Zealand is so committed to nonobjectivity in science teaching that the lecture-demonstration tables have been removed from all the science classrooms in the country. . . . This is to prevent teachers from claiming to know more than their students, thus unduly influencing how the students construct their own knowledge." But is that not carrying ideology too far? Surely the motive for having lecture tables and demonstrations is not to let teachers coerce students' beliefs about nature. Rather, the purpose of those tables and the appropriate activities occurring upon them is to enable chemicals and rocks and frogs to coerce students' and teachers' beliefs alike! The tension between teachers' and students' perspectives should be resolved by noticing that both lose to a real and objective nature that holds us all on a short leash.

Yager (1991), an enthusiastic constructivist, concluded that once their thinking had taken hold, all would see that "It is no longer possible to cling to the notion that a given problem has only one solution. It is also difficult to justify conceptions of right and wrong answers."

But then what about a typical little science problem, such as this: A baseball is dropped from a bridge and hits the ground in 1.5 seconds.

Approximately how high is the bridge? Are there really no right or wrong answers? If one student gets the wrong answer, how does it benefit that student to say that his answer is just fine too, as if he had some special excuse for not dealing with reality, whereas if another student bothers to get the right answer, how does she benefit from seeing that all answers receive the same approval? Science cannot really be for truth without also being against error, just as one cannot really be for freedom without also being against slavery. This is not intolerance, but rather sincerity. Reality need not be considered a threat to children's dignity!

Surely, creative teachers can affirm children while they also reject errors: "The subject matter and the tradition of scientific investigation need to be respected, and so do students. However, respect for students is consistent with telling them they are wrong, that they have misunderstood the meaning of certain key concepts, and that they have to work and think harder to master certain subject matters. A capable teacher can do all of this in a way that is not disrespectful, that is consistent with [a] student's intellectual attainment and capacity, and is not counterproductive to learning. What is less than respectful is to occupy students with tasks that, even if good fun, contribute little to intellectual growth" (Matthews 2000:344).

Certainly much of the public has the perception that feel-good psychology and soft thinking all too often go to school right along with the students. Figure 11.3 shows a cartoon that appeared in the 2 November 1998 issue of *U.S. News & World Report*. It is an exaggeration, but its point is that relativistic philosophy and feel-good psychology can reach such proportions as to subvert and displace subject-matter instruction. It reflects the ongoing negotiation between the public and educators over the philosophy and psychology that go to school with the students.

In contrast, Hoffmann et al. (1996) have described the real situation in a single sentence: "If we had to operate under an equal opportunity clause for every concept that was ever espoused, we would have such an impossibly complex and self-contradictory description of Nature, that we could never feel that we were making progress in understanding or utilizing our environment." Science education must operate in an environment in which all students are valuable, but some ideas are right, and others wrong. Anything else disadvantages everyone, because all people must deal with physical reality.

Incidentally, scientists need to be warned that constructivists sometimes bend common words, such as "knowledge," to serve their purposes, giving them meanings other than those that are customary among scientists. For instance, Golinski (1998:7) claims that "Constructivism was inaugurated by a determination to explain the formation of natural knowledge without engaging in assessment of its truth or validity.... It accepts as 'science' for

Figure 11.3. Educational fads and feel-good psychology can displace proper education. (This drawing by Tom Toles is reproduced with kind permission from the Universal Press Syndicate. TOLES © 1998 *The Buffalo News*, all rights reserved.)

the purpose of study what has passed for such in the context under discussion.... It is best understood as a pragmatic or methodological deployment of 'relativism' in the service of a comprehensive study of human knowledge." But that "knowledge," isolated from rationality and truth, is neither the justified true belief of Aristotle nor the knowledge that is spoken of in scientific and technological discourse. Likewise, the distance between constructivist "truth" and scientific truth, with and without scare quotes, is as vast as intergalactic distances.

Turning finally to the ugly in constructivism, something as important as children's education should not be treated too abstractly. Accordingly, this analysis of science education moves along to the reflections of concerned parents.

A review of Cromer's book (1997), by Douglas R. O. Morrison in *Scientific American* [1997, 277(5):114–118], begins with some parental concerns about the currently dominant paradigm in science education, constructivism: "I began to wonder some years ago why my children were learning science in such a crazy fashion. Teachers told them to do lab experiments but gave them no textbooks or notes to explain why they were doing those experiments or what they meant – evidently, the students were supposed to work it all out for themselves. At a P.T.A. [Parents' and Teachers' Association] meeting, I protested and was told that this was the new fashion in education. None of the other parents, I was informed, had made any complaint, except the ones who were scientists. . . . But it is becoming increasingly common knowledge that a harmful vision of science has been steadily taking over education in schools and universities. . . . This interpretation would imply that science is a subjective human construction, like art or music."

If what Morrison reported occurred in just one science classroom in all of America, that would be unfortunate. The message of the book by Cromer (1997) that Morrison was reviewing was that constructivism had become the dominant paradigm in science education and was taught to millions of children every year. Cromer (1997:10) did not mince words: "*Constructivism* is a postmodern antiscience philosophy." Cromer's historical perspective and sharp criticisms deserve careful attention from anyone involved in science education. Likewise, Matthews (1994:9) has written an important book that counters constructivism with "the three Rs – Reason, Realism and Rationality." Also see Matthews (1998a) and the review of that book by Irzik (2001). Morrison's observations illustrate how students privileged to have scientist parents often can compensate for constructivist excesses at school. However, Noretta Koertge (in Koertge 1998:257–271) has argued that many constructivist programs are counterproductive, especially disadvantaging women and minorities.

In fairness to educators, it must be emphasized that not all constructivists follow the ideas and agenda that Cromer and Morrison found so disturbing. But in fairness to students and parents, it must be said that many constructivists do: "For instance, a former New Zealand colleague marked a group of 35 constructivist-based Bachelor of Education students' practice teaching lessons on goldfish. Of the 35 lesson plans, only two were concerned with knowledge outcomes in their classes. The rest were all pleasant enough activities – imagine you are a goldfish, make a goldfish collage, brainstorm goldfish – but not the things for which parents primarily send their children to school, or for which governments fund education. Only two of the lessons were concerned to inform pupils about what food goldfish ate, what temperature water they survive in, and whether they are fresh or saltwater fish" (Matthews 2000:379–380).

The current situation places new demands on science educators: "Previously the science teacher had to master his or her subject matter and the techniques of making it interesting and intelligible to students. . . . Good teachers tried to master something of the orthodox history and philosophy of their subject. All of this was demanding enough. But now, in addition, teachers need to understand and evaluate the postmodern challenges to orthodoxy. This is a hard call, but it cannot be avoided. University students, and increasingly school students as well, encounter challenges to the legitimacy and objectivity of science.... Science teachers should be able to say something informed and intelligent about these challenges. At least they should be able to say what the strong, weak and disputed points are, and how they might bear upon the curriculum topics" (Matthews 2000:325). "To do otherwise is professionally irresponsible" (Matthews 1998b).

From the perspective of scientists observing the education community, it is perplexing indeed that constructivism motivates some educators to give students lots of hands-on experience and motivates other educators to remove lecture tables and demonstration materials. Constructivism moves some science educators to emphasize scientific method, and moves others to deny that such exists. Constructivism prompts some educators to value truth, and others to abandon truth. Perhaps this paradigm has been so successful, with nearly all educators wearing the name "constructivism," that this label is no longer informative. Anyway, as long as valuable constructivist pedagogy is associated with unfortunate philosophical commitments, constructivism will continue to draw fire from scandalized scientists and concerned parents.

"Education is notoriously subject to fashions and fads, especially ones imported from psychology and philosophy. They usually pass if they do not sustain the core educational business of teaching, learning and, hopefully, growing in wisdom. It is perhaps too early to judge constructivism, but there are clear signs that it too will eventually be seen as a passing fad – at least the excesses of the position will be so seen" (Matthews 2000:343).

Perhaps the most important point is that the current situation for school administrators and university professors is nevertheless extremely positive. They need not accept materials with questionable philosophical content that subverts science just because those materials also feature attractive and effective pedagogy. Rather, it is a big wide world out there, with many offerings. There is plenty of opportunity for administrators and professors to choose curriculum materials that combine substantial science, sensible philosophy, and first-rate pedagogy. That is the choice that can best serve the students' interests.

CONSTRUCTIVISM IN THE THIRD WORLD

This section examines constructivist science education in the Third World. It is important if for no other reason than that the Third World is where most of the world's people live. The positive pedagogical aspects of constructivism, as discussed earlier, can be enormously beneficial, but the concern here is with harmful philosophical alignments. A penetrating and poignant analysis has been offered by Meera Nanda, a native of India (Koertge 1998:286–311). She provides detailed documentation, especially on India, Pakistan, and China, but only an overview can be included here.

She feels that constructivists, primarily in the United States and England, have promulgated their view of science in the Third World with great energy and good intent. Nevertheless, she wants to return the gift: "The gift that I want to return is the cluster of theories that forbids outsiders from evaluating the truth or falsity of any beliefs of other people in other cultures from the vantage point of what is scientifically known about the world and, conversely, allows the insiders to reject as ethnocentric and imperialistic any truth claim that does not use locally accepted metaphysical categories and rules of justification.... This gift has many names, many givers, and many presumed beneficiaries. It is variously called *ethnoscience*, situated knowledge, anti-Northern Eurocentric, or *postcolonial* science – labels that derive their force from their parental rubric of social constructivist theories of science.... Social constructivist theories hold as their first principle that the standards of evaluation of truth, rationality, success, and progressiveness are relative to a culture's assumptions and that the ways of seeing further vary with gender, class, race, and caste in any given culture" (pp. 286–287).

Although Meera Nanda refuses the gift of constructivism or ethnoscience, she explains its appeal to many Third World intellectuals: "Such relativization has some 'liberating' effects for both the sponsors and the intended beneficiaries of ethnosciences. To begin with, it frees both of them from ever having to say that others, or they themselves, though acting rationally by local standards, could be holding false beliefs and that these beliefs can or should be corrected" (p. 287). The experience of colonial exploitation and disrespect has prompted anti-Western sentiments that extend to Western science.

Regarding that appealing gift of constructivism and ethnoscience to Third World intellectuals, she then asks "Why should anyone want to refuse so generous a gift, least of all someone like myself whose own native Indian culture was berated for so long by the British rulers as irrational, mystical, and superstitious? How can anyone urge ex-colonial people to refuse this poultice of relativism when they are still so obviously smarting from the indignities of colonialism?" (p. 288).

"My reason for urging a rejection of ethnoscience is this: What from the perspective of Western liberal givers looks like a tolerant, nonjudgmental, therapeutic 'permission to be different' appears to some of us 'others' as a condescending act of charity. This epistemic charity dehumanizes us by denying us the capacity for a reasoned modification of our beliefs in the light of better evidence made available by the methods of modern science. This kind of charity, moreover, enjoins us to stop struggling against the limits that our cultural heritage imposes on our knowledge and our freedoms.... By defining the very nature of rationality and truth as internal to social practices, social constructivist theories do indeed give the natives their 'permission' to be different – but, then, so did apartheid" (pp. 288–289).

"So the reason that we – the dogged advocates of scientific temper in non-Western societies – must reject the privilege of having our traditional knowledge considered at par with science is clear: the project of different and equal sciences for different people completely negates our project of science for all people. We prefer our much-maligned universalistic project because we are not interested in a supposed cognitive equality of different cultures but, rather, in substantive equality for all people in terms of healthier, fuller, and freer lives. We prefer the cold, objective facts of science to the comfortable, situated knowledge of our ancestors for the simple reason that we refuse to subordinate what is good to what is ours.... Insofar as cultures and traditions have a cognitive component – that is, to the extent they justify themselves on the basis of some theory of the nature of things – they will remain open to competition from other explanatory theories in the same domain" (pp. 299, 302).

"The constructivist giveth and the constructivist taketh away. The Strong Programme (SP) endorses the rational unity of humankind but denies as 'myth' the idea that there could be a universality of knowledge; that is, it denies that there could be facts about the natural world that knowers in different times and places could recognize as more rational to hold on the grounds that they are based on better evidence, are better approximations of the truth, or have withstood serious attempts to refute them. The constructivists, then, confer on the 'Others' the ability to think, but then take away the ability to choose, on occasion, the knowledge of aliens over the knowledge of ancestors" (p. 295).

That call from the Third World that they not be marginalized as tribal "others" communicates a wisdom that can benefit those in the First World. The heart of the matter is expressed with compelling simplicity by Simon, as quoted by Matthews (2000:339): For science educators "to start from the standpoint of individual differences is to start from the wrong position. To develop effective pedagogy means starting from the standpoint of what children have in common as members of the human species; to establish the general principles of teaching and, in the light of them, to determine what modifications of practice are necessary to meet specific individual needs."

Gálvez et al. (2000) have explained some of the hardships that hinder Third World science educators, and Matthews (2000:346) has offered a word of perspective and encouragement to those who teach science in hard circumstances: "Learning and education have flourished in circumstances blighted by desperate poverty," as his several specific examples illustrate. As for wealthy nations, however, "the Western educational crisis is deeper than funds and equipment." Modest resources accompanied by sensible philosophy, effective pedagogy, and a love of learning often can outperform lavish resources burdened with crazy philosophy, trendy pedagogy, and a love of entertainment. Poor philosophy is just as detrimental as poor microscopes.

Inevitably, all manner of ideas will flow to Third World nations from the outside world. It is to be hoped that as Third World scientists and scholars decide which of those intellectual gifts to accept and which to return, their motives will dictate that precedence be given to tangible benefits, rather than philosophical games and ideological trinkets.

A MODEST EXPERIMENT

Given the shifting sands of educational philosophies and fads, as well as disagreements about the NOS and scientific method, a bit of back-to-basics thinking can help us to get our bearings. This brief section constitutes my reply to the unrealistic and foolish ideas that have gripped some educators, and as such it may seem simple and even foolish. But as Machamer (1998) has noted in a different context, "The simplest answer is also the best." There is no remedy for fundamental errors other than basic truths. Accordingly, a modest experiment is proposed here.

Envision a science class, at the high-school or college level, performing some routine experiment regarding plant photosynthesis, frog physiology, or whatever. For the sake of concreteness, envision a teacher with a dozen students who are using an electric current to separate water into hydrogen and oxygen (and perhaps are subsequently combining the gases to reconstitute water). Thereby, they replicate some of the familiar evidence showing that water is composed of hydrogen and oxygen.

Suppose that this simple experiment is followed immediately by a discussion of the NOS, specifically the reliability of scientific findings. As a point of departure, the teacher presents official views from international standards for science education (William F. McComas and Joanne K. Olson, in McComas 1998:41–52). Those documents reflect great unanimity that science is tentative. For instance, "Scientific knowledge is not absolute; rather, it is tentative, approximate, and subject to revision," and "Our scientific understanding of the world is continually changing, sometimes incrementally, sometimes in gigantic, revolutionary bounds." The lack of qualifiers in those statements causes the implicit domain to be all scientific knowledge. Yet other documents are quite explicit: "But science *never* commits itself irrevocably to any fact or theory, no matter how firmly it appears to be established in the light of what is known," and "Because all scientific ideas depend on experimental and observational confirmation, *all* scientific knowledge is, in principle, subject to change as new evidence becomes available" (italics added).

Obviously, if all scientific knowledge is tentative and revisable, then necessarily that includes the finding that water is composed of hydrogen and oxygen. Is that finding really insecure, tentative, and revisable? Might tomorrow's scientists instead believe that water is composed of carbon and aluminum, or perhaps much more excitedly, show that it is composed of silver and gold? And what else is up for grabs next? Does the earth orbit around Neptune instead of the sun? Do humans not need vitamin A for good nutrition and health?

That clash between simple experiments and science standards might cause some perplexity. Some students, and even some teachers, might feel uncertain that science can deliver at least some certain and final truth. If a unanimous verdict about the water experiment is not forthcoming, then all educators know what to do: Try something easier.

Accordingly, one might propose a more modest experiment: The teacher and the dozen students stand in a circle, with everyone in plain view for everyone. The teacher displays a cookie jar containing several cookies, somewhat fewer than the number of students. The teacher passes the jar, with the instruction to take one cookie (until they run out) and to continue passing the jar around until it returns to the teacher, who then displays the empty cookie jar. Thereupon, the teacher raises the question "Who took the last cookie?"

Can the class handle this? If not, then these high-school or college students need to visit a kindergarten class to see that human beings are capable of performing this modest inquiry! But if so, then these students should be invited to reflect on what it is about the physical world and human beings that enables them to know who took the last cookie. With that small success under their belts, they can then discuss whether or not there is something fundamentally or impossibly harder about the water experiment.

The lesson here from comparing the water and cookie experiments is that philosophically, historically, psychologically, and pedagogically there is an inexorable connection between science and common sense. Therefore, the claim that "all scientific knowledge is tentative and uncertain" is unsupportable apart from the broader and bolder claim that "all knowledge is tentative and uncertain," which is simply skepticism. It is not some trendy or sophisticated educational philosophy; it is simply skepticism.

Clearly, the position papers from major scientific organizations strive for clarity and balance regarding science's reach, but many educators are concerned that those goals have not been achieved. A survey of eight international science documents listed 38 points on which there was wide consensus, the first two being that "Scientific knowledge is stable" and "Scientific knowledge is tentative" (William F. McComas and Joanne K. Olson, in McComas 1998:41–52). Similarly, another survey identified 14 points of consensus, of which the first was that "Scientific knowledge while durable, has a tentative character" (William F. McComas et al., in McComas 1998:3–39).

Many educators are concerned that such polarized statements can encourage selective and extreme readings, or simply confused readings, rather than balanced views. For instance, "To say that 'scientific ideas are tentative and open to change,' a statement that accurately portrays one facet of the nature of science, fails to consider the fact that 'most scientific ideas are not likely to change greatly in the future'" (Good and Shymansky 2001; also see Loving 1997).

Perhaps it might be suggested that a single balanced statement could express science's complexity better than several polar statements. For instance, one could say that "Some scientific knowledge and findings are certainly true, some are probably true, and some are rather speculative, and usually scientists have clear guidelines and substantial consensus regarding which category fits a given knowledge claim." Of course, that same balanced verdict applies equally to common sense, law, and any other kind of human inquiry aimed at finding the truth. So there is nothing wildly peculiar or desperately precarious here.

Perhaps it might also be suggested that any common-sense, unequivocal statement about the aim and reach of science simply *must* use the big t-word. For instance, many documents acknowledge that science is "tentative," but then seek counterpoint by adding that it is also "durable." However, what does the latter mean, and how does it rescue science? If a knowledge claim is true, then being durable is praiseworthy; but if it is false, then being persistent is blameworthy. Therefore, durable knowledge imparts no value to science apart from its being true knowledge. So science should take the high road and talk about truth, rather than expecting weasel words to do the job of communicating science's value.

FUTURE PROSPECTS

The ground already gained by science educators positions their community for exciting advances in the near future. Four possible directions merit special mention.

(1) Broader Vision for the NOS. The survey of eight science standards by William F. McComas and Joanne K. Olson (in McComas 1998:41–52) encompassed only documents in English from Western cultures. But their paper concluded with a vision for a broader survey: "As in any ongoing project, the conclusions presented here will be strengthened or contradicted only by examining additional standards documents and add[ing] their NOS elements to the existing data set. We look forward to including a review of science education standards in languages other than English and particularly from non-Western cultures. Not only would such an approach permit us to make firmer generalizations, but if as we suspect, our conclusions are upheld, we would be able to present evidence that a single scientific tradition unites humankind."

I might offer hopeful speculations about such a broader survey. Science standards from Western cultures often exhibit a tiresome rhetoric of polarized and even skeptical statements that seem to be overreactions to overly zealous positions from the past. But other cultures with a shorter history of scientific inquiry may be freer to lead the way in producing refreshingly balanced standards. Also, diversity and equity issues have vexed Western scholars and educators, but collegial interactions with scholars from other cultures can promote authentic solutions. A notable contribution is that from Meera Nanda (in Koertge 1998:286–311).

(2) Acculturating Students to Science. The AAAS (1989:111) combines historical accuracy with warm inclusiveness in saying that Western science "happened to develop in Europe during the last 500 years" and that it is now "a tradition to which people from all cultures contribute." The people who teach science, however, are the first to recognize that Dewey (1916:189) was right when he observed long ago that science is a learned and acquired art, not a spontaneous and native one. Learners from various cultures bring to science diverse assets and liabilities.

Recently an intriguing strategy has been developed to help acculturate learners to science. It deliberately and deeply integrates science and the NOS, reflecting an explicit understanding that both need to be communicated. For instance, a study unit can cover both "important understandings about the mechanisms and basic requirements of plant growth" *and* "the nature [of] and relationship between evidence and explanation," and "This successful balancing act results in acculturating students to mainstream science, while at the same time incorporating the unique perspectives of their prior knowledge and interest" (Loving 1997). That strategy also sits comfortably with the AAAS (1989, 1990) position that the NOS should be an integral component in science education, not an optional add-on.

(3) Effective Pedagogy for NOS Instruction. Many of the articles and books cited in this chapter offer good ideas for teaching the NOS effectively, especially those by Matthews (1994, 2000), McComas (1998), and Bevilacqua et al. (2001). One suggestion, however, is particularly intriguing: Marbach-Ad and Sokolove (2000) collected students' questions in an introductory biology course for biology majors and then organized them into several categories, beginning with low-level factual questions and progressing to high-level questions, including questions about the NOS and scientific method, especially hypothesis testing. That taxonomy of questions was introduced to the students partway through the semester. Comparing the students' questions before and after that taxonomy was presented, they found a highly significant increase in the proportion of high-level questions: "We believe that good science begins with good questions, and one of our objectives is to encourage students to recognize 'good' questions. ... However, to help students evaluate questions, we need to provide them with appropriate criteria together with examples of different types of questions. Only then can they begin to recognize which are high-level and which are low-level questions."

Similarly, Machamer (1998) suggested that "From a pedagogical point of view... asking students to reflect upon their activities when engaging in science, or studying science, is a way to enable them to understand themselves and their motivations more clearly. Having them ask – at whatever level – many of the questions that philosophers of science ask, actively engages them in the process of inquiry and challenges them to increase understanding of what they are doing."

More research into stimulating good questions would be worthwhile. It does seem, however, that effective NOS instruction must begin with helping students to discern and ask high-level questions. Although Marbach-Ad and Sokolove (2000) divided their questions into eight categories, perhaps a simpler taxonomy might suffice. For instance, one might distinguish low-level factual questions about the observational and experimental data, mid-level questions about interpretation of the data, and high-level methodological questions about the methods used in science to discover, predict, explain, test, and confirm.

(4) Education of Science Professionals. The literature in science education, with hundreds of books and dozens of journals, is simply enormous. Matthews (1994) and Bevilacqua et al. (2001) have provided magisterial overviews of teaching science with historical and philosophical insight, and Bell et al. (2001) have provided a recent bibliography.

It must be observed, however, that most of that literature addresses K–12 science education, with less attention to undergraduate and graduate education and still less to continuing education for science professionals. Notable exceptions include documents by the AAAS (1990), NSF (1996), and NRC (1999) at the undergraduate level and the NAS (1995) at the graduate level, although those are broadly conceived documents with limited attention to the NOS.

The importance of science literacy is widely appreciated. But Moore (1998), then president of the scientific research society Sigma Xi, asked the important question "But what kind of scientific literacy is important?" His reply was that "The basic concepts measured by survey questions may have little to do with the kinds of knowledge that are needed for competitiveness in the technological world," so it would be regrettable if "Basic literacy and numeracy take precedence over more sophisticated scientific understanding in the priorities of education." In other words, the very real value of basic scientific literacy for "simple good citizenship in a technological world" should not be allowed to supersede or dwarf the equally real value of advanced scientific literacy for science professionals. Neither elementary nor advanced education should be neglected.

Accordingly, a promising direction for further research is to explore NOS instruction for persons who already are, or who soon will become, science professionals. The education community could provide the scientific community with rigorous answers to several vital questions: What do practicing scientists need to know about the NOS in general, or scientific method in particular? Are the most critical elements of the NOS somewhat different for the physical, biological, and social sciences? When scientists with and without NOS instruction are compared, what differences emerge regarding research, teaching, and administrative abilities? Which will produce more creative and more productive scientists, 21 specialized courses or 20 specialized courses plus 1 course on the general principles of scientific method? Would a better balance between specialized and general knowledge result from 5% to 7% of the science curriculum being devoted to science's method, logic, philosophy, history, and ethics, as contrasted with the current practice of probably less than 1%?

Finally, this section about science education's future prospects will conclude with an expression of science education at its very best (Matthews 1994:1–3, 6). By "liberal," Matthews simply means encompassing the humanities:

At its most general level the liberal tradition in education embraces Aristotle's delineation of truth, goodness, and beauty as the ideals that people ought to cultivate in their appropriate spheres of endeavor. That is, in intellectual matters truth should be sought, in moral matters goodness, and in artistic and creative matters beauty. Education is to contribute to these ends: it is to assist the development of a person's knowledge, moral outlook and behavior, and aesthetic sensibilities and capacities. For a liberal, education is more than the preparation for work. . . . The liberal tradition seeks to overcome intellectual fragmentation. Contributors to the liberal tradition believe that science taught from such a perspective, and informed by the history and philosophy of the subject, can engender understanding of nature, the appreciation of beauty in both nature and science, and the awareness of ethical issues unveiled by scientific knowledge and created by scientific practice. . . . The liberal tradition maintains that science education should not just be an education or training *in* science, although of course it must be this, but also an education *about* science. Students educated in science should have an appreciation of scientific methods, their diversity and their limitations. They should have a feeling for methodological issues, such as how scientific theories are evaluated and how competing theories are appraised, and a sense of the interrelated role of experiment, mathematics and religious and philosophical commitment in the development of science.... The liberal approach requires a great deal from teachers; this needs to be recognized and provided for by those who educate and employ teachers.

SUMMARY

Science educators typically place scientific method within a broader context called the nature of science (NOS). From hundreds of empirical studies, they have found six benefits from properly focused, explicit instruction in the NOS: improvements in comprehension, adaptability, interest, realism, researchers, and teachers.

Currently the prevalent paradigm in science education is constructivism, which has many variants. Constructivists have achieved some excellent pedagogical reforms. However, one variant, radical constructivism, is of great concern, fundamentally because it would disregard or even deny the role that an observable and comprehensible nature has in constraining scientific beliefs, and it also would reject the idea that scientists and technologists can pursue and find truth. Radical constructivism and skepticism are outside the mainstream in philosophy and science alike. Such ideas are especially harmful in Third World countries. A modest experiment with cookies has been employed to expose its absurd excesses.

Four possible tasks for future research in science education merit special mention: a worldwide survey of science education standards for the NOS, techniques for acculturating students to science, effective pedagogy for NOS instruction, and a suitable NOS curriculum for science professionals. Regarding that fourth task, two questions merit particular attention: When scientists with and without NOS instruction are compared, what differences emerge regarding research, teaching, and administrative abilities? Which will produce more creative and more productive scientists, 21 specialized courses or 20 specialized courses plus 1 course on the general principles of scientific method?

CHAPTER TWELVE

CONCLUSIONS

In two words, the business of scientific method is *theory choice* – the choice of what to believe about the physical world. The presuppositions, evidence, and logic that comprise the scientific method have been explored in earlier chapters. Here a brief overview will draw much from the insightful and concise account of scientific method by Box et al. (1978:1–15).

Figure 12.1 shows the basic elements of data collection and analysis in scientific research. Starting at the top of this figure, the object under study is some part of the natural world. The scientist's objective is to find the truth about this physical thing, as emphasized in Chapter 2. The currently favored hypothesis H_i regarding this object will influence the design of a relevant experiment to generate empirical observations. When designing an experiment, a scientist must consider the prospective value of its data for discriminating between the competing hypotheses, together with the experiment's cost and risk. Informative experiments concentrate attention on situations for which the competing hypotheses predict different outcomes.

Naturally, different scientists may choose different research strategies: "Notice that, on this view of scientific investigation, we are not dealing with a *unique* route to problem solution. Two equally competent investigators presented with the same problem would typically begin from different starting points, proceed by different routes, and yet could reach the same answer. What is sought is not uniformity but convergence" (Box et al. 1978:5). To illustrate that point, they considered the game of "twenty questions," in which no more than twenty questions, each having only a yes or no answer, were allowed for determining what object someone was thinking of. Supposing that the object to be guessed was Abraham Lincoln's stovepipe hat, they then listed the questions and answers as the game was played by two different teams. One team asked 11 questions, and the other asked 8 mostly different questions, but both teams reached the same, right answer. A unique path to discovery is not required, so there is room for creativity, but

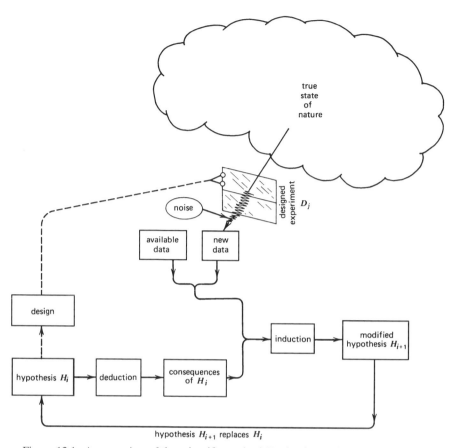

Figure 12.1. An overview of the scientific method. Deduction and induction are used to bring data to bear on theory choice. The goal is for theory choice to converge on the true state of nature. (Reproduced from Box et al., 1978:4, with kind permission from John Wiley & Sons.)

some subject-matter knowledge and strategy are needed. In science, efficient experimental designs and aggressive data analyses are the means for asking and answering questions posed to nature.

The scientist's window on the world is a designed experiment. The experiment is not perfect, however, for its data contain some noise. Good designs minimize the deleterious effects of experimental errors. Besides the new data being generated in the current experiment, usually there are additional data available from previous research. A particular strength of Bayesian analysis is its explicit formalism for combining prior and new information in inductive inferences, as elaborated in Chapter 7. Particularly because the data are noisy, parsimonious models are favored, as explained in Chapter 8. Good inductive methods will extract as much usable information from the data as possible, minimize distortions due to noise, and expose any problems that would indicate that new hypotheses and paradigms need to be considered.

Moving to the bottom left of Figure 12.1, there is a competition between two hypotheses, the currently favored H_i and the newly modified H_{i+1}. Deductive logic can be used to derive the consequences or predictions of the competing hypotheses, as explored in Chapter 5. Then inductive logic can be used to compare the various hypotheses' predictions against the data to determine which hypothesis is true or most likely to be true. In the example shown here, the evidence favors the new H_{i+1}, which then replaces H_i as the currently favored hypothesis.

Regarding the PEL model, this figure does mention experimental evidence and deductive and inductive logic, but it does not mention presuppositions. Nevertheless, presuppositions are in play and are indispensable. Presuppositions enable scientists to focus on a small roster of common-sense hypotheses, such as H_i and H_{i+1}, without hindrance from an endless roster of other wild hypotheses, as explained in Chapter 4. Presuppositions are needed to render any hypothesis sensible, but they do not render any hypothesis more (or less) credible than others, as does evidence, so scientists routinely present evidence but ignore presuppositions. But to give a scientific conclusion with full disclosure, the presuppositions must also be exhibited.

Figure 12.1 depicts a common situation with just two hypotheses, the currently favored hypothesis and a newer one. But sometimes there are multiple working hypotheses, such as H_A, H_B, and H_C (Platt 1964). Also, the hypotheses may be competing on more or less equal footing, with none decidedly favored by the scientific community. In any case, whether there are two or many hypotheses, and whether or not one is favored, the overall scheme remains basically the same. Deduction generates the predicted outcomes for each hypothesis, experiments determine actual outcomes, and induction weighs hypotheses in light of the data. But the size of the roster of hypotheses does affect how much work will be needed to rule out all the false alternatives, and a scientist's hunch about which hypothesis is true does affect which experiments will be tried first and how quickly the answer will be reached.

Sometimes scientists will win secrets from nature gradually, so over the years or centuries H_i will give way to H_{i+1}, then H_{i+2}, then H_{i+3}, and so on indefinitely. In other cases, definitive and accurate truth will be found, and with success the research can stop, leaving a settled theory. Interest then can move on to other topics that are still open. Sometimes researchers will specify a predetermined stopping rule to intervene when an experiment has generated enough data to support a definitive or adequate result. Knowing when to stop is important if the data are expensive, if research conclusions are needed to guide pressing applications, or if a scientific specialty has numerous other interesting questions that have not yet been researched.

In this figure's overview of scientific method, the key feature is that the hypotheses are confronted by data, leading to convergence on the truth about physical reality. Although the conjectured state of nature – the hypothesis or theory – may be inexact or even false, "the data themselves are generated by the true state of nature" (Box et al. 1978:5). Thereby nature constrains theory. That is how science finds truth. That is why science works.

REFERENCES

AAAS (American Association for the Advancement of Science). 1989. *Science for All Americans: A Project 2061 Report on Literacy Goals in Science, Mathematics, and Technology*. Washington, DC: AAAS.

1990. *The Liberal Art of Science: Agenda for Action*. Washington, DC: AAAS.

1993. *Benchmarks for Science Literacy*. Oxford University Press.

2000. *Designs for Science Literacy*. Oxford University Press.

Abd-El-Khalick, F., Bell, R. L., and Lederman, N. G. 1998. The nature of science and instructional practice: Making the unnatural natural. *Science Education* 82:417–436.

Achinstein, P. (ed.). 1983. *The Concept of Evidence*. Oxford University Press.

Ackoff, R. L. 1962. *Scientific Method: Optimizing Applied Research Decisions*. New York: Wiley.

Adams, M. M. 1987. *William Ockham* (2 vols.). Notre Dame, IN: University of Notre Dame Press.

Adams, P. D., Pannu, N. S., Read, R. J., and Brünger, A. T. 1997. Cross-validated maximum likelihood enhances crystallographic simulated annealing refinement. *Proceedings of the National Academy of Sciences USA* 94:5018–5023.

Adler, M. J. 1978. *Aristotle for Everybody: Difficult Thought Made Easy*. New York: Macmillan.

Alters, B. J. 1997. Whose nature of science? *Journal of Research in Science Teaching* 34:39–55.

Anderson, R. D., Anderson, B. L., Varanka-Martin, M. A., Romagnano, L., Bielenberg, J., Flory, M., Mieras, B., and Whitworth, J. 1994. *Issues of Curriculum Reform in Science, Mathematics and Higher Order Thinking across the Disciplines*. Washington, DC: U.S. Department of Education.

Annas, J., and Barnes, J. 1985. *The Modes of Scepticism: Ancient Texts and Modern Interpretations*. Cambridge University Press.

1994. *Sextus Empiricus: Outlines of Scepticism*. Cambridge University Press.

Annicchiarico, P. 1992. Cultivar adaptation and recommendation from alfalfa trials in northern Italy. *Journal of Genetics & Breeding* 46:269–278.

1997. Additive main effects and multiplicative interaction (AMMI) analysis of genotype–location interaction in variety trials repeated over years. *Theoretical and Applied Genetics* 94:1072–1077.

1998. Yield gain from recommendation of lucerne varieties based on mega-environment identifications by AMMI analysis. In: *Proceedings of the Fifth Congress of the European Society for Agronomy*, vol. 2, pp. 121–122. Nitra: Slovak Agricultural University.

2000. Variety × location interaction and its implications on breeding of lucerne: A case study. In: *Proceedings XIII EUCARPIA Medicago spp Group Meeting*, pp. 35–43. Università di Perugia.

Annicchiarico, P., Bertolini, M., and Mazzinelli, G. 1995. Analysis of genotype-environment interactions for maize hybrids in Italy. *Journal of Genetics & Breeding* 49:61–68.

Annicchiarico, P., and Perenzin, M. 1994. Adaptation patterns and definition of macro-environments for selection and recommendation of common-wheat genotypes in Italy. *Plant Breeding* 113:197–205.

Anscombe, G. E. M., and von Wright, G. H. (eds.). 1969. *On Certainty* (by Ludwig Wittgenstein). New York: Harper & Row.

Apostle, H. G. 1969. *Aristotle's Physics*. Bloomington: Indiana University Press.

Ashby, D. (ed.). 1993. Methodological and ethical issues in clinical trials. *Statistics in Medicine* 12:1373–1534.

Asmis, E. 1984. *Epicurus' Scientific Method*. Ithaca, NY: Cornell University Press.

Audi, R. (ed.). 1999. *The Cambridge Dictionary of Philosophy*, 2nd ed. Cambridge University Press.

Azzalini, A., and Diggle, P. J. 1994. Prediction of soil respiration rates from temperature, moisture content and soil type. *Applied Statistics* 43:505–526.

Ball, W. W. R., and Coxeter, H. S. M. 1987. *Mathematical Recreations and Essays*, 13th ed. New York: Dover.

Bandyopadhyay, P. S., and Boik, R. J. 1999. The curve fitting problem: A Bayesian rejoinder. *Philosophy of Science (Proceedings)* 66:S390–402.

Banner, M. C. 1990. *The Justification of Science and the Rationality of Religious Belief*. Oxford University Press.

Barnard, G. A. 1958. Studies in the history of probability and statistics. IX. Thomas Bayes's essay towards solving a problem in the doctrine of chances. *Biometrika* 45:293–315.

Barnes, B. 1991. How not to do the sociology of knowledge. *Annals of Scholarship* 8:321–335.

Barnes, J. 1975. *Aristotle's Posterior Analytics*. Oxford University Press.

Barnett, V. 1982. *Comparative Statistical Inference*, 2nd ed. New York: Wiley.

Baumgardt, C. 1951. *Johannes Kepler: Life and Letters*. New York: Philosophical Library.

Bayes, T. 1763. An essay towards solving a problem in the doctrine of chances. *Philosophical Transactions of the Royal Society* 53:370–418.

Beauchamp, T. L. (ed.). 1999. *David Hume: An Enquiry Concerning Human Understanding*. Oxford University Press.

Becker, B. J. 2000. MindWorks: Making scientific concepts come alive. *Science & Education* 9:269–278.

Bell, R., Abd-El-Khalick, F., Lederman, N. G., McComas, W. F., and Matthews, M. R. 2001. The nature of science and science education: A bibliography. *Science & Education* 10:187–204.

Bell, R. L., Lederman, N. G., and Abd-El-Khalick, F. 1998. Implicit versus explicit nature of science instruction: An explicit response to Palmquist and Finley. *Journal of Research in Science Teaching* 35:1057–1061.

Ben-Chaim, D., Ron, S., and Zoller, U. 2000. The disposition of eleventh-grade science students toward critical thinking. *Journal of Science Education and Technology* 9:149–159.

Benjamin, A. S., and Hackstaff, L. H. 1964. *Saint Augustine: On Free Choice of the Will.* Indianapolis, IN: Bobbs-Merrill.

Benjamini, Y., and Hochberg, Y. 1995. Controlling the false discovery rate: A practical and powerful approach to multiple testing. *Journal of the Royal Statistical Society, Series B* 57:289–300.

Bennett, J. H. (ed.). 1965. *Experiments in Plant Hybridisation* (by Gregor Mendel). London: Oliver & Boyd.

Berger, J. O. 1985. *Statistical Decision Theory and Bayesian Analysis*, 2nd ed. New York: Springer-Verlag.

Berger, J. O., and Berry, D. A. 1988. Statistical analysis and the illusion of objectivity. *American Scientist* 76:159–165.

Berger, J. O., and Pericchi, L. R. 1996. The intrinsic Bayes factor for model selection and prediction. *Journal of the American Statistical Association, Theory and Methods* 91:109–122.

Bernardo, J. M., Berger, J. O., Dawid, A. P., and Smith, A. F. M. (eds.). 1992. *Bayesian Statistics 4: Proceedings of the Fourth Valencia International Meeting.* Oxford University Press.

Bernardo, J. M., DeGroot, M. H., Lindley, D. V., and Smith, A. F. M. (eds.). 1980. *Bayesian Statistics: Proceedings of the First International Meeting.* Valencia, Spain: University Press.

Berry, D. A. 1987. Interim analysis in clinical trials: The role of the likelihood principle. *American Statistician* 41:117–122.

(ed.). 1990. *Statistical Methodology in the Pharmaceutical Sciences.* New York: Marcel Dekker.

1996. *Statistics: A Bayesian Perspective.* Belmont, CA: Duxbury Press.

Bettoni, E. 1961. *Duns Scotus: The Basic Principles of his Philosophy.* Washington, DC: Catholic University of America Press.

Bevilacqua, F., Giannetto, E., and Matthews, M. R. (eds.). 2001. *Science Education and Culture: The Contribution of History and Philosophy of Science.* Dordrecht: Kluwer.

Bhattacharyya, M. K., Smith, A. M., Ellis, T. H. N., Hedley, C., and Martin, C. 1990. The wrinkled-seed character of pea described by Mendel is caused by a transposon-like insertion in a gene encoding starch-branching enzyme. *Cell* 60:115–122.

Blackburn, S. 1973. *Reason and Prediction.* Cambridge University Press.

1994. *The Oxford Dictionary of Philosophy.* Oxford University Press.

Blackwell, R. J., Spath, R. J., and Thirlkel, W. E. 1963. *Commentary on Aristotle's Physics by St. Thomas Aquinas.* London: Routledge & Kegan Paul.

Boehner, P. 1957. *Ockham Philosophical Writings.* New York: Nelson.

Bohm, D., and Hiley, B. J. 1993. *The Undivided Universe.* London: Routledge.

Bourgeois, D. 1999. New processing tools for weak and/or spatially overlapped macromolecular diffraction patterns. *Acta Crystallographica, Section D* 55:1733–1741.

Bourne, L. E., Dominowski, R. L., Loftus, E. F., and Healy, A. F. 1986. *Cognitive Processes*, 2nd ed. Englewood Cliffs, NJ: Prentice-Hall.

Box, G. E. P., Hunter, W. G., and Hunter, J. S. 1978. *Statistics for Experimenters: An Introduction to Design, Data Analysis, and Model Building*. New York: Wiley.

Box, G. E. P., and Tiao, G. C. 1973. *Bayesian Inference in Statistical Analysis*. Reading, MA: Addison-Wesley.

Boyd, R., Gasper, P., and Trout, J. D. (eds.). 1991. *The Philosophy of Science*. Cambridge, MA: MIT Press.

Brancourt-Hulmel, M., Denis, J. B., and Lecomte, C. 2000. Determining environmental covariates which explain genotype–environment interaction in winter wheat through probe genotypes and biadditive factorial regression. *Theoretical and Applied Genetics* 100:285–298.

Bränden, C.-I., and Jones, T. A. 1990. Between objectivity and subjectivity. *Nature* 343:687–689.

Bricogne, G., and Carter, C. W. (eds.). 1996. *Likelihood, Bayesian Inference and Their Application to the Solution of New Structures*. Buffalo, NY: American Crystallographic Association.

Broad, C. D. 1952. *Ethics and the History of Philosophy*. London: Routledge & Kegan Paul.

1968. *Induction, Probability, and Causation*. Dordrecht: Reidel.

Broad, W. J. 1979. Paul Feyerabend: Science and the anarchist. *Science* 206:534–537.

Brodley, C. E., Lane, T., and Stough, T. M. 1999. Knowledge discovery and data mining. *American Scientist* 87(1):54–61.

Brooke, J. H. 1991. *Science and Religion: Some Historical Perspectives*. Cambridge University Press.

Brown, G. E. 1993. *The Objectivity Crisis: Rethinking the Role of Science in Society*. Chairman's report to the Committee on Science, Space, and Technology, House of Representatives, 103rd Congress, first session, serial D. Washington, DC: U.S. Government Printing Office.

Brown, H. I. 1987. *Observation and Objectivity*. Oxford University Press.

Brown, S. C. (ed.). 1984. *Objectivity and Cultural Divergence*. Cambridge University Press.

Brünger, A. T. 1992. Free R value: A novel statistical quantity for assessing the accuracy of crystal structures. *Nature* 355:472–475.

1993. Assessment of phase accuracy by cross validation: The free R value. *Acta Crystallographica, Section D* 49:24–36.

1997. Free R value: Cross-validation in crystallography. *Methods in Enzymology* 277:366–396.

Bugg, C. E., Carson, W. M., and Montgomery, J. A. 1993. Drugs by design. *Scientific American* 269(6):92–98.

Bugliarello, G. 1992. Verifiable truths. *American Scientist* 80:306.

Burks, A. W. 1977. *Chance, Cause, Reason: An Inquiry into the Nature of Scientific Evidence*. University of Chicago Press.

Cajori, F. 1947. *Sir Isaac Newton's Mathematical Principles of Natural Philosophy and His System of the World*. Berkeley: University of California Press.

Caldin, E. F. 1949. *The Power and Limits of Science: A Philosophical Study*. London: Chapman & Hall.

Callebaut, W. 1993. *Taking the Naturalistic Turn, or How Real Philosophy of Science Is Done*. University of Chicago Press.

Callus, D. A. (ed.). 1955. *Robert Grosseteste, Scholar and Bishop: Essays in Commemoration of the Seventh Centenary of His Death*. Oxford University Press.

Carey, S. S. 1994. *A Beginner's Guide to Scientific Method*. Belmont, CA: Wadsworth.

Carnap, R., Hahn, H., and Neurath, O. 1929. *Wissenschaftliche Weltauffassung: Der Wiener Kreis*. Wien: Artur Wolf Verlag.

Carson, R. N. 1997. Science and the ideals of liberal education. *Science & Education* 6:225–238.

Cartier, J. L., and Stewart, J. 2000. Teaching the nature of inquiry: Further developments in a high school genetics curriculum. *Science & Education* 9:247–267.

Carus, P. 1902. *Immanuel Kant: Prolegomena to Any Future Metaphysics That Can Qualify as a Science*. La Salle, IL: Open Court.

Caspar, M. 1959. *Kepler*, trans. C. D. Hellman. New York: Abelard-Schuman.

Cederblom, J., and Paulsen, D. W. 1986. *Critical Reasoning: Understanding and Criticizing Arguments and Theories*, 2nd ed. Belmont, CA: Wadsworth.

Chakrabarti, K. K. 1995. *Definition and Induction: A Historical and Comparative Study*. Honolulu: University of Hawaii Press.

Charlton, W. 1970. *Aristotle's Physics Books I and II*. Oxford University Press.

Chatalian, G. 1991. *Epistemology and Skepticism: An Enquiry into the Nature of Epistemology*. Carbondale, IL: Southern Illinois University Press.

Chattopadhyaya, D. P. 1991. *Induction, Probability, and Skepticism*. Albany: State University of New York Press.

Cho, Y., Gorina, S., Jeffrey, P. D., and Pavletich, N. P. 1994. Crystal structure of a p53 tumor suppressor–DNA complex: Understanding tumorigenic mutations. *Science* 265:346–355.

Cleave, J. P. 1991. *A Study of Logics*. Oxford University Press.

Cohen, J. 1994. The earth is round ($p < 0.5$). *American Psychologist* 49:997–1003.

Collingwood, R. G. 1940. *An Essay on Metaphysics*. Oxford University Press.

Collins, H. M. 1981. Stages in the empirical programme of relativism. *Social Studies of Science* 11:3–10.

Collins, H., and Pinch, T. 1993. *The Golem: What Everyone Should Know About Science*. Cambridge University Press.

1998. *The Golem at Large: What You Should Know About Technology*. Cambridge University Press.

Copleston, F. 1953. *A History of Philosophy. Vol. III: Late Medieval and Renaissance Philosophy*. New York: Doubleday.

Cornelius, P. L., and Crossa, J. 1999. Prediction assessment of shrinkage estimators of multiplicative models for multi-environment cultivar trials. *Crop Science* 39:998–1009.

Cottingham, J. (ed.). 1986. *René Descartes Meditations on First Philosophy*. Cambridge University Press.

(ed.). 1992. *The Cambridge Companion to Descartes*. Cambridge University Press.

Cottingham, J., Stoothoff, R., and Murdoch, D. 1988. *Descartes: Selected Philosophical Writings*. Cambridge University Press.

Couvalis, G. 1997. *The Philosophy of Science: Science and Objectivity*. London: Sage.

Cowtan, K. 1999. Error estimation and bias correction in phase improvement calculations. *Acta Chrystallographica, Section D* 55:1555–1567.

Craig, E. (ed.). 1998. *Routledge Encyclopedia of Philosophy*, 10 vols. London: Routledge.

Crombie, A. C. 1962. *Robert Grosseteste and the Origins of Experimental Science.* Oxford University Press.

Cromer, A. 1997. *Connected Knowledge: Science, Philosophy, and Education.* Oxford University Press.

Curley, A. J. 1996. *Augustine's Critique of Skepticism: A Study of Contra Academicos.* New York: Peter Lang.

Dales, R. C. 1973. *The Scientific Achievement of the Middle Ages.* Philadelphia: University of Pennsylvania Press.

Dalgarno, M., and Matthews, E. (eds.). 1989. *The Philosophy of Thomas Reid.* Dordrecht: Kluwer.

Dampier, W. C. 1961. *A History of Science and Its Relations with Philosophy and Religion.* Cambridge University Press.

Daniels, N. 1989. *Thomas Reid's 'Inquiry.'* Stanford, CA: Stanford University Press.

Davenport, H. 1992. *The Higher Arithmetic: An Introduction to the Theory of Numbers.* Cambridge University Press.

Davis, W. A. 1986. *An Introduction to Logic.* Englewood Cliffs, NJ: Prentice-Hall.

Deeds, D. G., Allen, C. S., Callen, B. W., and Wood, M. D. 2000. A new paradigm in integrated math and science courses: Finding common ground across disciplines. *Journal of College Science Teaching* 30:178–183.

DeVito, S. 1997. A gruesome problem for the curve-fitting solution. *British Journal for the Philosophy of Science* 48:391–396.

Dewey, J. 1916. *Democracy and Education.* New York: Macmillan.

Diederichs, K., and Karplus, P. A. 1997. Improved R-factors for diffraction data analysis in macromolecular crystallography. *Nature Structural Biology* 4:269–275.

Ding, A. A., and Hwang, J. T. G. 1999. Prediction intervals, factor analysis models, and high-dimensional empirical linear prediction (HELP). *Journal of the American Statistical Association* 94:446–455.

Dirac, P. A. M. 1963. The evolution of the physicist's picture of nature. *Scientific American* 208(5):45–53.

Drake, S., and MacLachlan, J. 1975. Galileo's discovery of the parabolic trajectory. *Scientific American* 232(3):102–110.

Driver, R., Newton, P., and Osborne, J. 2000. Establishing the norms of scientific argumentation in classrooms. *Science Education* 84:287–312.

Duggan, T. J. 1978. Ayer and Reid: Responses to the skeptic. *The Monist* 61:205–219.

Earman, J. 2000. *Hume's Abject Failure: The Argument against Miracles.* Oxford University Press.

Easterbrook, G. 1997. Science and God: A warming trend? *Science* 277:890–893.

Ebdon, J. S., and Gauch, H. G. 2002. Additive main effect and multiplicative interaction analysis of national turfgrass performance trials. II. Cultivar recommendations. *Crop Science* 42:497–506.

Economist, The. 1999. 20 February, p. 72.

Edwards, P. (ed.). 1967. *The Encyclopedia of Philosophy*, 8 vols. New York: Macmillan.

Einstein, A. 1936. Physics and reality, trans. Jean Piccard. *Journal of the Franklin Institute* 221:349–382.

——— 1954. *Ideas and Opinions.* New York: Crown.

Ellis, B. 1990. *Truth and Objectivity*. Oxford: Blackwell.

Ezzell, C. 2002. Proteins rule. *Scientific American* 286(4):40–47.

Farnham, N. H., and Yarmolinsky, A. (eds.). 1996. *Rethinking Liberal Education*. Oxford University Press.

Faust, D. 1984. *The Limits of Scientific Reasoning*. Minneapolis: University of Minnesota Press.

Ferejohn, M. 1991. *The Origins of Aristotelian Science*. New Haven, CT: Yale University Press.

Fetzer, J. H. (ed.). 1984. *Principles of Philosophical Reasoning*. Totowa, NJ: Rowman & Allanheld.

Feynman, R. P., Leighton, R. B., and Sands, M. 1964. *The Feynman Lectures on Physics*, vol. II. Reading, MA: Addison-Wesley.

Finlay, K. W., and Wilkinson, G. N. 1963. The analysis of adaptation in a plant-breeding programme. *Australian Journal of Agricultural Research* 14:742–754.

Fisher, R. A. 1973. *Statistical Methods and Scientific Inference*, 3rd ed. New York: Macmillan. (Originally published 1956.)

Forman, P. 1997. Assailing the seasons. *Science* 276:750–752.

Forster, M. R. 1999. Model selection in science: The problem of language variance. *British Journal for the Philosophy of Science* 50:83–102.

Forster, M., and Sober, E. 1994. How to tell when simpler, more unified, or less *ad hoc* theories will provide more accurate predictions. *British Journal for the Philosophy of Science* 45:1–35.

Frank, P. G. (ed.). 1956. *The Validation of Scientific Theories*. Boston: Beacon Press.

    1957. *Philosophy of Science: The Link Between Science and Philosophy*. Englewood Cliffs, NJ: Prentice-Hall.

Frege, G. 1893. *Grundgesetze der Arithmetik*. Jena: Hermann Pohle.

Freireich, E. J., Gehan, E., Frei, E., Schroeder, L. R., Wolman, I. J., Anbari, R., Burgert, E. O., Mills, S. D., Pinkel, D., Selawry, O. S., Moon, J. H., Gendel, B. R., Spurr, C. L., Storrs, R., Haurani, F., Hoogstraten, B., and Lee, S. 1963. The effect of 6-mercaptopurine on the duration of steroid-induced remissions in acute leukemia: A model for evaluation of other potentially useful therapy. *Blood* 21:699–716.

French, S. 1986. *Decision Theory: An Introduction to the Mathematics of Rationality*. New York: Wiley.

Friedman, K. S. 1990. *Predictive Simplicity: Induction Exhum'd*. Elmsford Park, NY: Pergamon.

Friedman, M. 1992. *Kant and the Exact Sciences*. Cambridge, MA: Harvard University Press.

    1999. *Reconsidering Logical Positivism*. Cambridge University Press.

Fuller, S. 2000. *Thomas Kuhn, A Philosophical History for Our Times*. University of Chicago Press.

Galison, P., and Stump, D. J. (eds.). 1996. *The Disunity of Science: Boundaries, Contexts, and Power*. Stanford, CA: Stanford University Press.

Galton, F. 1894. A plausible paradox in chances. *Nature* 49:365–366, 413.

Gálvez, A., Maqueda, M., Martínez-Bueno, M., and Valdivia, E. 2000. Scientific publication trends and the developing world. *American Scientist* 88: 526–533.

Gärdenfors, P. (ed.). 1992. *Belief Revision*. Cambridge University Press.

Gauch, H. G. 1982. Noise reduction by eigenvector ordinations. *Ecology* 63: 1643–1649.

1988. Model selection and validation for yield trials with interaction. *Biometrics* 44:705–715.

1992. *Statistical Analysis of Regional Yield Trials: AMMI Analysis of Factorial Designs*. New York: Elsevier.

1993. Prediction, parsimony, and noise. *American Scientist* 81:468–478, 507–508.

Gauch, H. G., and Zobel, R. W. 1997. Identifying mega-environments and targeting genotypes. *Crop Science* 37:311–326.

Gerstein, M., and Levitt, M. 1998. Simulating water and the molecules of life. *Scientific American* 279(5):100–105.

Gibbs, W. W. 2001. Cybernetic cells. *Scientific American* 265(2):52–57.

Giere, R. N. 1984. *Understanding Scientific Reasoning*, 2nd ed. New York: Holt, Rinehart & Winston.

1997. Scientific inference: Two points of view. *Philosophy of Science* (*Proceedings*) 64:S180–184.

Gillispie, C. C. (ed.). 1970. *Dictionary of Scientific Biography*, 18 vols. New York: Scribner.

Glyer, D., and Weeks, D. L. (eds.). 1998. *The Liberal Arts in Higher Education: Challenging Assumptions, Exploring Possibilities*. Lanham, MD: University Press of America.

Goldman, M. 1983. *The Demon in the Aether: The Story of James Clerk Maxwell*. Edinburgh: Paul Harris Publishing.

Golinski, J. 1998. *Making Natural Knowledge: Constructivism and the History of Science*. Cambridge University Press.

Good, R., and Shymansky, J. 2001. Nature-of-science literacy in *Benchmarks* and *Standards*: Post-Modern/Relativist or Modern/Realist? *Science & Education* 10:173–185.

Goodman, N. 1983. *Fact, Fiction, and Forecast*, 4th ed. Cambridge, MA: Harvard University Press.

Gottfried, K., and Wilson, K. G. 1997. Science as a cultural construct. *Nature* 386:545–547.

Gower, B. 1997. *Scientific Method: An Historical and Philosophical Introduction*. London: Routledge.

Grinnell, F. 1987. *The Scientific Attitude*. Boulder, CO: Westview Press.

Gross, P. R., and Levitt, N. 1994. *Higher Superstition: The Academic Left and Its Quarrels with Science*. Baltimore: Johns Hopkins University Press.

Gross, P. R., Levitt, N., and Lewis, M. W. (eds.). 1996. *The Flight from Science and Reason*. Baltimore: Johns Hopkins University Press.

Gruender, D. 2001. A new principle of demarcation: A modest proposal for science and science education. *Science & Education* 10:85–95.

Gunther, S. 1896. *Kepler und Galilei*. Berlin: E. Hofmann & Co.

Gustason, W. 1994. *Reasoning from Evidence: Inductive Logic*. New York: Macmillan.

Guyer, P. (ed.). 1992. *The Cambridge Companion to Kant*. Cambridge University Press.

Hacking, I. 1965. *Logic of Statistical Inference*. Cambridge University Press.

Hald, A. 1990. *A History of Probability and Statistics and Their Applications Before 1750*. New York: Wiley.

Halloun, I. A., and Hestenes, D. 1985a. The initial knowledge state of college physics students. *American Journal of Physics* 53:1043–1055.

1985b. Common sense concepts about motion. *American Journal of Physics* 53:1056–1065.

Hamilton, A. G. 1978. *Logic for Mathematicians*. Cambridge University Press.

Hamilton, W. (ed.). 1872. *The Works of Thomas Reid, D.D.*, 7th ed. Edinburgh: Maclachlan & Stewart.

Hansen, H. V., and Pinto, R. C. (eds.). 1995. *Fallacies: Classical and Contemporary Readings*. University Park: Pennsylvania State University Press.

Haskins, C. H. 1923. *The Rise of Universities*. Ithaca, NY: Cornell University Press.

Hayes, B. 1999. The web of words. *American Scientist* 87(2):108–112.

Herbert, W. 1995. The PC assault on science. *U.S. News & World Report* 118(7): 64, 66.

Himsworth, H. 1986. *Scientific Knowledge and Philosophic Thought*. Baltimore: Johns Hopkins University Press.

Hoefer, C., and Rosenberg, A. 1994. Empirical equivalence, underdetermination, and systems of the world. *Philosophy of Science* 61:592–607.

Hoffmann, R., Minkin, V. I., and Carpenter, B. K. 1996. Ockham's razor and chemistry. *Bulletin de la Société chimique de France* 133(2):117–130.

Hogan, K. 1999. Relating students' personal frameworks for science learning to their cognition in collaborative contexts. *Science Education* 83:1–32.

Hogan, K., and Maglienti, M. 2001. Comparing the epistemological underpinnings of students' and scientists' reasoning about conclusions. *Journal of Research in Science Teaching* 38:663–687.

Holton, G. 1993. *Science and Anti-science*. Cambridge, MA: Harvard University Press.

Horgan, J. 1991. Profile: Thomas S. Kuhn, reluctant revolutionary. *Scientific American* 264(5):40, 49.

1992. Profile: Karl R. Popper, the intellectual warrior. *Scientific American* 267(5):38–44.

1993. Profile: Paul Karl Feyerabend, the worst enemy of science. *Scientific American* 268(5):36–37.

1996. *The End of Science: Facing the Limits of Knowedge in the Twilight of the Scientific Age*. Reading, MA: Addison-Wesley.

Howson, C. 1997a. A logic of induction. *Philosophy of Science* 64:268–290.

1997b. Error probabilities in error. *Philosophy of Science* (*Proceedings*) 64:S185–194.

2000. *Hume's Problem: Induction and the Justification of Belief*. Oxford University Press.

Howson, C. and Urbach, P. 1993. *Scientific Reasoning: The Bayesian Approach*, 2nd ed. La Salle, IL: Open Court.

Hübner, K. 1983. *Critique of Scientific Reason*. University of Chicago Press.

Hunter, G. 1971. *Metalogic: An Introduction to the Metatheory of Standard First Order Logic*. Berkeley: University of California Press.

Hurley, P. J. 1994. *A Concise Introduction to Logic*, 5th ed. Belmont, CA: Wadsworth.

Hyman, A., and Walsh, J. J. (eds.). 1973. *Philosophy in the Middle Ages*, 2nd ed. Indianapolis: Hackett.

Irwin, A. R. 2000. Historical case studies: Teaching the nature of science in context. *Science Education* 84:5–26.

Irwin, T. H. 1988. *Aristotle's First Principles*. Oxford University Press.

Irzik, G. 2001. Back to basics: A philosophical critique of constructivism. *Studies in Philosophy and Education* 20:152–175.

Isaac, R. 1995. *The Pleasures of Probability*. Berlin: Springer-Verlag.

Jaki, S. L. 1966. *The Relevance of Physics*. University of Chicago Press.

Jefferys, W. H., and Berger, J. O. 1992. Ockham's razor and Bayesian analysis. *American Scientist* 80:64–72, 116, 212–214.

Jeffrey, R. C. 1983. *The Logic of Decision*, 2nd ed. University of Chicago Press.

——— 1991. *Formal Logic: Its Scope and Limits*, 3rd ed. New York: McGraw-Hill.

Jeffreys, H. 1973. *Scientific Inference*, 3rd ed. Cambridge University Press.

——— 1983. *Theory of Probability*, 3rd ed. Oxford University Press.

Johnson, M. A., and Lawson, A. E. 1998. What are the relative effects of reasoning ability and prior knowledge on biology achievement in expository and inquiry classes? *Journal of Research in Science Teaching* 35:89–103.

Joyce, J. M. 1999. *The Foundations of Causal Decision Theory*. Cambridge University Press.

Judson, L. (ed.). 1991. *Aristotle's Physics: A Collection of Essays*. Oxford University Press.

Kadane, J. B. (ed.). 1996. *Bayesian Methods and Ethics in a Clinical Trial Design*. New York: Wiley.

Kadane, J. B., Schervish, M. J., and Seidenfeld, T. 1999. *Rethinking the Foundations of Statistics*. Cambridge University Press.

Kang, M. S., and Gauch, H. G. (eds.). 1996. *Genotype-by-Environment Interaction*. Boca Raton, FL: CRC Press.

Kearns, J. T. 1988. *The Principles of Deductive Logic*. Albany: State University of New York Press.

Keeports, D., and Morier, D. 1994. Teaching the scientific method: Assessing a seminar on science and pesudoscience at Oakland's Mills College. *Journal of College Science Teaching* 24:45–50.

Kelly, K. T. 2000. The logic of success. *British Journal for the Philosophy of Science* 51:639–666.

Kemeny, J. G. 1959. *A Philosopher Looks at Science*. New York: Van Nostrand.

Keynes, J. M. 1962. *A Treatise on Probability*. New York: Harper & Row.

Kieseppä, I. A. 1997. Akaike information criterion, curve-fitting, and the philosophical problem of simplicity. *British Journal for the Philosophy of Science* 48:21–48.

——— 2001a. Statistical model selection criteria and Bayesianism. *Philosophy of Science (Proceedings)* 68:S141–152.

——— 2001b. Statistical model selection criteria and the philosophical problem of underdetermination. *British Journal for the Philosophy of Science* 52:761–794.

King, P. 1985. *Jean Buridan's Logic*. Dordrecht: Reidel.

——— 1995. *Augustine: Against the Academicians; The Teacher*. Indianapolis: Hackett.

Kleywegt, G. J., and Brünger, A. T. 1996. Checking your imagination: Applications of the free R value. *Structure* 4:897–904.

Kleywegt, G. J., and Jones, T. A. 1995. Where freedom is given, liberties are taken. *Structure* 3:535–540.

Kneale, W., and Kneale, M. 1986. *The Development of Logic*. Oxford University Press.

Koenig, H. G., McCullough, M. E., and Larson, D. B. 2001. *Handbook of Religion and Health*. Oxford University Press.

Koertge, N. (ed.). 1998. *A House Built on Sand: Exposing Postmodernist Myths about Science*. Oxford University Press.

Kolmogorov, A. N. 1933. *Grundbegriffe der Wahrscheinlichkeitsrechnung*. Berlin: Julius Springer.

1956. *Foundations of the Theory of Probability*, 2nd ed. New York: Chelsea.

Kornblith, H. 1993. *Inductive Inference and Its Natural Ground: An Essay in Naturalistic Epistemology*. Cambridge, MA: MIT Press.

Koulaidis, V., and Ogborn, J. 1995. Science teachers' philosophical assumptions: How well do we understand them? *International Journal of Science Education* 17:273–283.

Kretzmann, N., and Stump, E. (eds.). 1993. *The Cambridge Companion to Aquinas*. Cambridge University Press.

Kuhn, T. S. 1970. *The Structure of Scientific Revolutions*, 2nd ed. University of Chicago Press. (Originally published 1962.)

Kukla, A. 1995. Forster and Sober on the curve-fitting problem. *British Journal for the Philosophy of Science* 46:248–252.

1996. Does every theory have empirically equivalent rivals? *Erkenntnis* 44:137–166.

2001. Theoreticity, underdetermination, and the disregard for bizarre scientific hypotheses. *Philosophy of Science* 68:21–35.

Kyburg, H. E. 1970. *Probability and Inductive Logic*. New York: Macmillan.

Lakatos, I. (ed.). 1968. *The Problem of Inductive Logic*. Amsterdam: North-Holland.

Lakatos, I., and Musgrave, A. (eds.). 1968. *Problems in the Philosophy of Science*. Amsterdam: North-Holland.

(eds.). 1970. *Criticism and the Growth of Knowledge*. Cambridge University Press.

Lakin, S., and Wellington, J. 1994. Who will teach the 'nature of science'? Teachers' views of science and their implications for science education. *International Journal of Science Education* 16:175–190.

Larochelle, M., Bednarz, N., and Garrison, J. (eds.). 1998. *Constructivism and Education*. Cambridge University Press.

Larson, E. J., and Witham, L. 1999. Scientists and religion in America. *Scientific American* 281(3):88–93.

Laudan, L. 1977. *Progress and Its Problems: Towards a Theory of Scientific Growth*. Berkeley: University of California Press.

1996. *Beyond Positivism and Relativism: Theory, Method, and Evidence*. Boulder, CO: Westview.

Lavoie, D. R. 1999. Effects of emphasizing hypothetico-predictive reasoning within the science learning cycle on high school students' process skills and conceptual understandings in biology. *Journal of Research in Science Teaching* 36:1127–1147.

Lawson, A. E. 1990. Use of reasoning to a contradiction in grades three to college. *Journal of Research in Science Teaching* 27:541–551.

1999. What should students learn about the nature of science and how should we teach it? Applying the 'If-and-then-therefore' pattern to develop students' theoretical reasoning abilities in science. *Journal of College Science Teaching* 28:401–411.

Lawson, A. E., Clark, B., Cramer-Meldrum, E., Falconer, K. A., Sequist, J. M., and Kwon, Y.-J. 2000. Development of scientific reasoning in college biology: Do two levels of general hypothesis-testing skills exist? *Journal of Research in Science Teaching* 37:81–101.

Lawson, A. E., and Weser, J. 1990. The rejection of nonscientific beliefs about life: Effects of instruction and reasoning skills. *Journal of Research in Science Teaching* 27:589–606.

Leblanc, H., and Wisdom, W. A. 1976. *Deductive Logic*, 2nd ed. Boston: Allyn & Bacon.

Lederman, N. G. 1992. Students' and teachers' conceptions of the nature of science: A review of the research. *Journal of Research in Science Teaching* 29:331–359.

Lefevre, R., Escaut, A., and Bouldoires, B. 1997. Students' views of science and technology when entering university. *Journal of Science Education and Technology* 6:285–296.

Lehrer, K. 1978. Reid on primary and secondary qualities. *The Monist* 61:184–191.
 1989. *Thomas Reid*. London: Routledge.

Leplin, J. (ed.). 1984. *Scientific Realism*. Berkeley: University of California Press.
 1997. *A Novel Defense of Scientific Realism*. Oxford University Press.

Lester, D. R., Ross, J. J., Davies, P. J., and Reid, J. B. 1997. Mendel's stem length gene (Le) encodes a gibberellin 3β-hydroxylase. *The Plant Cell* 9:1435–1443.

Levi, I. 1986. *Hard Choices: Decision Making under Unresolved Conflict*. Cambridge University Press.
 1991. *The Fixation of Belief and Its Undoing: Changing Beliefs through Inquiry*. Cambridge University Press.

Lewis, R. J., and Berry, D. A. 1994. Group sequential clinical trials: A classical evaluation of Bayesian decision-theoretic designs. *Journal of the American Statistical Association, Theory and Methods* 89:1528–1534.

Lide, D. R. (ed.). 1995. *CRC Handbook of Chemistry and Physics*, 76th ed. Boca Raton, FL: CRC Press.

Lindberg, D. C. 1992. *The Beginnings of Western Science: The European Scientific Tradition in Philosophical, Religious, and Institutional Context, 600 B.C. to A.D. 1450*. University of Chicago Press.

Losee, J. 1993. *A Historical Introduction to the Philosophy of Science*, 3rd ed. Oxford University Press.

Loving, C. C. 1997. From the summit of truth to its slippery slopes: Science education's journey through positivist-postmodern territory. *American Educational Research Journal* 34:421–452.

Lucas, J. R. 2000. *The Conceptual Roots of Mathematics: An Essay on the Philosophy of Mathematics*. London: Routledge.

Lynch, S. 2001. 'Science for all' is not equal to 'One size fits all': Linguistic and cultural diversity and science education reform. *Journal of Research in Science Teaching* 38:622–627.

McAllister, J. W. 1996. *Beauty and Revolution in Science*. Ithaca, NY: Cornell University Press.
 1998. Is beauty a sign of truth in scientific theories? *American Scientist* 86:174–183.

McCloskey, M. 1983. Intuitive physics. *Scientific American* 248(4):122–130.

McComas, W. F. (ed.). 1998. *The Nature of Science in Science Education: Rationales and Strategies*. Dordrecht: Kluwer.

McGinn, M. 1989. *Sense and Certainty: A Dissolution of Scepticism*. Oxford: Blackwell.

Machamer, P. 1998. Philosophy of science: An overview for educators. *Science & Education* 7:1–11.

Macilwain, C. 1997. Campuses ring to a stormy clash over truth and reason. *Nature* 387:331–333.

MacKay, D. J. C. 1992. Bayesian interpolation. *Neural Computation* 4:415–447.

McKay, T. J. 1989. *Modern Formal Logic*. New York: Macmillan.

McKeon, R. 1941. *The Basic Works of Aristotle*. New York: Random House.

McManners, J. (ed.). 1993. *The Oxford Illustrated History of Christianity*. Oxford University Press.

McNeill, P. M. 1993. *The Ethics and Politics of Human Experimentation*. Cambridge University Press.

Marbach-Ad, G., and Sokolove, P. G. 2000. Good science begins with good questions: Answering the need for high-level questions in science. *Journal of College Science Teaching* 30:192–195.

Marrone, S. P. 1983. *William of Auvergne and Robert Grosseteste: New Ideas of Truth in the Early Thirteenth Century*. Princeton, NJ: Princeton University Press.

1985. *Truth and Scientific Knowledge in the Thought of Henry of Ghent*. Cambridge, MA: Medieval Academy of America.

Martin, J. 1992. *Francis Bacon, the State, and the Reform of Natural Philosophy*. Cambridge University Press.

Matthews, M. R. 1994. *Science Teaching: The Role of History and Philosophy of Science*. London: Routledge.

(ed.). 1998a. *Constructivism in Science Education: A Philosophical Examination*. Dordrech: Kluwer.

1998b. In defense of modest goals when teaching about the nature of science. *Journal of Research in Science Teaching* 35:161–174.

2000. *Time for Science Education: How Teaching the History and Philosophy of Pendulum Motion Can Contribute to Science Literacy*. Dordrecht: Kluwer.

Mauldin, R. F., and Lonney, L. W. 1999. Scientific reasoning for nonscience majors: Ronald N. Giere's approach. Analyzing scientific press reports to teach nonmajors the value of research. *Journal of College Science Teaching* 28:416–421.

Mayo, D. G. 1996. *Error and the Growth of Experimental Knowledge*. University of Chicago Press.

1997. Error statistics and learning from error: Making a virtue of necessity. *Philosophy of Science (Proceedings)* 64:S195–212.

Medawar, P. B. 1969. *Induction and Intuition in Scientific Thought*. Philadelphia: American Philosophical Society.

1984. *The Limits of Science*. New York: Harper & Row.

Megill, A. (ed.). 1994. *Rethinking Objectivity*. Durham, NC: Duke University Press.

Mendel, G. 1865. Versuche über Pflanzen-Hybriden. *Verhandlungen des Naturforschenden Vereins in Brünn* 4:3–47.

Mendenhall, W., and Beaver, R. J. 1994. *Introduction to Probability and Statistics*, 9th ed. Belmont, CA: Duxbury Press.

Meyling, H. 1997. How to change students' conceptions of the epistemology of science. *Science & Education* 6:397–416.

Midgley, M. 1992. *Science as Salvation: A Modern Myth and Its Meaning*. London: Routledge.

Miller, R. W. 1987. *Fact and Method: Explanation, Confirmation and Reality in the Natural and the Social Sciences*. Princeton, NJ: Princeton University Press.

Mintzes, J. J., Wandersee, J. H., and Novak, J. D. (eds.). 1998. *Teaching Science for Understanding: A Human Constructivist View*. Orlando, FL: Academic Press.

Misak, C. J. 1995. *Verificationism: Its History and Prospects*. London: Routledge.

Mooers, C. A. 1921. The agronomic placement of varieties. *Journal of the American Society of Agronomy* 13:337–352.

Moore, J. H. 1998. Public understanding of science – and other fields. *American Scientist* 86:498.

Moreno-Gonzalez, J., and Crossa, J. 1998. Combining genotype, environment and attribute variables in regression models for predicting the cell-means of multi-environment cultivar trials. *Theoretical and Applied Genetics* 96:803–811.

Morrison, D. R. O. 1997. Bad science, bad education. *Scientific American* 277(5): 114–118.

Moss, D. M., Abrams, E. D., and Robb, J. 2001. Examining student conceptions of the nature of science. *International Journal of Science Education* 23:771–790.

Mueller, I. 1981. *Philosophy of Mathematics and Deductive Structure in Euclid's Elements*. Cambridge, MA: MIT Press.

Mukerjee, M. 1998. Undressing the emperor. *Scientific American* 278(3):30, 32.

Mulaik, S. A. 2001. The curve-fitting problem: An objectivist view. *Philosophy of Science* 68:218–241.

Nachit, M. M., Sorrells, M. E., Zobel, R. W., Gauch, H. G., Fischer, R. A., and Coffman, W. R. 1992a. Association of morpho-physicological traits with grain yield and components of genotype–environment interaction in durum wheat. *Journal of Genetics and Breeding* 46:363–368.

   1992b. Association of environment variables with sites' mean grain yield and components of genotype–environment interaction in durum wheat. *Journal of Genetics and Breeding* 46:369–372.

Nagel, E. 1950. *Philosophy of Scientific Method, John Stuart Mill*. New York: Hafner.

NAS (National Academy of Sciences). 1995. *Reshaping the Graduate Education of Scientists and Engineers*. Washington, DC: National Academy Press.

Nash, L. K. 1963. *The Nature of the Natural Sciences*. Boston: Little, Brown.

Natter, W., Schatzki, T. R., and Jones, J. P. (eds.). 1995. *Objectivity and Its Other*. New York: Guilford Press.

NCEE (National Commission on Excellence in Education). 1983. *A Nation at Risk: The Imperative for Educational Reform*. Washington, DC: U.S. Department of Education.

Neurath, M., and Cohen, R. S. (eds.). 1973. *Otto Neurath: Empiricism and Sociology*. Dordrecht: Reidel.

Newell, R. W. 1986. *Objectivity, Empiricism and Truth*. London: Routledge & Kegan Paul.

Newton, I. 1964. *The Mathematical Principles of Natural Philosophy*. New York: Citadel Press.

Newton-Smith, W. H. 1981. *The Rationality of Science*. London: Routledge & Kegan Paul.

   1999. *A Companion to the Philosophy of Science*. Oxford University Press.

Nicolaou, K. C., and Boddy, C. N. C. 2001. Behind enemy lines. *Scientific American* 282(5):54–61.

Nielsen, D. R. 1992. Presidential address: Global agronomic opportunities. *Agronomy Journal* 84:131–132.

Niemann, J. T., Cairns, C. B., Sharma, J., and Lewis, R. J. 1992. Treatment of prolonged ventricular fibrillation: Immediate countershock versus high-dose epinephrine and CPR preceding countershock. *Circulation* 85:281–287.

Niven, I., Zucherman, H. S., and Montgomery, H. L. 1991. *An Introduction to the Theory of Numbers*, 5th ed. New York: Wiley.

Nola, R. 1997. Constructivism in science and science education: A philosophical critique. *Science & Education* 6:55–83.

Norton, D. F. (ed.). 1993. *The Cambridge Companion to Hume*. Cambridge University Press.

Nott, M., and Wellington, J. 1998. Eliciting, interpreting and developing teachers' understandings of the nature of science. *Science & Education* 7:579–594.

Novak, J. D., and Gowin, D. B. 1984. *Learning How to Learn*. Cambridge University Press.

NRC (National Research Council). 1996. *National Science Education Standards*. Washington, DC: National Academy Press.

1997. *Science Teaching Reconsidered: A Handbook*. Washington, DC: National Academy Press.

1999. *Transforming Undergraduate Education in Science, Mathematics, Engineering, and Technology*. Washington, DC: National Academy Press.

NSF (National Science Foundation). 1996. *Shaping the Future: New Expectations for Undergraduate Education in Science, Mathematics, Engineering, and Technology*. Arlington, VA: National Science Foundation.

NSTA (National Science Teachers Association). 1995. *A High School Framework for National Science Education Standards*. Arlington, VA: National Science Teachers Association.

O'Hear, A. 1989. *An Introduction to the Philosophy of Science*. Oxford University Press.

Pais, A. 1982. *Subtle Is the Lord: The Science and the Life of Albert Einstein*. Oxford University Press.

Palmer, H. 1985. *Presupposition and Transcendental Inference*. New York: St. Martin's Press.

Parry, W. T., and Hacker, E. A. 1991. *Aristotelian Logic*. Albany: State University of New York Press.

Patterson, R. 1995. *Aristotle's Modal Logic: Essence and Entailment in the Organon*. Cambridge University Press.

Peace, K. E. (ed.). 1988. *Biopharmaceutical Statistics for Drug Development*. New York: Marcel Dekker.

Pearson, E. S. 1947. The choice of statistical tests illustrated on the interpretation of data classed in a 2×2 table. *Biometrika* 34:139–169.

Peltonen, M. (ed.). 1996. *The Cambridge Companion to Bacon*. Cambridge University Press.

Pérez-Ramos, A. 1988. *Francis Bacon's Idea of Science and the Maker's Knowledge Tradition*. Oxford University Press.

Phillips, D. C. 1995. The good, the bad, and the ugly: The many faces of constructivism. *Educational Researcher* 24(7):5–12.

Piburn, M. D. 1990. Reasoning about logical propositions and success in science. *Journal of Research in Science Teaching* 27:887–900.

Pinborg, J. (ed.). 1976. *The Logic of John Buridan.* Copenhagen: Museum Tusculanum.

Pinnick, C. L. 1992. Cognitive commitment and the strong program. *Social Epistemology* 6:289–298.

Pirie, M. 1985. *The Book of the Fallacy: A Training Manual for Intellectual Subversives.* London: Routledge & Kegan Paul.

Platt, J. R. 1964. Strong inference. *Science* 146:347–353.

Polanyi, M. 1962. *Personal Knowledge: Towards a Post-Critical Philosophy.* University of Chicago Press.

Pomerance, C. 1982. The search for prime numbers. *Scientific American* 247(6): 136–147.

Popper, K. R. 1945. *The Open Society and Its Enemies.* London: Routledge & Kegan Paul.

1968. *The Logic of Scientific Discovery.* New York: Harper & Row. (Originally published 1959.)

1974. *Conjectures and Refutations: The Growth of Scientific Knowledge,* 5th ed. New York: Harper & Row.

1979. *Objective Knowledge: An Evolutionary Approach.* Oxford University Press.

Potter, M. 2000. *Reason's Nearest Kin: Philosophies of Arithmetic from Kant to Carnap.* Oxford University Press.

Press, S. J. 1989. *Bayesian Statistics: Principles, Models, and Applications.* New York: Wiley.

Quine, W. V. 1975. On empirically equivalent systems of the world. *Erkenntnis* 9: 313–328.

Quinton, A. 1980. *Francis Bacon.* Oxford University Press.

Raftery, A. E., Tanner, M. A., and Wells, M. T. (eds.). 2002. *Statistics in the 21st Century.* London: Chapman & Hall.

Reif, F. 1995. Understanding and teaching important scientific thought processes. *Journal of Science Education and Technology* 4:261–282.

Rescher, N. 1980. *Induction: An Essay on the Justification of Inductive Reasoning.* Pittsburgh, PA: University of Pittsburgh Press.

1983. *Risk: A Philosophical Introduction to the Theory of Risk Evaluation and Management.* Lanham, MD: University Press of America.

(ed.). 1985. *The Heritage of Logical Positivism.* Lanham, MD: University Press of America.

1998. *Predicting the Future: An Introduction to the Theory of Forecasting.* Albany: State University of New York Press.

1999. *The Limits of Science,* rev. ed. Pittsburgh, PA: University of Pittsburgh Press.

Rips, L. J. 1994. *The Psychology of Proof: Deductive Reasoning in Human Thinking.* Cambridge, MA: MIT Press.

Ritchie, A. D. 1923. *Scientific Method: An Inquiry into the Character and Validity of Natural Laws.* New York: Harcourt, Brace.

Robert, C. P. 1994. *The Bayesian Choice: A Decision-Theoretic Motivation.* Berlin: Springer-Verlag.

Robinson, J. T. 1998. Science teaching and the nature of science. *Science & Education* 7:617–634. (Reprint of 1965 article.)

Romagosa, I., Fox, P. N., Garcia-del-Moral, L. F., and Ramos, J. M. 1993. Integration and statistical and physiological analyses of adaptation of near-isogenic barley lines. *Theoretical and Applied Genetics* 86:822–826.

Romagosa, I., Ullrich, S. E., Han, F., and Hayes, P. M. 1996. Use of the additive main effects and multiplicative interaction model in QTL mapping for adaptation in barley. *Theoretical and Applied Genetics* 93:30–37.

Rose, L. E. 1968. *Aristotle's Syllogistic*. Springfield, IL: Charles C Thomas.

Rosenberg, A. 2000. *Philosophy of Science: A Contemporary Introduction*. London: Routledge.

Rosenthal-Schneider, I. 1980. *Reality and Scientific Truth: Discussions with Einstein, von Laue, and Planck*. Detroit, MI: Wayne State University Press.

Ross, A. (ed.). 1996. *Science Wars*. Durham, NC: Duke University Press.

Ross, S. 1994. *A First Course in Probability*, 4th ed. New York: Macmillan.

Royall, R. M. 1997. *Statistical Evidence: A Likelihood Paradigm*. London: Chapman & Hall.

Rubinstein, M. F. 1986. *Tools for Thinking and Problem Solving*. Englewood Cliffs, NJ: Prentice-Hall.

Russo, J. E., and Schoemaker, P. J. H. 1989. *Decision Traps: Ten Barriers to Brilliant Decision-Making and How to Overcome Them*. New York: Doubleday.

Ryder, J., and Leach, J. 1999. University science students' experiences of investigative project work and their images of science. *International Journal of Science Education* 21:945–956.

Salmon, W. C. 1967. *The Foundations of Scientific Inference*. Pittsburgh, PA: University of Pittsburgh Press.

1984. *Logic*, 3rd ed. Englewood Cliffs, NJ: Prentice-Hall.

1999. The spirit of logical empiricism: Carl G. Hempel's role in twentieth-century philosophy of science. *Philosophy of Science* 66:333–350.

Scheffler, I. 1982. *Science and Subjectivity*, 2nd ed. Indianapolis: Hackett.

Schield, M. 1996. Using Bayesian inference in classical hypothesis testing. In: *Proceedings of the Section on Statistical Education*, pp. 274–279. Alexandria, VA: American Statistical Association.

Schilpp, P. A. (ed.). 1951. *Albert Einstein: Philosopher-Scientist*. New York: Tudor Publishing.

(ed.). 1974. *The Philosophy of Karl Popper*, 2 vols. Evanston, IL: Library of Living Philosophers.

Schrödinger, E. 1928. *Collected Papers on Wave Mechanics*, trans. J. F. Shearer. London: Blackie & Son.

Schum, D. A. 1994. *The Evidential Foundations of Probabilistic Reasoning*. New York: Wiley.

Segerstråle, U. (ed.). 2000. *Beyond the Science Wars: The Missing Discourse about Science and Society*. Albany: State University of New York Press.

Shafer, G. 1976. *A Mathematical Theory of Evidence*. Princeton, NJ: Princeton University Press.

Shankar, N. 1994. *Metamathematics, Machines, and Gödel's Proof*. Cambridge University Press.

Shiland, T. W. 1998. The atheoretical nature of the National Science Education Standards. *Science Education* 82:615–617.

Siegmund, D. 1985. *Sequential Analysis: Tests and Confidence Intervals*. Berlin: Springer-Verlag.

Simon, H. A. 1962. The architecture of complexity. *Proceedings of the American Philosophical Society* 106:467–482.

Smith, M. U., Lederman, N. G., Bell, R. L., McComas, W. F., and Clough, M. P. 1997. How great is the disagreement about the nature of science? A response to Alters. *Journal of Research in Science Teaching* 34:1101–1103.

Smith, M. U., and Scharmann, L. C. 1999. Defining versus describing the nature of science: A pragmatic analysis for classroom teachers and science educators. *Science Education* 83:493–509.

Snedecor, G. W., and Cochran, W. G. 1989. *Statistical Methods*, 8th ed. Ames: Iowa State University Press.

Sneller, C. H., and Dombek, D. 1995. Comparing soybean cultivar ranking and selection for yield with AMMI and full-data performance estimates. *Crop Science* 35:1536–1541.

Sneller, C. H., Kilgore-Norquest, L., and Dombek, D. 1997. Repeatability of yield stability statistics in soybean. *Crop Science* 37:383–390.

Snow, C. P. 1993. *The Two Cultures and the Scientific Revolution*. Cambridge University Press. (Originally published 1959.)

Sober, E. 1975. *Simplicity*. Oxford University Press.

Sokal, A. 1996. Transgressing the boundaries: Toward a transformative hermeneutics of quantum gravity. *Social Text* 46/47:217–252.

Sokal, A., and Bricmont, J. 1998. *Fashionable Nonsense: Postmodern Intellectuals' Abuse of Science*. New York; Picador.

Souders, T. M., and Stenbakken, G. N. 1991. Cutting the high cost of testing. *IEEE Spectrum* 28(3):48–51.

Stalker, D. (ed.). 1994. *Grue! The New Riddle of Induction*. La Salle, IL: Open Court.

Stein, C. 1955. Inadmissibility of the usual estimator for the mean of a multivariate normal distribution. *Proceedings of the Third Berkeley Symposium on Mathematical Statistics and Probability* 1:197–206.

Stenmark, M. 1995. *Rationality in Science, Religion, and Everyday Life: A Critical Evaluation of Four Models of Rationality*. Notre Dame, IN: University of Notre Dame Press.

Stephenson, B. 1987. *Kepler's Physical Astronomy*. Berlin: Springer-Verlag.

Stewart, I. 1996. The interrogator's fallacy. *Scientific American* 275(3):172–175.

Stirzaker, D. 1994. *Elementary Probability*. Cambridge University Press.

Stove, D. C. 1973. *Probability and Hume's Inductive Scepticism*. Oxford University Press.

1982. *Popper and After: Four Modern Irrationalists*. Elmsford Park, NY: Pergamon.

Swinburne, R. 1973. *An Introduction to Confirmation Theory*. London: Methuen.

(ed.). 1974. *The Justification of Induction*. Oxford University Press.

(ed.). 1989. *Miracles*. New York: Macmillan.

1997. *Simplicity as Evidence of Truth*. Milwaukee, WI: Marquette University Press.

Tarantola, A. 1987. *Inverse Problem Theory: Methods for Data Fitting and Model Parameter Estimation*. Amsterdam: Elsevier.

Terwilliger, T. C., and Berendzen, J. 1996a. Bayesian difference refinement. *Acta Crystallographica Section D* 52:1004–1011.

1996b. Bayesian weighting for macromolecular crystallographic refinement. *Acta Crystallographica Section D* 52:743–748.

Theocharis, T., and Psimopoulos, M. 1987. Where science has gone wrong. *Nature* 329:595–598.

Thornburn, W. M. 1918. The myth of Occam's razor. *Mind* 27:345–353.

Tobias, S. 1990. *They're Not Dumb, They're Different: Stalking the Second Tier.* Tucson, AZ: Research Corporation.

Tobin, K. (ed.). 1993. *The Practice of Constructivism in Science Education.* Hillsdale, NJ: Lawrence Erlbaum.

Todhunter, I. 1933. *Euclid's Elements, Books I–VI, XI and XII.* London: J. M. Dent.

Trachtman, L. E., and Perrucci, R. 2000. *Science under Siege? Interest Groups and the Science Wars.* Lanham, MD: Rowman & Littlefield.

Treagust, D. F., Duit, R., and Fraser, B. J. (eds.). 1996. *Improving Teaching and Learning in Science and Mathematics.* New York: Teachers College Press, Columbia University.

Trigg, R. 1980. *Reality at Risk: A Defence of Realism in Philosophy and the Sciences.* Totowa, NJ: Barnes & Noble.

1993. *Rationality and Science: Can Science Explain Everything?* Oxford: Blackwell.

Troyer, A. F. 1996. Breeding widely adapted, popular maize hybrids. *Euphytica* 92: 163–174.

1999. Background of U.S. hybrid corn. *Crop Science* 39:601–626.

Trundle, R. 1999. *Medieval Modal Logic and Science: Augustine on Scientific Truth and Thomas on Its Impossibility without a First Cause.* Lanham, MD: University Press of America.

Urbach, P. 1987. *Francis Bacon's Philosophy of Science.* La Salle, IL: Open Court.

Velleman, D. J. 1994. *How To Prove It: A Structured Approach.* Cambridge University Press.

Vier, P. C. 1951. *Evidence and Its Function According to John Duns Scotus.* St. Bonaventure, NY: Franciscan Institute.

Vilmorin-Andrieux, M. 1885. *The Vegetable Garden.* London: John Murray.

Wade, N. 1977. Thomas S. Kuhn: Revolutionary theorist of science. *Science* 197: 143–145.

Ward, S. C. 1996. *Reconfiguring Truth: Postmodernism, Science Studies, and the Search for a New Model of Knowledge.* Lanham, MD: Rowman & Littlefield.

Warren, B., Ballenger, C., Ognowski, M., Rosebery, A. S., and Hudicourt-Barnes, J. 2001. Rethinking diversity in learning science: The logic of everyday sense-making. *Journal of Research in Science Teaching* 38:529–552.

Watkins, J. 1984. *Science and Scepticism.* Princeton, NJ: Princeton University Press.

Weaver, R. M. 1948. *Ideas Have Consequences.* University of Chicago Press.

Weisheipl, J. A. (ed.). 1980. *Albertus Magnus and the Sciences.* Toronto: Pontifical Institute of Mediaeval Studies.

Whitehead, A. N. 1925. *Science and the Modern World.* Cambridge University Press.

Whitehead, A. N., and Russell, B. 1910–13. *Principia Mathematica,* 3 vols. Cambridge University Press.

Wilkinson, D. (ed.). 1991. *Report of the Forty-sixth Annual Corn and Sorghum Research Conference.* Washington, DC: American Seed Trade Association.

Williams, L. P., and Steffens, H. J. 1978. *The History of Science in Western Civilization. Vol. II: The Scientific Revolution.* Lanham, MD: University Press of America.

Williams, M. 1991. *Unnatural Doubts: Epistemological Realism and the Basis of Scepticism.* Oxford: Blackwell.

Wilson, E. B. 1952. *An Introduction to Scientific Research.* New York: McGraw-Hill.

Wilson, E. O. 1998. *Consilience: The Unity of Knowledge.* New York: Knopf.

Woolhouse, R. S. 1988. *The Empiricists.* Oxford University Press.

Wright, J. N. 1991. *Science and the Theory of Rationality.* Aldershot, UK: Avebury.

Wyckoff, S. 2001. Changing the culture of undergraduate science teaching: Shifting from lecture to interactive engagement and scientific reasoning. *Journal of College Science Teaching* 30:306–312.

Yager, R. E. 1991. The constructivist learning model. *The Science Teacher* 58(6):52–57.

Yates, F., and Cochran, W. G. 1938. The analysis of groups of experiments. *Journal of Agricultural Science, Cambridge* 28:556–580.

Zeidler, D. L. 1997. The central role of fallacious thinking in science education. *Science Education* 81:483–496.

Zhao, R., Pevear, D. C., Kremer, M. J., Giranda, V. L., Kofron, J. A., Kuhn, R. J., and Rossmann, M. G. 1996. Human rhinovirus 3 at 3.0 Å resolution. *Structure* 4:1205–1220.

Zivin, J. A. 2000. Understanding clinical trials. *Scientific American* 282(4):69–75.

Zoller, U., Ben-Chaim, D., Ron, S., Pentimalli, R., and Borsese, A. 2000. The disposition toward critical thinking of high school and university science students: An inter-intra Israeli-Italian study. *International Journal of Science Education* 22:571–582.

INDEX